Advanced Course on FAIRSHAPE

Edited by
Prof. Dr. Josef Hoschek
University of Technology Darmstadt
and Prof. Dr. Panagiotis Kaklis
National Technical University of Athens

B. G. Teubner Stuttgart 1996

Die Deutsche Bibliothek – CIP-Einheitsaufnahme

Advanced Course on FAIRSHAPE < 1996, Lambrecht >:
Advanced Course on FAIRSHAPE / ed. by Josef Hoschek and
Panagiotis Kaklis. – Stuttgart : Teubner, 1996
   ISBN 3-519-02634-1
NE: Hoschek, Josef [Hrsg.]

Das Werk einschließlich aller seiner Teile ist urheberrechtlich geschützt. Jede Verwertung außerhalb der engen Grenzen des Urheberrechtsgesetzes ist ohne Zustimmung des Verlages unzulässig und strafbar. Das gilt besonders für Vervielfältigungen, Übersetzungen, Mikroverfilmungen und die Einspeicherung und Verarbeitung in elektronischen Systemen.

© B. G. Teubner Stuttgart 1996
Printed in Germany
Gesamtherstellung: Präzis-Druck GmbH, Karlsruhe

1   PREAMBLE

Assembled here is a collection of articles presented at the *Advanced Course on "Automatic Fairing and Shape-Preserving Methodologies for CAD/CAM"* held at Pfalz Akademie, Lambrecht (near Kaiserslautern), Germany, during the period of March 25th to 29th, 1996. The course has been organized by the Network **FAIRSHAPE**, a two-year EU-project, which is economically supported by the XII Directorate General of the European Commission in the context of the Framework Programme: **H**uman **C**apital and **M**obility (**HCM**, 1992-1994).

The objectives of FAIRSHAPE can be classified into two major categories, namely the *general* objectives and the *technical* ones. The general objectives, which are of strategic character, can be summarized as follows:

1. Initiate and establish close interaction between European Organizations, which are involved in the CAGD/CAD/CAM activity by doing either basic or applied research or by exploiting the results of applied research.
2. Provide the academic partners of the network with CAGD problems, which are of current industrial interest.
3. Reduce the time needed for the results of basic research in CAGD to be disseminated to the CAD/CAM end-users and tested in the context of a real CAD/CAM environment.
4. Exchange methods and expertise between partners belonging to different industrial sectors, more specifically, the automobile and marine sector.

The major technical goals of FAIRSHAPE are the following:

1. Create a knowledge base containing the state-of-the-art on *fairing* and *shape-preserving* methods for curves and surfaces.
2. Compare the methods, classified in (1), from a theoretical point of view and evaluate the best of them by means of benchmark tests prepared by the industrial partners of FAIRSHAPE.
3. Based on the results obtained in (2), develop new automatic algorithms for constructing fair or shape-preserving interpolating curves/surfaces.
4. Formulate and solve practical problems, in which the designer needs to perform fairing and shape-preservation, simultaneously.
5. Establish criteria for shape-quality control and derive computationally efficient implementations of them. Check the so-obtained criteria in real CAD/CAM environment prepared by the industrial partners of FAIRSHAPE.

Besides *the Exchange and Training of Young Researchers*, which is the principal activity of the network, the Steering Committee of FAIRSHAPE is also organizing *Internal Workshops*, for monitoring the materialization of the afore-mentioned strategic and technical goals. So far, we have organized two internal workshops, the first in Athens (March '95) and the second in Berlin (August '95). Furthermore, and in accordance with the Work Programme adopted by the European Commission, the activities of FAIRSHAPE should also include the organization of

1. An **Advanced Course** and
2. A *Benchmarking Workshop*.

The purpose of the Lambrecht Advanced Course was two-fold. First, provide an, as complete as possible, description of the current status of the basic research done in the areas of *fairing* and *shape-preserving interpolation*. Secondly, report and submit to the criticism of a broader audience (46 participants) the so-far obtained results of the bench-marking activities of FAIRSHAPE in the areas of *fairing, shape-preserving interpolation* and *constrained approximation*. We believe that both purposes have been fully materialized and hope that this is reflected in the diversity and quality of the papers contained in the present volume. On this occasion, I would like to express, also on the behalf of my colleagues in FAIRSHAPE, my thanks to the members of the Differential-Geometry-and-Kinematics-Group of the Technical University of Darmstadt, and especially Professor J. Hoschek, for their contribution in organizing the whole event at Lambrecht and editing the articles of the collection in hand.

Let me conclude this preamble by making a reference to the internal structure of our network. FAIRSHAPE comprises partners from both academia and industry. More specifically, the partners of FAIRSHAPE can be classified into three major groups:

1. *The academic group, which is involved in basic research in the area of CAGD.* This group consists of:
   (a) Differential Geometry and Kinematics Group, Department of Mathematics, Technische Hochschule Darmstadt (DE), *senior scientist: Prof. J. Hoschek*,
   (b) Numerical Analysis Section, Department of Energetics, University of FLorence (IT), *senior scientists: Prof. F. Fontanella and Prof. P. Costantini*,
   (c) Department of Applied Mathematics, University of Zaragoza (SP), *senior scientists: Prof. M. Gasca and Prof. J. Carnicer*,
   (d) Department of Applied Mathematical Sciences, School of Mathematics, University of Leeds (UK), *senior scientists: Prof. M. Bloor and Prof. M. Wilson*, and
   (e) Department of Mathematical Sciences, The University, Dundee, *senior scientist: Prof. T.N.T. Goodman*.

2. *The academic group, which is involved in applied research in the areas of CAGD and CAD/CAM.* This group consists of:
   (a) Division of Ship Design, Department of Naval Architecture & Marine Engineering, Technische Universität Berlin (DE), *senior scientist: Prof. H. Nowacki*, and
   (b) Ship-Design Laboratory, Department of Naval Department of Naval Architecture & Marine Engineering, National Technical University of Athens (GR), *senior scientist: Assoc. Prof. P.D. Kaklis*.

3. *The industrial group*, consisting of:
   (a) The SYRKO-CAD/CAM Development Group, Mercedes-Benz AG, *senior scientist: Dr. E. Kaufmann*, and
   (b) KCS (Kockums Computer Systems) UK, Ltd., *senior scientists: D. Catley and A. Ives-Smith*.

P.D. Kaklis,
Project-coordinator.

Contents

Preamble 3

Contents 5

I: Fairing and Shape Preserving of Curves 7

Horst Nowacki, Justus Heimann, Elefterios Melissaratos, Sven-Holm Zimmermann:
   Experiences in Curve Fairing 9

J. M. Carnicer, M. S. Floater:
   Co-convexity preserving Curve Interpolation 17

Tim Goodman:
   Shape Preserving Interpolation by Planar Curves 29

Tim Goodman, Boon-Hua Ong:
   Shape Preserving Interpolation by Curves in Three Dimensions 39

Andrew Ives-Smith:
   A coparative study of two curve fairing methods in Tribon Initial Design . . . 47

II: Fairing Curves and Surfaces 57

Jan Hadenfeld:
   Fairing of B–Spline Curves and Surfaces 59

Marc Daniel:
   Declarative Modeling of fair shapes: An additional approach to curves and
   surfaces computations 77

III: Shape Preserving of Curves and Surfaces 86

Paolo Costantini:
   Shape-preserving interpolation with variable degree polynomial splines 87

IV: Fairing of Surfaces 115

Malcolm I.G. Bloor, Michael J. Wilson:
   Functional Aspects of Fairness 117

Roger Andersson:
   Surface design based on brightness intensity or isophotes–theory and practice 131

Alain Massabo:
   Fair surface blending, an overview of industrial problems 145

Hartmut Prautzsch:
   Multivariate Splines with Convex B-Patch Control Nets are Convex 153

V: Shape Preserving of Surfaces 160

Malcolm I.G. Bloor, Michael J. Wilson:
    Parametrizing Wing Surfaces using Partial Differential Equations 161

J. M. Carnicer, M. S. Floater:
    Algorithms for convexity preserving interpolation of scattered data 175

Paolo Costantini:
    Abstract schemes for functional shape–preserving interpolation 185

B. Mulansky, J. W. Schmidt, M. Walther:
    Tensor Product Spline Interpolation subject to Piecewise Bilinear Lower and Upper Bounds 201

Bert Jüttler:
    Construction of Surfaces by Shape Preserving Approximation of Contour Data 217

Ulrich Dietz:
    B–Spline Approximation with Energy Constraints 229

Günther Greiner:
    Curvature approximation with application to surface modeling 241

Ron Pfeifle, Hans-Peter Seidel:
    Scattered Data Approximation with Triangular B-Splines 253

VI: Benchmarks 264

Kaklis D. Panagiotis and Gabrielides C. Nikolaos:
    Benchmarking in the Area of Planar Shape-Preserving Interpolation 265

Steffen Wahl:
    Benchmark Process in the Area of Shape-Constrained Approximation 283

Fairing and Shape Preserving of Curves

Experiences in Curve Fairing

Horst Nowacki, Justus Heimann, Elefterios Melissaratos, Sven-Holm Zimmermann

Technical University of Berlin, Division of Ship Design
Salzufer 17-19, D–10587 Berlin, Germany
E–mail: nowacki@cadlab.tu-berlin.de

Abstract: The task of fairing a curve interpolating a given point data set and potentially given end conditions by minimizing an explicit fairness measure is discussed from the viewpoint of the resulting curve quality. Results are compared for different choices of a fairness criteria applied to a variety of data sets. The improvements achievable by going from integer to rational cubic B–spline curves are examined in particular. Fairness quality can be raised both by lessening the constraints and by increasing the freedoms in curve representation.

1 Introduction

One of the objectives in the FAIRSHAPE project (HCM-Network CHRX-CT94-0522) is the fairing of curves with given interpolation constraints. The main interest is in comparing curve shapes when different fairness criteria and varying curve representations are applied. The objective is to improve the fairness quality of curves as measured by some explicit fairness criterion.

The special case of cubic spline interpolation of a given point set is a well-known classical problem of curve fitting. It is well understood that this interpolation problem is uniquely defined except for two missing conditions, e.g., end conditions [1]. The missing conditions can be replaced by optimality requirements, e.g., by minimizing a fairness measure [2]. This has been done in practice, however the cubic spline interpolation problem is highly constrained by the point data set and the quality of curve thus is usually not very sensitive to the fairing process.

A contrasting situation exists when a sufficient surplus of freedoms is introduced into the curve representation, e.g. by raising the polynomial degrees, increasing the number of segments etc., so that the problem statement becomes more strongly underdeterminate. In this case the shape of the curve becomes more responsive to the fairing objectives. Meier [3] and Meier, Nowacki [4] have drawn the attention to this option and have documented the results of shape improvements by a curve fairing process with interpolation.

The current investigation resumes the basic premise of this earlier work. The working hypothesis is that an improved fairness quality can be achieved by introducing additional degrees of freedom in the curve representation. This will be illustrated by several examples.

In principle the curve representation may be extended to offer additional flexibility by:
- Raising the polynomial degrees [3],
- Increasing the number of curve segments,
- Changing the parametrization of the curve knot vector, especially from uniform to non-uniform,
- Moving from integer to rational representations, thus bringing in free weighting factors, e.g., by going from NUBS to NURBS curves.

Although several of these options were explored in our current work, we will mainly report results from our comparisons between NUBS and NURBS. They are indicative of the potential of shape improvement in such fairing processes.

A second main interest was in the influence of the choice of fairness measure upon curve shape, a main theme of the FAIRSHAPE project. A variety of fairness criteria was applied to several data sets and the effects on fairness measures and curve shape were recorded.

The results reported here are part of an ongoing activity whose goal it is to find a proper balance between data fidelity and shape quality in fairing process.

2   Fairness Criteria

In the current context the fairness of a curve is considered to be a global property of the curve that can be measured and evaluated by some suitable integral criterion taken over the curve range. The choice of the criterion is somewhat subjective and should depend on the purpose of the fairing process. In fact, in some applications the criterion may also be a functional, non-geometric integral property of the shape, like the viscous drag of a foil.

Whatever the choice of fairness criterion may be, it is intended to reflect the design objective and does result in a single measure of merit, a "fairness number", as a measure of shape quality.

In the current paper we have considered the following fairness criteria:

$$E_1 = \int_0^1 |\dot{\overline{Q}}(u)|^2 du = \text{first order parametric criterion}$$

$$E_2 = \int_0^1 |\ddot{\overline{Q}}(u)|^2 du = \text{second order parametric criterion}$$

$$E_3 = \int_0^1 |\dddot{\overline{Q}}(u)|^2 du = \text{third order parametric criterion}$$

$$E_k = \int_0^1 k^2(s) ds = \text{curvature criterion}$$

$$E_{k'} = \int_0^1 |dk(s)/ds|^2 ds = \text{change of curvature criterion}$$

where

$\overline{Q}(u)$ = parametric curve
$\dot{\overline{Q}}(u), \ddot{\overline{Q}}(u), \dddot{\overline{Q}}(u)$ = First, second, third parametric derivatives of $\overline{Q}(u)$
$k(s)$ = curvature as a function of arc length s

3 Fairing Process

A fairing process has a data set and an assumed curve representation as inputs and is defined by a fairness criterion and constraints (Fig. 1).

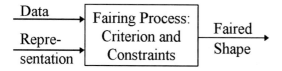

Fig. 1 Definition of Fairing Process

We use the following **test data sets** in this report:

- Meier's test data points [3]
- A tanker afterbody section
- A historical frigate midship section

All test examples are for planar curves.

The **curve representations** are:

- Non-uniform integer and rational B–splines (NUBS and NURBS)
- with a polynomial degree of 3 (potentially higher)
- and with variable segmentation

The **fairness criteria** are those recorded in Section 2.

Constraints consist of free or fixed end conditions (with or without end tangent vectors).

A workbench software environment was created at TU Berlin in order to do systematic variations on this set of assumptions in the statement of the fairing process. Existing software components were integrated into the workbench (Fig. 2). The point data are initially interpolated to generate a NUBS curve representation. This can readily be rewritten as a NURBS curve with unit weights, which forms the input to the optimization process. The result of the fairing process is a NURBS curve with optimized weights and its associated fairness criteria. The results are displayed by the visualization component of the workbench.

The idea of fairing NURBS curves by optimizing on the weights was first proposed by Hohenberger and Reuding [5], who for approximation of point data used a global change of curvature criterion as their fairness measure. In the current context the approach is extended to interpolation and any desired fairness criterion.

The optimization process of the weights is here implemented in a simplified way to save computer time. The weights w_i associated with the control points $\overline{P}_i$ of the defining polygon of the NURBS curve are constrained to be varied according to a cubic Bézier curve relationship

$$w_i = k_0 B_0\left(\frac{i}{n}\right) + k_1 B_1\left(\frac{i}{n}\right) + k_2 B_2\left(\frac{i}{n}\right) + k_3 B_3\left(\frac{i}{n}\right)$$

where
$B_j(u)$ = cubic Bernstein basis functions, j = 0, ... ,3
and
k_0, k_1, k_2, k_3 = free optimization parameters in curve range

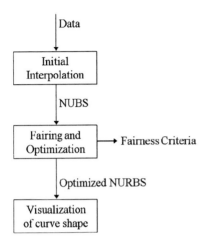

Fig. 2 Fairing Process in Workbench

This limits the number of free parameters to four per curve even when the curve has many control points.

The numerical optimization algorithm used here to minimize the fairness functional is an unconstrained minimization technique, viz., the well known simplex method of downhill search by Nelder and Mead [6]. It converges rather reliably in reasonable computing time.

4 Results

Figs. 3 to 7 show several results of the NURBS fairing tests. The curves are represented by piecewise cubic NURBS with non-uniform chord length parametrization and tangential end conditions throughout these examples. The initial curve with unit weights (NURBS curve) was optimized by varying the weights. The fairness criterion was systematically varied including E_1, E_2, E_3 and E_k. The optimization succeeded in all cases to reduce the fairness criterion compared to the initial status although in some cases only slightly.

Fig. 3 shows an interpolating cubic NURBS curve for Meier's data set with given end tangent vectors and chord length parametrization. The curvature criterion E_k is minimized. The faired curve differs in the middle segments and exhibits reduced curvature (flatness). This is typical for curvature or strain energy minimization. The criterion was reduced by about 13 percent.

Fig. 4 presents the HSVA tanker cross section with 8 non-equidistant data points and end tangent constraints. The E_1 criterion is applied. It reflects the intention to reduce arc length and does result in greater tautness of the curve, although numerically it changes only slightly. Note that the (non-optimal) E_2 decreases while E_k rises. Hence the curvature and the second parametric derivative criteria do not necessarily show equivalent trends.

Figs. 5 to 6 give corresponding results for the E_2, E_3 and E_k criteria. In all cases there are certain improvements in the optimized measure of merit whereas the others may or may not be favourably affected. Table 1 summarizes these results. It shows what penalty must be paid in other criteria when one of the criteria is optimized.

Calculated Criterion	Optimized Criterion			
	E_1	E_2	E_3	E_k
E_1	204.0	205.0	205.0	206.1
E_2	11413.	4139.	6216.	53557.
E_3	$1.25 \cdot 10^8$	$0.92 \cdot 10^8$	$0.75 \cdot 10^8$	$14.6 \cdot 10^8$
E_k	0.980	0.945	0.929	0.924

Tab. 1 Fairness Criteria for HSVA Tanker Section

Although the choice of criterion remains subjective, there are many applications where the E_3 or E_k fairness measure may be favoured because they produce the most gentle changes in curvature. The E_k criterion is still under investigation.

Finally in Fig. 7 the midship section curve of a historical frigate is faired by the curvature criterion which is improved by only 2 percent.

Yet the spike visualization of radii of curvature demonstrates a certain curvature integral reduction in this example.

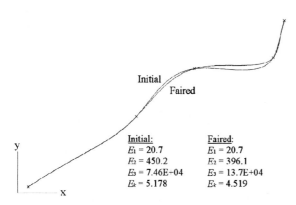

Initial:
E_1 = 20.7
E_2 = 450.2
E_3 = 7.46E+04
E_k = 5.178

Faired:
E_1 = 20.7
E_2 = 396.1
E_3 = 13.7E+04
E_k = 4.519

Fig. 3 Initial and faired curve for Meier's test data set. The E_k criterion is minimized.

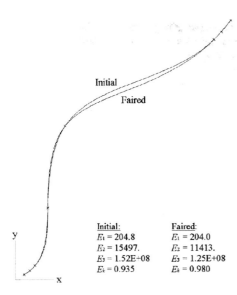

Fig. 4 Initial and faired curve for HSVA Tanker (aft section) data set. The E_1 criterion is minimized.

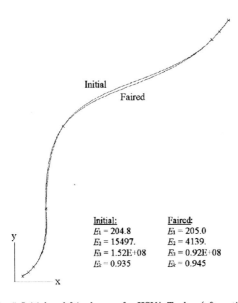

Fig. 5 Initial and faired curve for HSVA Tanker (aft section) data set. The E_2 criterion is minimized.

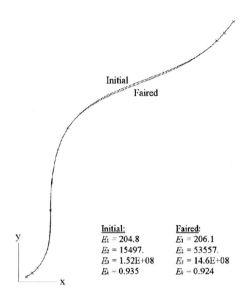

Fig. 6 Initial and faired curve for HSVA Tanker (aft section) data set. The E_k criterion is minimized.

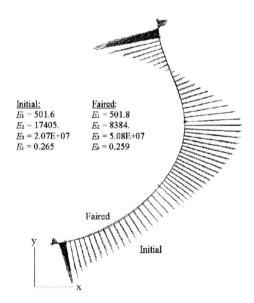

Fig. 7 Initial and faired curve for the historical frigate (midship section) data set. The E_k criterion is minimized.

5 Conclusions

A fairing process can be applied to improve the quality of a shape in terms of some chosen fairness measure. The fairness criterion is essential for the shape character of the resulting curve and should be selected so as to reflect the design objectives.

The curve to be faired is as responsive to the fairing process as the number of free parameters permits. Many curve interpolation problems are tightly constrained by the given data sets and end conditions and thus do not leave much room to fairing improvements.

For demonstrable success in curve fairness quality it is essential to avoid too many constraints and to a allow sufficient surplus freedoms in curve representation. This can be achieved in several ways, i.e., by modifying polynomial degrees, number of curve segments, curve parametrization, and by rational forms with free weights.

In the present article examples were tested where the additional freedoms consisted of the weights of the defining polygon vertices in going from integer to rational non-uniform B–spline curves. This provides only a limited measure of shape control. Despite that curve shapes were responsive to this fairing process. The influence of choosing different fairness criteria could also be demonstrated again.

Further work is needed to draw more systematic conclusions on the influence of various fairness criteria and different methods for providing extra freedom upon shape character and shape fidelity.

Acknowledgement

The help provided by Stefan Harries in setting up the workbench environment is gratefully acknowledged.

References

[1] Holladay, J. C.: *Smoothest Curve Approximation.* Math. Tables Aids Computations **11** (1957).

[2] Theilheimer, F., Starkweather: *The Fairing of Ship Lines on a High-Speed Computer.* David Taylor Model Basin Report No. 1474, Bethesda, MD (1961).

[3] Meier, H.: *Differential Geometry Based Design and Analytic Representation of Curvature Continuous Ship Surfaces and Similar Free Form Surfaces.* Diss., TU Berlin, VDI-Verlag, Reihe Rechnerunterstützte Verfahren, **20** (5), in German, Düsseldorf (1987).

[4] Meier, H., Nowacki, H.: *Interpolating Curves with Gradual Changes in Curvature.* Computer Aided Geometric Design **4** (1987), XX–YY.

[5] Hohenberger, W., Reuding, T.: *Smooth rational B–Spline curves using the weights in an optimization procedure.* Computer Aided Geometric Design **12** (1995) 8, 837–848.

[6] Nelder, J.A., Mead, R.: *A simplex method for function minimization.* Computer Journal **7** (1965), 308–313.

Co-convexity preserving Curve Interpolation

J. M. Carnicer, M. S. Floater

Departamento de Matemática Aplicada,
Universidad de Zaragoza,
Spain

Abstract: An algorithm is described for constructing C^1 co-convexity preserving interpolants to arbitrary sequences of points and possible tangents. The interpolants can be chosen to consist of cubic and parabolic pieces. The main tool is the numerical computation of C^2 convexity preserving interpolants to convex data arising from the minimization of certain functionals.

Key words: Convexity, Shape preserving interpolation.

1 Introduction

This paper is devoted to the problem of interpolating a sequence of points and optionally tangents, with smooth functions which preserve co-convexity, that is to say, functions which, roughly speaking, are convex in regions where the data is convex, and concave in regions where the data is concave.

The basic approach is to split the domain into regions of three types, according to whether the data is locally convex, concave, or neither. For regions of convex (resp. concave) data, a convexity (resp. concavity) preserving interpolant is found numerically by solving a nonlinear system. The theory behind the existence of the interpolant was developed in [1] and [4]. Cubic Hermite polynomials are employed when the data is neither locally convex nor locally concave.

Section 2 describes convexity preserving interpolation, Section 3 co-convexity interpolation and Sections 4 deals with numerical results.

2 A convexity preserving C^2 interpolant

Given a sequence of points in $[a, b]$,

$$a = x_1 = x_2 < x_3 \leq \ldots \leq x_{n-2} < x_{n-1} = x_n = b \quad \text{with} \quad x_i < x_{i+2}, \tag{2.1}$$

and a sequence of values

$$y_1, \ldots, y_n \in \mathbb{R}, \tag{2.2}$$

we may pose a Hermite interpolation problem

$$L_i(f) = y_i, \tag{2.3}$$

where L_i are the corresponding Hermite functionals, defined by

$$L_i(f) := \begin{cases} f'(x_i), & \text{if } x_{i-1} = x_i, \\ f(x_i), & \text{otherwise.} \end{cases}$$

Let us define

$$m_i := \begin{cases} y_{i+1}, & \text{if } x_i = x_{i+1}, \\ \dfrac{y_{i+1} - y_i}{x_{i+1} - x_i}, & \text{otherwise,} \end{cases} \quad i = 1, \ldots, n-1, \quad (2.4.\text{a})$$

$$d_i := m_{i+1} - m_i, \quad i = 1, \ldots, n-2. \quad (2.4.\text{b})$$

The interpolation data (x_i, y_i), $i = 1, \ldots, n$ is said to be strictly convex, if there exists a strictly convex function $f : [a, b] \to \mathbb{R}$ such that $L_i(f) = y_i$, $i = 1, \ldots, n$.

In Theorem 2.2 of [4], it was shown that the data is strictly convex if and only if $m_1 < \cdots < m_{n-1}$ or equivalently if $d_i > 0$. Furthermore, it was shown that any set of strictly convex data has a C^∞ interpolant whose second derivative is bounded below by a positive number, that is $f'' \geq \varepsilon$, for some $\varepsilon > 0$.

Now, let us pose the following convexity preserving interpolation problem. Given strictly convex data (2.1), (2.2) find a function

$$f \in W^2_\infty[a, b] = \left\{ f \in C^1[a, b] \,|\, f'' \in L^\infty[a, b] \right\},$$

such that (2.3) holds and f is *strongly convex*, that is $f'' \geq \varepsilon$ (a.e.) for some $\varepsilon > 0$.

In order to select only one solution of this problem among all possible strongly convex interpolants, we will minimize the functional

$$J_p(f) = \int_a^b F_p(f''(t))dt, \quad f \in W^2_\infty[a, b], \quad (2.5)$$

where $F_p : \mathbb{R}_+ \to \mathbb{R}$, $p \in [0, 1]$ is defined by

$$F_p(t) := \begin{cases} t \log t & \text{if } p = 1 \\ -t^p/p & \text{if } 0 < p < 1 \\ -\log t & \text{if } p = 0. \end{cases} \quad (2.6)$$

Let us reduce our interpolation problem to a problem in $L^\infty[a, b]$. If we call $u = f''$, the interpolation conditions take the form

$$\int_a^b u(x) N_i(x) dx = d_i \quad i = 1, \ldots, n-2, \quad (2.7)$$

where d_i is given by (2.4) and $N_i(x) := N(x | x_i, x_{i+1}, x_{i+2})$ denotes the linear B-spline or *hat function* given by

$$N(x | x_i, x_{i+1}, x_{i+2}) := \begin{cases} (x - x_i)/(x_{i+1} - x_i), & \text{if } x \in [x_i, x_{i+1}], \\ (x_{i+2} - x)/(x_{i+2} - x_{i+1}), & \text{if } x \in [x_i, x_{i+1}], \\ 0, & \text{otherwise.} \end{cases} \quad (2.8)$$

Let S be the set of all functions $u \in L^\infty[a, b]$ such that (2.7) holds and are strongly positive, that is $u \geq \varepsilon$ (a.e.) for some $\varepsilon > 0$. The set S is nonempty provided that the $d_i > 0$.

Now, from the discussion in section 5 of [1], we deduce that there exists a unique function $u_0 \in S$ such that
$$\int_a^b F_p(u_0(x))dx \le \int_a^b F_p(u(x))dx, \quad \forall u \in S.$$
Furthermore $u_0(x) = \Phi'_p(\sum_{i=1}^{n-2} \alpha_i N_i(x))$, where $\Phi_p : D_p \to \mathbb{R}$,
$$D_p := \begin{cases} \mathbb{R}, & \text{if } p = 1, \\ (0, \infty), & \text{otherwise}, \end{cases}$$
$$\Phi_p(t) := \begin{cases} \exp(t-1) & \text{if } p = 1, \\ (1-p)/(pt^{p/(1-p)}) & \text{if } 0 < p < 1, \\ -\log t & \text{if } p = 0 \end{cases} \tag{2.9}$$
and $(\alpha_1, \ldots, \alpha_{n-2}) \in D_p^{n-2}$ is the unique solution of the nonlinear system
$$\int_a^b \Phi'_p\left(\sum_{i=1}^{n-2} \alpha_i N_i(x)\right) N_j(x) dx = d_j, \quad j = 1, \ldots, n-2. \tag{2.10}$$

The previous known facts of [1] can be restated as follows:

Theorem 2.1. *Let $p \in [0, 1]$ and let (2.1), (2.2) be a set of strictly convex data. Then there exists a unique strongly convex interpolant $f \in W^2_\infty[a, b]$ minimizing the functional J_p given by (2.6). Furthermore, f is characterized by the property*
$$f''(x) = \Phi'_p\left(\sum_{i=1}^{n-2} \alpha_i N_i(x)\right),$$
where $\alpha_1, \ldots, \alpha_{n-2}$ ($\alpha_i > 0$ for all i, if $p \ne 1$) is the unique solution of the nonlinear system (2.10)

Let us remark that the interpolant mentioned in the previous theorem is a piecewise function with knots at the interpolation points. The function is globally C^1 and has smoothness C^2 on each interval not containing a double knot $x_i = x_{i+1}$. Specifically, the function in each subinterval $[x_i, x_{i+1}]$, $x_i < x_{i+1}$ has the form
$$a_i x + b_i + H_s(c_i x + d_i),$$
where $s = (1 - 2p)/(1 - p) \in [-\infty, 1]$ and
$$H_s(x) := \begin{cases} \exp x, & \text{if } s = -\infty, \\ \frac{x^s}{s(s-1)}, & \text{if } s \in (-\infty, 0) \cup (0, 1), \\ -\log x, & \text{if } s = 0, \\ x \log x, & \text{if } s = 1, \end{cases} \tag{2.11}$$
and, in the case of $p \ne 1$, $c_i x + d_i$ has to be strictly positive on $[x_i, x_{i+1}]$.

The case $p = 1$, $s = -\infty$ has been dealt with in detail in the last section of [1]. In the next section we shall describe how to make the computations in practice for the remaining cases $p \in [0, 1)$.

Two cases are of special interest because they produce functions whose graph is formed by pieces which are conic sections. The case $p = 1/3$, $s = 1/2$ produces in each piece an arc of a parabola and the case $p = 2/3$, $s = -1$ will produce in each piece an arc of a hyperbola. For related information the reader is referred to [2],[3].

Let us remark that the C^2 (functional) parabolic solution for $p = 1/3$ is a particular case of the geometric construction given by Schaback in [6]. In this case the solution maximizes the functional $\int_a^b (f''(x))^{1/3} dx$. This can be stated equivalently in the following geometric terms: if $\kappa(s)$, $s \in [0, l]$ is the curvature of the corresponding curve in terms of the arc length s and l is the length of the curve:

$$\int_0^l (\kappa(s))^{1/3} ds = \int_a^b (f''(x))^{1/3} dx.$$

Therefore the solution can be identified as the convex interpolating curve maximizing $\int_0^l (\kappa(s))^{1/3} ds$. We conjecture that this characterization can be extended to the nonfunctional setting and be applied to the geometric (nonfunctional) construction of [6].

Another case which can be analyzed from another point of view is the case in which each knot is a double knot. In this case the interpolation problem can be separated into several Hermite interpolation problems at two points. These kinds of problems have been described in more detail in [4] but from the point of view of differential equations. In fact, the differential equations

$$y^{(IV)} = (2 - p) \frac{(y''')^2}{y''}, \quad y'' > 0,$$

analyzed in section 5 of [4] are the Euler-Lagrange equations corresponding to the problem of minimizing J_p.

3 The numerical solution of the convexity preserving interpolation problem

The main goal of this section is to describe how to compute the convex interpolant of Theorem 2.1 in the case $p \in [0, 1)$. This interpolant is of the form

$$f''(x) = \left(\sum_{i=1}^{n-2} \alpha_i N_i(x) \right)^{s-2}, \quad s = (1 - 2p)/(1 - p),$$

$\alpha_i \in (0, \infty)$, $i = 1, \ldots, n - 2$. There is a unique choice of $\alpha \in \mathbb{R}_+^{n-2}$ such that f agrees with the interpolation data: take $\alpha_i > 0$ as the unique solution of the system of equations

$$\int_a^b \left(\sum_{i=1}^{n-2} \alpha_i N_i(x) \right)^{s-2} N_j(x) dx = d_j \qquad (3.1)$$

where d_j is given by (2.4).

Let us define

$$I_s(\alpha) := \int_a^b H_s' \left(\sum_{i=1}^{n-2} \alpha_i N_i(x) \right) dx, \quad \alpha \in \mathbb{R}_+^{n-2}$$

where H_s is the function given in (2.11). Taking into account that $H_s''(x) = x^{s-2}$, we obtain

$$\frac{\partial I_s}{\partial \alpha_j}(\alpha) = \int_a^b \left(\sum_{i=1}^{n-2} \alpha_i N_i(x)\right)^{s-2} N_j(x)dx, \quad j = 1,\ldots,n-2.$$

On the other hand

$$I_s(\alpha) = \sum_{i=2}^{n-2} \int_{x_r}^{x_{i+1}} H_s'\left(\sum_{i=1}^{n-2} \alpha_i N_i(x)\right) dx = \sum_{i=2}^{n-2}(x_{i+1} - x_i)[\alpha_{i-1}, \alpha_i]H_s, \quad (3.2)$$

and differentiating formula (3.2) we obtain an explicit expression for the system (3.1), namely

$(x_3 - x_2) \cdot [\alpha_1, \alpha_1, \alpha_2]H_s = d_1,$
$(x_{j+1} - x_j) \cdot [\alpha_j, \alpha_j, \alpha_{j-1}]H_s + (x_{j+2} - x_{j+1}) \cdot [\alpha_j, \alpha_j, \alpha_{j+1}]H_s = d_j, \quad j = 2,\ldots,n-3,$
$(x_{n-1} - x_{n-2}) \cdot [\alpha_{n-2}, \alpha_{n-2}, \alpha_{n-3}]H_s = d_{n-2}.$

So, the computation of the system of equations can be reduced to the computation of $[x,x,y]H_s$.

Let us consider first the general case $s \neq 0$, $s \neq 1$. Then we have

$$[x,x,y]H_s = \frac{1}{s(s-1)} \cdot \frac{(y^s - x^s)/(y-x) - sx^{s-1}}{y-x} = x^{s-2}\left[\frac{(y/x)^s - 1 - s(y/x - 1)}{s(s-1)(y/x - 1)^2}\right].$$

Now taking $h = y/x - 1$ and defining $\varphi_s : (0, \infty) \to \mathbb{R}$ by

$$\varphi_s(h) := \frac{(1+h)^s - 1 - sh}{s(s-1)h^2},$$

we arrive at the formula

$$[x,x,y]H_s = x^{s-2}\varphi_s(y/x - 1).$$

Note that for small values of h we can approximate $\varphi_s(h)$ by the partial sums of the expansion

$$\varphi_s(h) = \frac{1}{2} + \frac{s-2}{6}h + \frac{(s-2)(s-3)}{24}h^2 + \cdots + \frac{(s-2)\cdots(s-1-k)}{(k+1)!}h^k + \cdots$$

Now, let us consider the special cases. In the case $s = 1$ we have

$$[x,x,y]H_1 = \frac{1}{x}\varphi_1(y/x - 1), \quad \varphi_1(h) = \frac{1}{h^2}[(1+h)\log(1+h) - h],$$

and in the case $s = 0$ we may write

$$[x,x,y]H_0 = \frac{1}{x^2}\varphi_0(y/x - 1), \quad \varphi_0(h) = \frac{1}{h^2}[\log(1+h) - h].$$

For small values of h, the functions $\varphi_0(h)$ and $\varphi_1(h)$ can be evaluated using the corresponding Taylor's expansions.

The system (3.1) can be written in terms of the function φ_s which can be evaluated following the general and special rules shown above:

$$\alpha_1^{s-2}(x_3 - x_2)\varphi_s\left((\alpha_2 - \alpha_1)/\alpha_1\right) = d_1,$$

$$\alpha_j^{s-2}\left[(x_{j+1} - x_j)\varphi_s\left((\alpha_{j-1} - \alpha_j)/\alpha_j\right) + (x_{j+2} - x_{j+1})\varphi_s\left((\alpha_{j+1} - \alpha_j)/\alpha_j\right)\right] = d_j,$$
$$(j = 2, \ldots, n-3)$$

$$\alpha_{n-2}^{s-2}(x_{n-1} - x_{n-2})\varphi_s\left((\alpha_{n-3} - \alpha_{n-2})/\alpha_{n-2}\right) = d_{n-2}.$$

In order to compute the solution α we propose using Newton's method. Another way of computing the solution α is to use the following iterative formula

$$\alpha_1^{(k+1)} = \left\{(x_3 - x_2)\varphi_s(1 - \alpha_2^{(k)}/\alpha_1^{(k)})/d_1\right\}^{\frac{1}{2-s}}$$

$$\alpha_j^{(k+1)} = \left\{\left[(x_{j+1} - x_j)\varphi_s(1 - \alpha_{j-1}^{(k+1)}/\alpha_j^{(k)}) + (x_{j+2} - x_{j+1})\varphi_s(1 - \alpha_{j+1}^{(k)}/\alpha_j^{(k)})\right]/d_j\right\}^{\frac{1}{2-s}}$$
$$(j = 2, \ldots, n-3)$$

$$\alpha_{n-2}^{(k+1)} = \left\{(x_{n-1} - x_{n-2})\varphi_s(1 - \alpha_{n-3}^{(k+1)}/\alpha_{n-2}^{(k)})/d_{n-2}\right\}^{\frac{1}{2-s}}.$$

Numerical experience has confirmed that these kinds of iterations converge to the unique solution of the system. However a proof of convergence has not been found.

Finally, once the α_i's are computed, we can use the following formula to evaluate f:

$$f(x) = \frac{x_{i+1} - x}{x_{i+1} - x_i}y_i + \frac{x - x_i}{x_{i+1} - x_i}y_{i+1} - [\alpha_i, \bar{\alpha}(x), \alpha_{i+1}]H_s \cdot (x - x_i)(x_{i+1} - x),$$

where

$$\bar{\alpha}(x) := \frac{x_{i+1} - x}{x_{i+1} - x_i}\alpha_{i-1} + \frac{x - x_i}{x_{i+1} - x_i}\alpha_i, \quad x \in [x_i, x_{i+1}].$$

4 A Co-convexity preserving C^1 interpolant

In this section we show how the method discussed in the previous sections can be applied to the task of interpolating arbitrary data with co-convexity preserving functions.

Suppose as before we are given the data (x_i, y_i) for $i = 1, \ldots, n$ as in (2.1) and (2.2).

Definition 4.1. Let $f \in C^1([a,b], \mathbb{R})$ be such that $L_i(f) = y_i$, $i = 1, \ldots, n$. We will say that f is a co-convexity preserving interpolant if, for all $i = 2, \ldots, n-2$ for which $x_i < x_{i+1}$,

(i) $f|_{[x_i, x_{i+1}]}$ is strictly convex whenever the four data $(x_{i-1}, y_{i-1}), \ldots, (x_{i+2}, y_{i+2})$ are strictly convex, and

(ii) $f|_{[x_i, x_{i+1}]}$ is strictly concave whenever the data $(x_{i-1}, y_{i-1}), \ldots, (x_{i+2}, y_{i+2})$ are strictly concave.

We remark that the four data $(x_{i-1}, y_{i-1}), \ldots, (x_{i+2}, y_{i+2})$ are: strictly convex iff $d_{i-1} > 0$ and $d_i > 0$; strictly concave iff $d_{i-1} < 0$ and $d_i < 0$; and otherwise $d_{i-1}d_i \le 0$.

In the following we will describe an algorithm for constructing a co-convexity preserving interpolant which applies the method presented in Sections 2 and 3 to regions of convex and concave data respectively. Thus the important part of the algorithm is the decomposition of the data into convex, concave, and inflective subsets. This algorithm will be easiest to explain by defining it recursively.

First of all, if there is any internal derivative condition among the interpolation conditions, we can simplify the problem.

Lemma 4.2. *Suppose there is some $i \in \{3, \ldots, n-3\}$ such that $x_i = x_{i+1}$. Let $f_1 : [a, x_i] \to \mathbb{R}$ be a C^1 co-convexity preserving interpolant to the data (x_k, y_k), $k = 1, \ldots, i+1$ and let $f_2 : [x_i, b] \to \mathbb{R}$ be a C^1 co-convexity preserving interpolant to the data (x_k, y_k), $k = i, \ldots, n$. Then*

$$f(x) = \begin{cases} f_1(x), & a \leq x \leq x_i, \\ f_2(x), & x_i \leq x \leq b, \end{cases}$$

is a C^1 co-convexity preserving interpolant to the data (x_k, y_k), $k = 1, \ldots, n$.

Proof: Indeed, since both f_1 and f_2 match the data (x_i, y_i) and (x_{i+1}, y_{i+1}) we find that $f_1(x_i) = f_2(x_i) = y_i$ and $f_1'(x_i) = f_2'(x_i) = y_{i+1}'$ and so f is C^1. Furthermore, if the data $(x_{j-1}, y_{j-1}), \ldots, (x_{j+2}, y_{j+2})$ are strictly convex (resp. concave), for any $2 \leq j < j+1 \leq i$, then $f_1|_{[x_{j-1}, x_j]}$ will be strictly convex (resp. concave) and consequently $f|_{[x_{j-1}, x_j]}$ will be strictly convex (resp. concave). For intervals $[x_j, x_{j+1}]$ where $i+1 \leq j < j+1 \leq n-1$ a similar argument using f_2 establishes that f is co-convexity preserving. ∎

Due to Lemma 4.2 we only need to consider data satisfying $x_j < x_{j+1}$ for $j = 2, \ldots, n-2$. In this case the data is Lagrange except for the two end points $a = x_1 = x_2$ and $b = x_{n-1} = x_n$. Now the basic idea is to identify intervals $[x_i, x_{i+1}]$ for which $d_{i-1} d_i \leq 0$. In such an interval, we wish to make two local derivative estimates μ_i and μ_{i+1} respectively, ensuring that we preserve, where necessary, strict convexity or strict concavity of the data to the left and right of the interval. However it is possible that there are two consecutive intervals in which $d_{j-1} d_j \leq 0$, $j = i-1, i$ and then estimates for the slope at x_i will be taken into consideration two times as endpoint of both intervals $[x_{i-1}, x_i]$, $[x_i, x_{i+1}]$. So, in order to avoid ambiguity we identify those $i \in \{3, \ldots, n-2\}$ for which either

$$d_{i-2} d_{i-1} \leq 0 \quad \text{or} \quad d_{i-1} d_i \leq 0, \tag{4.1}$$

(or both) and make a single derivative estimate m_i. If (4.1) holds we will say that x_i is an *inflection point*.

We have implemented two different methods for estimating a derivative at an inflection point x_i. In Method I, we always take the derivative from the Catmull-Rom spline [5]:

$$\mu_i = \frac{y_{i+1} - y_{i-1}}{x_{i+1} - x_{i-1}}. \tag{4.2}$$

In Method II, if $d_{i-2} d_{i-1} > 0$, we set the Bessel end condition from the left:

$$\mu_i = g'(x_i), \quad g \in \pi_2, \quad L_j(g) = y_j, \quad j = i-2, i-1, i, \tag{4.3}$$

and otherwise the Catmull-Rom derivative as in (4.2). Similarly if $d_i d_{i+1} > 0$, we set the Bessel end condition [5] from the right:

$$\mu_i = g'(x_i), \quad g \in \pi_2, \quad L_j(g) = y_j, \quad j = i, i+1, i+2, \tag{4.4}$$

and otherwise the Catmull-Rom derivative as in (4.2).

A little thought shows that both in Methods I and II, the possible strict convexity or strict concavity of any four consecutive data $(x_{j-1}, y_{j-1}), \ldots, (x_{j+2}, y_{j+2})$, with $j = i - 1$ or $j = i$ is preserved if we augment the data set with the data item (x_i, μ_i) between (x_i, y_i) and (x_{i+1}, y_{i+1}). We have thus established the following.

Lemma 4.3. *Suppose that $x_j < x_{j+1}$ for $j = 2, \ldots, n-2$. Let $x_{i_1} \ldots, x_{i_m}$, with $2 < i_1 < i_2 < \cdots < i_m < n-1$, be the inflection points. Further, define $i_0 = 1$, $\mu_1 = y_2$, and $i_{m+1} = n-1$, $\mu_{n-1} = y_n$. Then for each $k = 0, 1, \ldots, m$, let $f_k|_{[x_{i_k}, x_{i_{k+1}}]}$ be a C^1 co-convexity preserving interpolant to the data (x_{i_k}, y_{i_k}), (x_{i_k}, μ_{i_k}), (x_{i_k+1}, y_{i_k+1}), $(x_{i_k+2}, y_{i_k+2}), \ldots, (x_{i_{k+1}}, y_{i_{k+1}})$, $(x_{i_{k+1}}, \mu_{i_{k+1}})$. Then the piecewise function*

$$f(x) = f_k(x), \quad x \in [x_{i_k}, x_{i_{k+1}}], \quad k = 0, 1, \ldots, m,$$

is a C^1 co-convexity preserving interpolant to the data (x_j, y_j), $j = 1, \ldots, n$.

By using cubic polynomials in intervals whose endpoints are both inflection points and elsewhere either convex or concave interpolants, we now establish the following theorem.

Theorem 4.4. *Let (x_i, y_i), $i = 1, \ldots, n$ be a set of data as defined in (2.1) and (2.2). Then there is an interpolant $f : [a, b] \to \mathbb{R}$ which is C^1 and co-convexity preserving.*

Proof: Due to Lemmas 4.2 and 4.3 it is sufficient to construct a co-convexity preserving interpolant f when the data is Lagrange except for the end points ($x_j < x_{j+1}$ for $j = 2, \ldots, n-2$) and either

(i) $d_j > 0$ for all $j = 1, \ldots, n-2$, or
(ii) $d_j < 0$ for all $j = 1, \ldots, n-2$, or
(iii) $n = 4$, $x_1 = x_2 < x_3 = x_4$ and $d_1 d_2 \leq 0$.

In case (i) we let $p \in [0, 1]$ and f be the strict convexity preserving interpolant of Theorem 2.1. In case (ii) we let $p \in [0, 1]$ and g be the strict convexity preserving interpolant of Theorem 2.1 to the inverted data $(x_i, -y_i)$, and then let $f = -g$. Finally, in case (iii) we set $f \in \pi_3$ be the unique cubic Hermite polynomial interpolating the data $(x_1, y_1), \ldots, (x_4, y_4)$. In all three cases, f is clearly a co-convexity preserving interpolant to the data. ∎

When implementing an algorithm, the interpolant f can be generated in two sequential steps.

Algorithm 4.5

1. For $i = 3, \ldots, n-2$, if $x_{i-1} < x_i < x_{i+1}$, check whether either $d_{i-2} d_{i-1} \leq 0$ or $d_{i-1} d_i \leq 0$. If so, compute a derivative estimate μ_i at x_i from (4.2), (4.3), or (4.4). Then augment the interpolation data by inserting each new data item (x_i, μ_i) between (x_i, y_i) and (x_{i+1}, y_{i+1}).

2. Working with the augmented data set, identify each subsequence of the form $x_i = x_{i+1} < x_{i+2} < \cdots < x_j = x_{j+1}$, $1 \leq i < j \leq n$. ¿From the data $(x_i, y_i), \ldots, (x_{j+1}, y_{j+1})$ let $f|_{[x_i, x_j]}$ be

   2a. a convexity preserving interpolant if $d_k > 0$ for all $k = i, \ldots, j-1$, or
   2b. a concavity preserving interpolant if $d_k < 0$ for all $k = i, \ldots, j-1$, or
   2c. otherwise, the cubic Hermite polynomial interpolant, since in this case $j = i + 2$.

Co-convexity preserving Curve Interpolation

Finally let us present some examples of data sets to test our algorithm. For the convex and concave pieces we used the method of convexity preserving interpolation described in the previous sections with the choice $p = 1/3$. Therefore the function is piecewise parabolic or cubic. The interpolation points at which both the value and slope were given and also those in which the algorithm augmented the data are marked with a + sign. In a neighbourhood of those the interpolant is C^1. The remaining interpolation points are marked by a $\Diamond$. In a neighbourhood of those, the interpolant is C^2. If the derivatives at the endpoints of the interval x_1 or x_n are not given we estimate them using the Bessel end criterion.

In Examples 1 to 4, derivatives were estimated by both Methods I (solid lines) and II (dashed lines).

Examples.

1. Akima data:

x	0.0	1.0	2.0	5.0	6.0	8.0	9.0	11.0	12.0	14.0	15.0
y	10.0	10.0	10.0	10.0	10.0	10.0	10.5	15.0	50.0	60.0	85.0

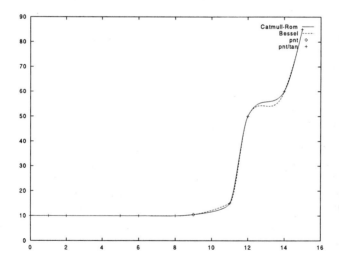

Fig. 1 Akima data

2. Titanium heat data:

x	595.	635.	695.	795.	855.	875.	885.	895.	905.	915.	935.	985.	1035.	1075.
y	.644	.652	.644	.694	.907	1.336	1.881	2.169	2.075	1.598	.916	.607	.603	.608

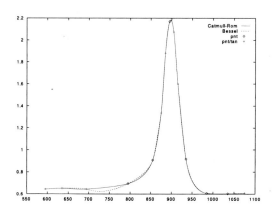

Fig. 2 Titanium heat data

3. Pruess data:

x	22.0	22.5	22.6	22.7	22.8	22.9	23.0	23.1	23.2	23.3	23.4	23.5	24.0
y	523.	543.	550.	557.	565.	575.	590.	620.	860.	915.	944.	958.	986.

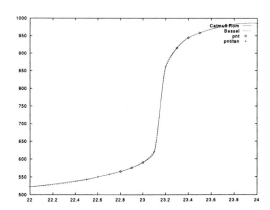

Fig. 3 Pruess data

4. Radiochemical data:

x	7.99	8.09	8.19	8.70	9.20	10.00	12.00	15.00	20.00
y	0.	2.76429×10^{-5}	4.37498×10^{-2}	.169183	.469428	.94374	.998636	.999919	.999994

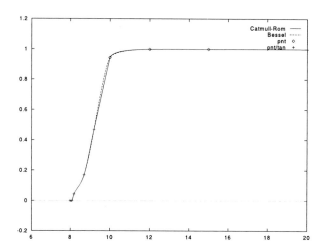

Fig. 4 Radiochemical data

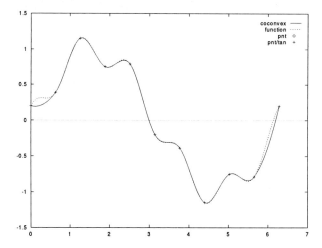

Fig. 5 The function $\sin x + \cos(5x)/5$ sampled at equidistant points

As a final example we sample the function $\sin x + \cos(5x)/5$ at 11 equidistant points of the interval $[0, 2\pi]$. It turns out that every internal point is an inflection point and so both

Methods I and II generate the same co-convexity preserving interpolant. We see from the figure that this interpolant closely mirrors the function except near the endpoints (due to using the Bessel end conditions at these points rather than using the first derivative of the function).

Acknowledgement: The authors wish to thank Prof. Constantini for providing the data used in the numercial examples.

Research partially supported by the EU Project FAIRSHAPE, CHRX-CT94-0522. The first author was also partially supported by DGICYT PB93-0310 Research Grant.

References

[1] Carnicer, J. M.: *On best constrained interpolation.* Numerical Algorithms **1** (1991), 155–176.

[2] Carnicer, J. M.: *Dual Bézier curves and convexity preserving interpolation.* Computer Aided Geometric Design **9** (1992), 435–445.

[3] Carnicer, J. M.: *Rational interpolation with a single variable pole.* Numerical Algorithms **3** (1992), 125–132.

[4] Carnicer, J. M., Dahmen, W.: *Characterization of local strict convexity preserving interpolation methods by C^1 functions.* Journal of Approximation Theory **77** (1994), 2–30.

[5] Farin, G: *Curves and Surfaces for Computer Aided Geometric Design.* Academic Press, Boston, 1988.

[6] Schaback, R: *Interpolation with piecewise quadratic visually C^2 Bézier polynomials.* Computer Aided Geometric Design **6** (1989), 219–233.

Authors

Prof. Dr. J.M. Carnicer
Dept. de Matematica Aplicada
Universidad de Zaragoza
Edificio de Matematicas
Ciudad Universitaria, Zaragoza
Spain
E–mail: carnicer@posta.unizar.es

Dr. Mikael Floater
Dept. de Matematica Aplicada
Universidad de Zaragoza
Ciudad Universitaria, Zaragoza
Spain
E–mail: floater@posta.unizar.es

Shape Preserving Interpolation by Planar Curves

Tim Goodman

University of Dundee, Dept. of Mathematics & Computer Science
DUNDEE DD1 4HN, Scotland
E–mail: tgoodman@mcs.dundee.ac.uk

Abstract: Requirements for shape preserving interpolation by planar curves are discussed. How these may be satisfied is then illustrated by four different schemes.

1   Introduction

We shall consider the following problem. Given data points in R^2,

$$I_i = (x_i, y_i), \quad i = 0, \ldots, N, \quad N \geq 3,$$

construct a continuous curve passing through them in order. Such a curve can be represented by a continuous function $r = (x, y) : [a, b] \to R^2$, in which case our interpolation condition is that there exist parameter values

$$a = t_0 < t_1 < \ldots < t_N = b,$$

for which

$$r(t_i) = I_i, \quad i = 0, \ldots, N.$$

It may be that we require a closed curve. In this case it is convenient to extend the data cyclically to $I_i, i \in Z$, defined by $I_{i+N} = I_i, i \in Z$. Similarly the function r can be extended to R by defining $r(t) = r(t + b - a), t \in R$.

By an **interpolation scheme** we shall mean a method of assigning to each set of such data a curve satisfying the above properties. In any practical problem it will be required that the curve satisfy additional properties, and in this paper we focus on the requirement that the 'shape' of the curve is 'similar' to the 'shape' of the data. Here the terms 'shape' and 'similar' are vague and the first part of Section 2 is devoted to discussing ways in which they can be made more precise.

By the 'shape of the data' is usually meant the shape of the polygonal arc $I_0 \ldots I_N$, i.e. the curve given by

$$r(t) = \frac{I_{i-1}(t_i - t) + I_i(t - t_{i-1})}{t_i - t_{i-1}}, \quad t_{i-1} \leq t \leq t_i, \quad i = 1, \ldots, N.$$

So if preserving shape were the only requirement, this polygonal arc would give the optimal solution. However it is usual to require further properties such as additional smoothness of the curve and its 'fairness', i.e. that the curve is visually pleasing. In Section 2 we shall discuss smoothness and fairness as well as stability and invariance under various transformations. We shall finish the section by discussing various ways of representing the interpolating curve.

Section 3 is devoted to a brief description of four schemes for shape preserving interpolation by planar curves. The literature on this topic is so large and varied that we have not attempted to give a survey of methods. We do not claim that the four presented here are the best, nor do we attempt to rank their merits. We have simply chosen them to illustrate some of the different approaches to solving the problems involved. Since most situations for practical application prefer explicit representation by piecewise polynomials or rationals, we have chosen schemes of these types. Moreover they all give in general curves with continuous tangent directions and curvatures, though only two of them give parametric representations which are twice continuously differentiable. As the Fairshape project is preparing a literature guide, we have not attempted to give here a comprehensive list of references.

2 Properties

We first consider various ways in which the shape of the interplating curve can be similar to the shape of the data.

2.1 Monotonicity

For $v \in R^2$, we say the data is increasing in direction v if

$$v.I_0 < v.I_1 < \ldots < v.I_N.$$

Thus the data is increasing in direction $(1,0)$ if $x_0 < x_1 < \ldots x_N$. Similarly the curve is said to be increasing in direction v if it has a representation $r : [a,b] \to R^2$ for which $v.r$ is a strictly increasing function. For simplicity we may throughout the paper confuse a curve with a representation r.

We say an interpolation scheme is **monotonicity preserving** in direction v if data which are increasing in direction v are assigned a curve which is increasing in direction v. If a curve is increasing in direction v, it seems natural to take $v.r(t)$ as the parameter. For example, when $v = (1,0)$, we may express the curve in the form $r(x) = (x, f(x))$, $x_0 \leq x \leq x_N$. Such a representation is often called 'functional' and many such schemes are considered in the literature. This is discussed, for example, in the papers in this Proceedings: "Abstract schemes and polynomial splines of adaptive degree for functional shape-preserving interpolation" by P. Costantini and "Convexity-preserving curve interpolation" by J. Carnicer and M. Floater. So in this paper we shall not consider in detail this class of schemes.

We say an interpolation scheme is **comonotone** in direction v if for all i, $0 \leq i \leq N-1$,

$$v.I_i < \text{(respectively } \leq, > \text{ or } \geq \text{)} \ v.I_{i+1}$$

implies that for $t_i \leq s < t \leq t_{i+1}$,

$$v.r(s) < \text{(respectively } \leq, > \text{ or } \geq \text{)} \ v.r(t).$$

Shape Preserving Interpolation by Planar Curves

Clearly a scheme which is comonotone in direction v is also monotonicity preserving in direction v. However in many circumstances comonotonicity may be stronger than what is desired. For example it implies that if $v.I_{i-1} < v.I_i > v.I_{i+1}$, then $v.r$ has a local maximum at $t = t_i$, while if $v.I_i = v.I_{i+1}$, then the curve r is linear on $[t_i, t_{i+1}]$.

A weaker condition is that of **local monotonicity preserving** (l. m. p.) in direction v. Essentially this says that $r.v$ has the minimum number of local extrema consistent with the data. To make this precise we need the following conditions.

For all i, $1 \leq i \leq N-2$, if $(v.I_{i-1}, v.I_i, v.I_{i+1}, v.I_{i+2})$ is increasing (resp. strictly increasing, decreasing or strictly decreasing) then $v.r$ is increasing (resp. strictly increasing, decreasing or strictly decreasing) on $[t_i, t_{i+1}]$.

For open curves we also require the following conditions on the end segments.

$(v.I_0, v.I_1, v.I_2)$ (respectively $(v.I_{N-2}, v.I_{N-1}, v.I_N)$) increasing (resp. etc.)

implies that

$v.r$ is increasing (resp. etc.) on $[t_0, t_1]$ (resp. $[t_{N-1}, t_N]$).

Thus for $1 \leq i \leq N-1$, if $v.I_{i-1} = v.I_i = v.I_{i+1}$, then $v.r$ is constant on $[t_{i-1}, t_{i+1}]$. Otherwise we require that $v.r$ has at most one local exremum on $[t_{i-1}, t_{i+1}]$. Note that if $v.I_{i-1} \leq v.I_i = v.I_{i+1} \leq v.I_{i+2}$, then $v.r$ is constant on $[t_i, t_{i+1}]$.

Clearly for given v, a comonotone scheme is l. m. p. which in turn is monotonicity preserving.

2.2 Convexity

For u, v in R^2 we define $u \times v = u_1 v_2 - u_2 v_1$. We say a curve $r : [a, b] \to R^2$ is positively (resp. negatively) convex if for any $a \leq s_1 < s_2 < s_3 \leq b$,

$$(r(s_2) - r(s_1)) \times (r(s_3) - r(s_2)) \geq 0 \text{ (resp. } \leq 0).$$

We say r is locally positively (resp. negatively) convex at s in $[a, b]$ if r is positively (resp. negatively) convex on some neighbourhood of s.

If r is not locally convex at s (with either sign), we say r has an inflection at s.

We say a scheme is **local convexity preserving** (l. c. p.) if the following hold.

For any $0 \leq i \leq j-2 \leq N-2$, the polygonal arc $I_i \ldots I_j$ being positively (resp. negatively) convex (resp. locally convex) implies that r is positively (resp. negatively) convex (resp. locally convex) on $[t_{i+1}, t_{j-1}]$.

For open curves, when $i = 0$ we replace t_{i+1} by t_0, and when $j = N$ we replace t_{j-1} by t_N. The above implies that r is locally convex on $[t_0, t_1]$ and on $[t_{N-1}, t_N]$. For $1 \leq i \leq N-2$, we require that r has at most one inflection on $[t_i, t_{i+1}]$.

The above conditions are stronger than requiring that r has the minimum number of inflections consistent with the data. We also remark that the above conditions on local convexity do not in general imply the conditions on convexity, but they will do so if for $1 \leq i \leq N-1$, r has a tangent at t_i which is a positive combination of $I_i - I_{i-1}$ and $I_{i+1} - I_i$, and r has appropriate tangents at I_0 and I_N.

Note that the first condition with $i = j - 2$ implies that r cannot have an inflection at a data point. In particular this implies that if, for any $1 \leq i \leq N-1$, I_{i-1}, I_i and I_{i+1} are collinear, then r is linear on $[t_{i-1}, t_{i+1}]$. These conditions may seem unnecessarily restrictive and we could relax the first condition by requiring that $i \leq j - 3$. However the following conditions are still implied.

If, for any $1 \leq i \leq N - 2$, $I_{i-1}, \ldots, I_{i+2}$ are collinear, then r is linear on $[t_{i-1}, t_{i+2}]$. If, for any $2 \leq i \leq N - 2$, I_{i-1}, I_i, I_{i+1} are collinear and $I_{i-2} \ldots I_{i+2}$ is locally convex, then r is linear on $[t_{i-1}, t_{i+1}]$.

2.3 Monotone Curvature

Monotonicity and convexity are the main properties of shape considered in shape preserving interpolation. This does not preclude others and, as food for thought, we suggest here one other possibility. We shall assume that r is smooth enough for its curvature to be defined except at a finite number of points.

Suppose that for some i, $1 \leq i \leq N - 2$, $I_{i-1} \ldots I_{i+2}$ is locally convex and the magnitude of the curvature of the circle through I_{i-1}, I_i, I_{i+1} is $\leq$ (resp. $\geq$) that of the circle through I_i, I_{i+1}, I_{i+2}. Then we require that on $[t_i, t_{i+1}]$ the curve r is locally convex (with the same sign) and the magnitude of the curvature of r is increasing (resp. decreasing).

A motivation for this definition is mentioned briefly in Section 2.6. Note that if the points $I_{i-1}, \ldots, I_{i+2}$ lie in order on a circle, then the curve r must be a circular arc on $[t_i, t_{i+1}]$.

2.4 Invariance

We say a scheme is invariant under a transformation $T : R^2 \to R^2$ if whenever data (I_i) is assigned a curve r, then the data (TI_i) is assigned the curve Tr. We shall consider various choices of T. Whether invariance under T is desirable will depend on the problem to be considered.

(a) Shift
$$T(p) = p + a, \quad a \in R^2.$$

Invariance under T for all a in R^2 essentially means independence of the choice of an origin. It may not be appropriate if there is a natural choice of origin (e.g. when radially symmetric) or one or both variables represent a quantity which is naturally positive, e.g. length.

(b) Scaling
$$T(x,y) = (\lambda x, \mu y), \quad \lambda, \mu \in R.$$

Invariance under positive scaling is appropriate if the problem is independent of the units chosen. For example, if x and y are measured independently (e.g. time and distance), then it would be apprpriate for all $\lambda, \mu > 0$, while if they represent the same measurement (e.g. points in a plane), then it would be appropriate only for $\lambda, \mu > 0$. Choosing λ or μ equal to -1 will give preservation of certain symmetries.

(c) Rotation
$$T = \begin{bmatrix} \cos\theta & -\sin\theta \\ \sin\theta & \cos\theta \end{bmatrix}.$$

We say a scheme is rotation invariant if it is invariant under such T for all θ. In problems where data represent points in a plane it would seem appropriate to have rotation invariance as well as invariance under global scaling $T(x,y) = (\lambda x, \lambda y)$, $\lambda > 0$. Note that rotation invariance and rotation under scaling in one direction $T(x,y) = (\lambda x, y)$, all $\lambda > 0$, imply invariance under all linear transformations. If, in addition, we impose invariance under all shifts, then we have invariance under all affine transformations. Many schemes are affine invariant but this may not be appropriate in all situations as we now illustrate.

Suppose that a scheme is comonotone in a direction v and is also rotation invariant. It is then clearly comonotone in all directions. Take any i, $0 \leq i \leq N-1$, and choose u orthogonal to $I_i I_{i+1}$, so that $u.I_i = u.I_{i+1}$. Since the scheme is comonotone in direction u, the curve r must be linear on $[t_i, t_{i+1}]$. Thus the scheme must always give the polygonal arc $I_0 \ldots I_N$.

If a scheme is l. m. p. in some direction and rotation invariant, then it is also l. m. p. in all directions. If $I_i I_{i+1}$ is any 'inflection segment', i.e. $I_i \ldots I_{i+2}$ is not locally convex, then since the scheme is l. m. p. in a direction orthogonal to $I_i I_{i+1}$, the curve r is linear on $[t_i, t_{i+1}]$. Thus for data which are not locally convex, requiring both l. m. p. and rotation invariance seems overly restrictive. On the other hand, for locally convex data, it can be seen that l. c. p. implies l. m. p. in all directions. Thus for rotation invariant schemes, requiring l. m. p. seems either too restrictive or redundant.

We also note that monotone curvature is not affine invariant. Indeed it is not invariant under scaling in one direction, since such a transformation sends a circle into an ellipse.

To finish Section 2.4 we mention that it may be appropriate for a scheme to be invariant in the following sense. If data I_i, $i = 0, \ldots, N$, are assigned to a curve r on $[a,b]$, then the data I_{N-i}, $i = 0, \ldots, N$ are assigned to the curve $r(-t)$ on $[-b,-a]$.

2.5 Smoothness

We have already remarked that unless we require some extra smoothness for our interpolating curve, the problem of shape preserving interpolation is trivial. It is usual to call a continuous curve G^n for $n \geq 1$ if it has some representation by a C^n function $r : [a,b] \to R^2$ with $r'(t) \neq 0$, $a \leq t \leq b$. The curve is G^1 if and only if it has a continuous unit tangent vector and is G^2 if and only if it has, in addition, continuous curvature.

In practice an interpolation scheme will give the curve in terms of a particular function r. Even though the curve is G^n, the particular representation may not be C^n. If r is G^1 but not C^1, then a linear change of parameter on each interval $[t_i, t_{i+1}]$, $0 \leq i \leq N-1$, will give a new representation which is C^1, so that in practice there is no distinction between C^1 and G^1. For $n \geq 2$ a suitable change of parameter is not so simple. As the form of the representation r may be important (see Section 2.7) it is customary to distinguish between C^n and G^n for $n \geq 2$. We note that even if a curve is represented by a C^n function r, the curve may not be G^n if r' vanishes at some point.

Smoothness of the interpolating curve may not be compatible with shape preserving interpolation. For suppose we have non-collinear data points $I_{i-2} \ldots I_{i+2}$ such that I_{i-2}, I_{i-1}, I_i and I_i, I_{i+1}, I_{i+2} are collinear. Then any locally convex curve which interpolates the data must be linear on $[t_{i-2}, t_i]$ and on $[t_i, t_{i+2}]$ and so cannot be G^1 in a neighbourhood of t_i.

2.6 Fairness

It is usual to call a curve 'fair' if it is visually pleasing. This is obviously subjective but there are various objective criteria that have been suggested to be relevant, such as the magnitude of the curvature, the rate of change of the curvature and the monotonicity of the curvature. (It was the last of these which motivated Section 2.3.) Although it is common practice to approximate a given curve by a 'fairer' curve (see the articles on 'fairing' in this proceedings), it has not been usual to incorporate fairness criteria into shape preserving interpolation schemes. However there have been attempts to obtain fairness indirectly by the following two methods. The C^2 cubic spline interpolant (see Section 3.1) can be gained by minimising the integral of the square of the second derivative and, since the second derivative can be considered as a linearised curvature, it is generally regarded that this gives 'fair' curves. However this interpolation scheme is certainly not shape preserving in general. Hence one approach to shape preserving interpolation is to modify the C^2 cubic spline interpolant so that it satisfies the shape preserving criteria. Examples of such schemes will be given in Sections 3.1 and 3.3. An alternative approach is to allow more degrees of freedom in the choice of curve than is necessary for interpolation and to try and use the extra freedom to gain a fair curve, either through the algorithm itself or by allowing the user to modify the curve with certain 'shape parameters'. An example of such a scheme is given in Section 3.4.

2.7 Stability

A scheme is stable if the map from the data $(I_0, \ldots, I_N)$ to the interpolating curve r is continuous, i.e. a small change in the data gives a small change in the curve. Another form of stability is that a change in a single data point I_i does not affect the curve r on its full domain $[a, b]$ but only on a relatively small neighbourhood of t_i. A scheme satisfying this property we shall call **local**.

2.8 Representation

To finish this section we consider different ways in which schemes represent the interpolating curve.

(a) Subdivision

In subdivision an initial polygonal arc is recursively modified into polygonal arcs with larger numbers of smaller edges, which can provide an efficient and stable way of constructing a curve which is arbitrarily close to a smooth curve. While curves with explicit representation can sometimes be constructed in this manner, some subdivision algorithms produce curves with no simple explicit representation. As an example, see [3].

(b) Implicit

A planar curve can be defined as the zero set of a bivariate function $f : R^2 \to R$. For example algebraic curves, i.e. zero sets of polynomials, are used in [10].

(c) Explicit

The vast majority of schemes produce curves with an explicit representation $r : [a, b] \to R^2$. While other functions r are sometimes used, e.g. exponentials in [14], [11], by far the most popular are piecewise polynomials or piecewise rationals. These can be represented as linear combinations of B-splines, which can be evaluated by subdivision. For polynomial or rational segments this gives the ubiquitous Bézier representation. While piecewise polynomials have the advantage of simplicity, the use of piecewise rationals gives extra degrees of freedom which can be traded for lower degree, while they also allow projective invariance and reproduction of conics.

3 Some methods

To finish the paper we illustrate some different approaches to shape preserving interpolation by planar curves by describing four different schemes. These must be brief and we refer to the quoted papers for full details. The schemes are all affine invariant and use piecewise polynomials or piecewise rationals. They are C^2 or G^2 and l. c. p., except when dealing with consecutive collinear data for which we have explained in Section 2.5 these conditions may be incompatible. For brevity we shall not discuss end conditions, i.e. the modifications which must be made to the scheme near I_0 and I_N.

3.1 Polynomial C^2

This scheme is due to Kaklis and Sapidis [9], based on an earlier scheme for the functional case by Kaklis and Pandelis [8]. As mentioned in Section 2.6, it modifies the C^2 cubic spline interpolant. As with other such schemes one has initially a choice of the values (t_i) of the parameter at which to interpolate, e.g. one popular choice is the 'chord-length parameterisation' in which $t_{i+1} - t_i = |I_{i+1} - I_i|$, $0 \leq i \leq N-1$. As the choice of values does not otherwise affect the algorithm, we shall not mention it further here or in Section 3.3. Also in common with other such schemes, the idea is to modify those segments of the interpolating curve which have unnecessary inflections. If such an inflection occurs in the interval $[t_i, t_{i+1}]$, i.e. between I_i and I_{i+1}, then the curve is 'pulled' closer to the line segment $I_i I_{i+1}$ until the unwanted inflection is removed.

The technique used in this scheme to modify the segment is to increase its degree as follows. For $m \geq 3$, the curve between I_i and I_{i+1} is of the form

$$r(t) = a(1-s)^m + b(1-s) + cs + ds^m, \quad s = \frac{t - t_i}{t_{i+1} - t_i}, \quad t_i \leq t \leq t_{i+1},$$

where a, b, c, d are in R^2. This is equivalent to requiring the Bézier representation

$$r(t) = \sum_{i=0}^{m} b_i \binom{m}{i} s^i (1-s)^{m-i}, \quad s = \frac{t - t_i}{t_{i+1} - t_i}, \quad t_i \leq t \leq t_{i+1},$$

to satisfy

$$b_i = \frac{(m-1-i)b_1 + (i-1)b_{m-1}}{m-2}, \quad i = 1, \ldots, m-1.$$

By the interpolation conditions, $b_0 = I_i$ and $b_m = I_{i+1}$, and the curve lies in the convex hull of b_0, b_1, b_{m-1} and b_m. For any choices of m on the different segments $[t_i, t_{i+1}]$, $i = 0, \ldots, N-1$, there is a unique C^2 curve of this form interpolating the data (with suitable end conditions) which is determined by solving a global, tridiagonal, strictly diagonally dominant linear system. As $m \to \infty$ on a particular segment $[t_i, t_{i+1}]$, the curve will converge to the line segment $I_i I_{i+1}$. Thus, provided there are no consecutive collinear data, choosing the degrees m large enough will ensure the scheme is l. c. p. For 'nearly collinear' data, this may require very high degrees and so in this case the degrees may be chosen so that the l. c. p. conditions are satisfied within a tolerance, and this allows C^2 continuity to be maintained in all cases.

A variant of the scheme is given in [12] in which high degrees are avoided by taking the tangent directions and curvatures from the above scheme and interpolating them by a piecewise quintic G^2 curve which conserves the l. c. p. properties.

3.2 Polynomial G^2

This scheme is due to Schaback [13]. We shall first describe the case of locally convex data $I_0, \ldots, I_N$, with no three consecutive collinear data points. Then we can write

$$(I_i - I_{i-1}) \times (I_{i+1} - I_i) = |I_i - I_{i-1}||I_{i+1} - I_i| \sin \gamma_i, \quad 1 \leq i \leq N - 1,$$

where either $0 < \gamma_i < \pi$, $1 \leq i \leq N - 1$, or $-\pi < \gamma_i < 0$, $1 \leq i \leq N - 1$. Without loss of generality we assume the former. We shall need the further conditions that $\gamma_i + \gamma_{i+1} < \pi$, $1 \leq i \leq N - 2$. In this case it can be shown that there is a G^2 curve interpolating the data which is represented by a quadratic function between consecutive data points. If the stronger conditions $\gamma_i + \gamma_{i+1} < \frac{\pi}{2}$, $1 \leq i \leq N - 2$, are imposed, then the curve is unique (with suitable end conditions). Since we do not claim C^2 continuity, we can assume that $t_i = i$, $0 \leq i \leq N$, and so the curve is given, for $0 \leq i \leq N - 1$, by

$$r(t) = I_i(1-s)^2 + b_i 2s(1-s) + I_{i+1}s^2, \quad s = t - i, \quad i \leq t \leq i + 1,$$

for b_i in R^2. In practice one does not find directly the point b_i but the angle α_i between $I_i I_{i+1}$ and $I_i b_i$, i.e. the tangent direction at I_i. The curve is found by solving a global, tridiagonal, **non-linear** system for $\alpha_1, \ldots, \alpha_{N-1}$.

Now in general the data can be divided into sections as above. Adjacent sections will be divided by either an inflection segment $I_j I_{j+1}$ where

$$[(I_j - I_{j-1}) \times (I_{j+1} - I_j)][(I_{j+1} - I_j) \times (I_{j+2} - I_{j+1})] < 0,$$

or by collinear data $I_j, \ldots, I_{j+n}$, where $n \geq 2$. On an inflection segment the curve is defined by a cubic polynomial which is determined uniquely by the data points I_j, I_{j+1} and the tangent directions and curvatures at these points, under certain restrictions on the tangent directions. The collinear data $I_j, \ldots, I_{j+n}$ are interpolated by a straight line and the adjacent segments $I_{j-1} I_j$ and $I_{j+n} I_{j+n+1}$ are each defined by a cubic polynomial which at I_j and I_{j+n} have the same tangent direction and curvature (i.e. 0) as the line $I_j \ldots I_{j+n}$. The remaining locally convex sections are then interpolated by the above method.

3.3 Rational C^2

This scheme, due to Clements [2], following work in [1], is similar to that in Section 3.1 in that it modifies the C^2 cubic spline interpolant on segments with unwanted inflections. In this case, the curve between I_i and I_{i+1} is of the form

$$r(t) = \frac{a(1-s)^3}{ws+1} + b(1-s) + cs + \frac{ds^3}{w(1-s)+1}, \quad s = \frac{t-t_i}{t_{i+1}-t_i}, \quad t_i \leq t \leq t_{i+1},$$

where $w \geq 0$, and a, b, c, d in R^2 depend on i. For any choices of w on the different segments $[t_i, t_{i+1}]$, $i = 0, \ldots, N - 1$, there is a unique C^2 curve of this form interpolating the data (with suitable end conditions). As in Section 3.1, this is determined by solving a global, tridiagonal, strictly diagonally dominant linear system. For $w = 0$ on all segments this gives the usual cubic spline. As $w \to \infty$ on a particular segment $[t_i, t_{i+1}]$, the curve will converge to the linear segment $I_i I_{i+1}$. It is assumed that there are no consecutive collinear data and so choosing the 'weights' w large enough will ensure that the scheme is l. c. p.

3.4 Rational G^2

Our final scheme was originated by Goodman [4] and implemented and extended by Goodman and Unsworth [6], [7]. Whereas the previously described schemes required solving a global system of equations with, in general, a unique solution, this scheme does not solve any equations but specifies the curve explicitly. The scheme is local in that the curve segment on $[t_i, t_{i+1}]$ depends in general only on the data $I_{i-2}, \ldots, I_{i+3}$. There are also some degrees of freedom which can be used both in the construction of the algorithm and for the user to interactively modify the curve, if so desired. In [5] this freedom is utilised in constructing l. c. p. interpolating curves which are constrained to lie within certain regions bounded by straight lines. As in the scheme in Section 3.2, the l. c. p. conditions are satisfied directly rather than by increasing certain parameters. As in that scheme, we can assume that $t_i = i$, $0 \leq i \leq N$.

The first step in the scheme is to assign the tangent directions T_i and the curvature magnitudes κ_i at the data points I_i. As a help to obtain a fair curve, the scheme will in general reproduce circular arcs (see the end of Section 2.3). Therefore κ_i is chosen to be the curvature of the circle passing through the points I_{i-1}, I_i, I_{i+1}. The tangent direction is of the form

$$T_i = a_i(I_i - I_{i-1}) + b_i(I_{i+1} - I_i)$$

for some $a_i \geq 0$, $b_i \geq 0$, (see the comments on this in Section 2.2). In order to satisfy the l. c. p. conditions, it is required that $b_i = 0$ when I_{i-2}, I_{i-1}, I_i are collinear, and that $a_i = 0$ when I_i, I_{i+1}, I_{i+2} are collinear. In order to reproduce circular arcs, we need that when $I_{i-2}, \ldots, I_{i+2}$ lie in order on a circle, then T_i is the tangent to that circle, i.e. that a_i and b_i are in the ratio given by $a_i = |I_{i+1} - I_i|^2$, $b_i = |I_i - I_{i-1}|^2$. Both these requirements are satisfied by choosing

$$a_i = \kappa_{i+1}|I_{i+1} - I_i|^2, \quad b_i = \kappa_{i-1}|I_i - I_{i-1}|^2.$$

It now remains to define the curve between I_i and I_{i+1}. This is of the form

$$r(t) = \frac{I_i \alpha (1-s)^3 + Bs(1-s)^2 + Cs^2(1-s) + I_{i+1} \beta s^3}{\alpha(1-s)^3 + s(1-s)^2 + s^2(1-s) + \beta s^3}, \quad s = t - i, \quad i \leq t \leq i+1,$$

where $\alpha, \beta > 0$ and B, C in R^2 depend on i. As this curve has tangent directions T_i, T_{i+1} at I_i, I_{i+1} respectively, we must have

$$B = I_i + xT_i, \quad C = I_{i+1} - yT_{i+1},$$

for some $x, y > 0$. The values of x and y can be chosen, with some degree of freedom, so that the l. c. p. conditions are satisfied and circular arcs are reproduced, but we do not give details here. Having chosen B and C, α and β are uniquely determined explicitly by requiring the curvature at I_i and I_{i+1} to be κ_i and κ_{i+1} respectively.

Although the above scheme, like the other three, gives an automatic algorithm requiring no knowledge from the user, it also allows the user to interactively modify the curve by varying, for example, the tangent directions T_i and curvatures κ_i.

References

[1] Clements, J. C.: *Convexity-Preserving Piecewise Rational Cubic Interpolation.* SIAM J. Numer. Anal. **27** (1990), 1016–1023.

[2] Clemants, J. C.: *A Convexity-Preserving C^2 Parametric Rational Cubic Interpolation.* Numer. Math. **63** (1992), 165–171.

[3] Dyn, N., Levin, D., Liu, D.: *Interpolatory Convexity-Preserving Subdivision Schemes for Curves and Surfaces.* Computer-aided Design **24** (1992), 211–216.

[4] Goodman, T. N. T.: *Shape Preserving Interpolation by Parametric Rational Cubic Splines.* In R. P. Agarwal, Y. M. Chow and S. J. Wilson (eds.): Numerical Mathematics Singapore 1988, International Series of Numerical Mathematics 86, Birkhauser Verlag, Basel (1988), 149 –158.

[5] Goodman, T. N. T., Ong, B. H., Unsworth, K.: *Constrained Interpolation Using Rational Cubic Splines.* In G. Farin (ed.): NURBS for Curve and Surface Design, SIAM (1991), 59–74.

[6] Goodman, T. N. T., Unsworth, K.: *An Algorithm for Generating Shape Preserving Parametric Interpolating Curves Using Rational Cubic Splines.* University of Dundee Report CS 89/01 (1989).

[7] Goodman, T. N. T., Unsworth, K.: *Interactive Shape Preserving Interpolation by Curvature Continuous Rational Cubic Splines.* CAT Report 239 (1990).

[8] Kaklis, P. D., Pandelis, D. G.: *Convexity-Preserving Polynomial Splines of Non-Uniform Degree.* IMA J. Numer. Anal. **10** (1990), 223–234.

[9] Kaklis, P. D., Sapidis, N. S.: *Convexity-Preserving Interpolatory Parametric Splines of Non-Uniform Degree.* Computer Aided Geometric Design **12** (1995), 1–26.

[10] Levin, D., Nadler, E.: *Convexity Preserving Interpolation by Algebraic Curves and Surfaces.* To appear Advances in Comp. Math.

[11] Pruess, S.: *Properties of Splines in Tension.* J. Approx. Theory **17** (1976), 86–96.

[12] Sapidis, N. S., Kaklis, P. D.: *A Hybrid Method for Shape-Preserving Interpolation with Curvature-Continuous Quintic Splines.* Comp. Suppl. **10** (1995).

[13] Schaback, R.: *On Global GC^2 Convexity Preserving Interpolation of Planar Curves by Piecewise Bézier Polynomials.* In T. Lyche and L. L. Schumaker (eds.): Mathematical Methods in CAGD, Academic Press (1989), 539–548.

[14] Schweikert, D. G.: *Interpolatory Tension Splines with Automatic Selection of Tension Factors.* J. Math. Phys. **45** (1966), 312–317.

Shape Preserving Interpolation by Curves in Three Dimensions

Tim Goodman[1], *Boon-Hua Ong*[2]

[1] University of Dundee
[2] Universiti Sains Malaysia

Abstract: Requirements for shape preserving interpolation by space curves are discussed and earlier work is mentioned. Then we describe two new local schemes using rational cubics. Both ensure continuity of the tangent directions and magnitudes of the curvatures, while the second gives curves with continuous osculating planes.

1 Introduction

As in the paper [2] in this volume, we shall consider the problem of interpolating points I_i, $i = 0, \ldots, N$, by a continuous curve which has a similar 'shape' to the 'shape' of the data. The only difference is that now the points are in R^3 and are not constrained to lie in a plane. Thus the curve is also in general non-planar, sometimes called a 'space curve', and can be represented by a continuous function $r : [a, b] \to R^3$. As in [2], there exist parameter values

$$a = t_0 < t_1 < \ldots < t_N = b$$

for which

$$r(t_i) = I_i, \quad i = 0, \ldots, N.$$

In contrast to the planar case, this problem has been little considered. Perhaps this is partly because it is not so clear what is meant by the 'shape' of a space curve. We shall suggest solutions to this problem in Section 2, motivated by the only prior scheme that we are aware of, which is due to Kaklis and Karavelas [6]. Then in Section 3 we shall briefly describe two methods for shape preserving interpolation by space curves which are both closely related to the method sketched in Section 3.4 of [2]. As there, these are local schemes using piecewise cubic rational curves. Both methods ensure continuity of the tangent direction and magnitude of the curvature. That in Section 3.2 also ensures continuity of the direction of the curvature, i.e. it preserves continuity of the osculating plane. That in Section 3.1 does not ensure this continuity but it is simpler and gives curves which seem to be as visually pleasing.

2 Properties

In [6], Kaklis and Karavelas give a scheme for shape preserving interpolation by space curves. The scheme uses the same method as that in [5], which is briefly described in Section 3.1 of [2]: modify the C^2 cubic spline interpolant by increasing the degree of certain segments until the shape preserving conditions are satisfied. We shall therefore not consider the scheme further but shall discuss the shape preserving conditions that they impose and suggest some extensions.

2.1 Torsion

We recall that the sign of the torsion of a curve r at a point t where $r^{(3)}(t)$ exists is equal to the sign of the triple scalar product $[r'(t), r''(t), r^{(3)}(t)]$, i.e. the determinant $|r'(t)\ r''(t)\ r^{(3)}(t)|$, where elements of R^3 are denoted by column vectors. More generally we can say that a curve has torsion > 0 (respectively < 0, $= 0$) at t if for any $s_1 < s_2 < s_3 < s_4$ in some neighbourhood of t,

$$[r(s_2) - r(s_1), r(s_3) - r(s_2), r(s_4) - r(s_3)] > 0 \text{ (resp. } < 0,\ = 0).$$

In [6] is considered the following **torsion condition** for an interplation scheme, where we write $L_i = I_{i+1} - I_i$, $i = 0, \ldots, N - 1$.

For $1 \leq i \leq N - 2$, if $[L_{i-1}, L_i, L_{i+1}] > 0$ (resp. < 0, $= 0$), then the torsion of r is > 0 (resp. < 0, $= 0$) on (t_i, t_{i+1}).

We note that the case '= 0' requires that when $I_{i-1}, \ldots, I_{i+2}$ are coplanar, then the curve r on $[t_i, t_{i+1}]$ lies in the same plane. In [6] they only require this to be satisfied up to a given tolerance.

We also remark that the torsion condition implies that if

$$[L_{i-2}, L_{i-1}, L_i]\ [L_{i-1}, L_i.L_{i+1}] < 0,$$

then the torsion of the curve r changes sign at I_i. The scheme in [6] actually implies that the torsion is zero at all data points I_i.

The above torsion condition corresponds to the local convexity part of the l. c. p. condition given in Section 2.2 of [2] for the planar case. However this l. c. p. condition also considers (global) convexity of the curve r. The corresponding definition to convexity for a space curve $r : [a, b] \to R^3$ would be that for any $a \leq s_1 < s_2 < s_3 < s_4 \leq b$,

$$[r(s_2) - r(s_1), r(s_3) - r(s_2), r(s_4) - r(s_3)] \geq 0 \text{ (resp. } \leq 0).$$

Such a curve is called an **ascending (resp. descending) coil** by Labenski and Piper [7]. Clearly an ascending (resp. descending) coil has non-negative (resp. non-positive) torsion. This definition suggests the following **coil condition**.

For any $0 \leq i \leq j-3 \leq N-3$, the polygonal arc $I_i \ldots I_j$ being an ascending (resp. descending) coil implies that the curve r is an ascending (resp. descending) coil on $[t_{i+1}, t_{j-1}]$.

In particular for $j = i + 3$ this gives the following, which is closely related to the torsion condition.

For $1 \leq i \leq N-2$, if $[L_{i-1}, L_i, L_{i+1}] \geq 0$ (resp. ≤ 0), then the curve r is an ascending (resp. descending) coil on $[t_i, t_{i+1}]$.

2.2 Convexity

The concept of convexity does not apply directly to a non-planar curve. However, as pointed out in [1], a curve may look convex when viewed from a particular direction, i.e. we may discuss the convexity of orthogonal projections of a curve. For any non-zero vector V in R^3, we define an orthogonal projection $P_V : R^3 \to R^2$ as follows. If u, v are vectors in R^3 such that u, v, V are mutually orthogonal with $[u, v, V] > 0$, then for any w in R^3,

$$P_V w = (w.u, w.v).$$

Note that for $x, y \in R^3$,

$$\begin{aligned}[] [x, y, V] &= [(x.u)u + (x.v)v, (y.u)u + (y.v)v, V] \\ &= [u, v, V]\{(x.u)(y.v) - (x.v)(y.u)\} \\ &= [u, v, V]|P_V x\ P_V y|, \end{aligned}$$

where we have represented elements of R^2 as column vectors. Thus $[x, y, V]$ and $|P_V x\ P_V y|$ have the same sign.

Now for $1 \leq i \leq N - 1$, we denote by N_i the vector product $L_{i-1} \times L_i$. Suppose that $N_i \neq 0$. Since $[L_{i-1}, L_i, N_i] > 0$, we have $|P_{N_i} L_{i-1}\ P_{N_i} L_i| > 0$ and so the planar polygonal arc $P_{N_i}(I_{i-1} I_i I_{i+1})$ is positively convex.

Now take $1 \leq i \leq N - 2$, and suppose that $N_i.N_{i+1} > 0$. Since $[L_i, L_{i+1} N_i] > 0$, we see that $|P_{N_i} L_i\ P_{N_i} L_{i+1}| > 0$ and so $P_{N_i}(I_i I_{i+1} I_{i+2})$ is positively convex and so the polygonal arc $P_{N_i}(I_{i-1} \ldots I_{i+2})$ is locally positively convex. Similarly $P_{N_{i+1}}(I_{i-1} \ldots I_{i+2})$ is locally positively convex.

In [6] the following **local convexity condition** is imposed.

If $N_i.N_{i+1} > 0$, then $P_{N_i} r$ and $P_{N_{i+1}} r$ are locally positively convex on $[t_i, t_{i+1}]$.

It is pointed out in [3] that if $P_{N_i} r$ and $P_{N_{i+1}} r$ are locally positively convex on $[t_i, t_{i+1}]$, then so is $P_V r$ whenever

$$V = \lambda N_i + \mu N_{i+1}, \quad \lambda \geq 0, \quad \mu \geq 0, \quad V \neq 0.$$

This follows because for any $t_i \leq s_1 < s_2 < s_3 \leq t_{i+1}$,

$$[r(s_2) - r(s_1), r(s_3) - r(s_2), V] =$$

$$\lambda[r(s_2) - r(s_1), r(s_3) - r(s_2), N_i] + \mu[r(s_2) - r(s_1), r(s_3) - r(s_2), N_{i+1}] \geq 0.$$

Similarly if $P_{N_i} r$ and $P_{N_{i+1}} r$ are positively convex on $[t_i, t_{i+1}]$, then so is $P_V r$ for

$$V = \lambda N_i + \mu N_{i+1}, \quad \lambda \geq 0, \quad \mu \geq 0, \quad V \neq 0.$$

The above local convexity conditions can be strengthened to the following **convexity conditions**.

If $N_i.N_{i+1} > 0$, then $P_{N_i} r$ and $P_{N_{i+1}} r$ are positively convex on $[t_i, t_{i+1}]$.

We now consider the case $N_i.N_{i+1} < 0$, which is not considered in [6]. As before, $P_{N_i}(I_{i-1} I_i I_{i+1})$ is positively convex but thus time $P_{N_i}(I_i I_{i+1} I_{i+2})$ is negatively convex. Following [3] we propose the following **inflection conditions**.

If $N_i.N_{i+1} < 0$, then $P_{N_i} r$ and $P_{N_{i+1}} r$ are locally positively convex at t_i and t_{i+1} respectively and for

$$V = \lambda N_i + \mu N_{i+1}, \quad \lambda\mu \leq 0, \quad V \neq 0,$$

$P_V r$ has only one inflection on (t_i, t_{i+1}).

The only case not so far considered is when $N_i.N_{i+1} = 0$. Stability considerations suggest the following **linearity conditions**.

When $N_i.N_{i+1} = 0$, then r is linear on $[t_i, t_{i+1}]$.

In particular this implies that if $N_i = 0$, i.e. I_{i-1}, I_i, I_{i+1} are collinear, then r is linear on $[t_{i-1}, t_{i+1}]$. This final case is imposed within a tolerance in [6].

All the above conditions in Section 2.2 have considered only vectors N_i, N_{i+1}. More generally we could consider a sequence of vectors $N_i, \ldots, N_j$, for some $1 \leq i \leq j - 1 \leq N - 2$. For example we could require that if the polygonal arc $P_V(I_{i-1} \ldots I_{j+1})$ is locally positively convex for $V = N_i, \ldots, N_j$, then $P_V r$ is locally positively convex for $V = N_i, \ldots, N_j$. Of course this would also imply that $P_V r$ is locally convex for $V = \lambda_i N_i + \ldots + \lambda_j N_j$, for any $\lambda_i \geq 0, \ldots, \lambda_j \geq 0$.

To finish this section we consider again the case of coplanar data in the case when the interpolating curve has G^2 continuity. We call a continuous curve G^2 if it has a representation by a C^2 function $r : [a, b] \to R^3$ with $r'(t) \neq 0$, $a \leq t \leq b$. This is equivalent to the curve having a continuous unit tangent vector and continuous curvature. Continuity of the direction of the curvature will imply continuity of the osculating plane, when this is defined.

Suppose that the interpolating curve r is G^2 at t_i with a well-defined osculating plane. Then we can assume that the parameterisation is chosen so that r is C^2 at t_i and $r'(t_i)$ and $r''(t_i)$ are linearly independent. We shall assume that $I_{i-2}, \ldots, I_{i+1}$ lie in a plane P. Then we have seen that the torsion condition implies that r lies in P on $[t_{i-1}, t_i]$ and hence $r(t_i)$, $r'(t_i)$, $r''(t_i)$ lie in P. We shall now show that if on $[t_i, t_{i+1}]$ r is a coil and $P_{N_i} r$ is convex, then r must also lie in P on $[t_i, t_{i+1}]$.

We shall suppose that r does not lie in P on $[t_i, t_{i+1}]$ and reach a contradiction. Take any s, $t_i < s < t_{i+1}$, such that $r(s)$ does not lie in P. Now

$$\lim_{t \to 0^+} t^{-3}[r(t_i+t)-r(t_i), r(t_i+2t)-r(t_i+t), r(s)-r(t_i+2t)] = [r'(t_i), r''(t_i), r(s)-r(t_i)] \neq 0.$$

Also

$$\lim_{t \to 0^+} t^{-1}[r(t_i+t)-r(t_i), r(s)-r(t_i+t), r(t_{i+1})-r(s)] = [r'(t), r(s)-r(t_i), r(t_{i+1})-r(t_i)].$$

If r is a coil on $[t_i, t_{i+1}]$, then for $t_i < t_i + 2t < s < t_{i+1}$, $[r(t_i+t) - r(t_i), r(t_i+2t) - r(t_i+t), r(s) - r(t_i+2t)]$ and $[r(t_i+t) - r(t_i), r(s) - r(t_i+t), r(t_{i+1}) - r(s)]$ must have the same sign and so $[r'(t_i), r''(t_i), r(s) - r(t_i)]$ and $[r'(t_i), r(s) - r(t_i), r(t_{i+1}) - r(t_i)]$ must have the same sign,

However if $P_{N_i} r$ is convex, $[r'(t_i), r''(t_i), r(s) - r(t_i)]$ and $[r'(t_i), r(t_{i+1}) - r(t_i), r(s) - r(t_i)]$ must have strictly the same sign, which is a contradiction.

Similarly if r is G^2 at t_{i-1}, and on $[t_{i-2}, t_{i-1}]$ r is a coil and $P_{N_{i-1}} r$ is convex, then r must lie in P on $[t_{i-2}, t_{i-1}]$. Thus the torsion and convexity conditions imply that under many circumstances coplanar data $I_i, \ldots, I_j$, $j \geq i + 3$, must give an interpolating curve which lies in the same plane on $[t_i, t_j]$. This is a very strong restriction and implies, for example, that if $I_{i-2}, \ldots, I_{i+1}$ are coplanar and $I_i, \ldots, I_{i+3}$ lie in a different plane, then r is linear on $[t_i, t_{i+1}]$.

3 Two methods

In this section we describe briefly two methods for shape preserving interpolation by space curves. Both are based on the method for planar curves described in Section 3.4 of [2], in which we specify tangent directions and curvatures at the data points and then define the curve between consecutive data points as a rational cubic. As there we may take $t_i = i$, $i = 0, \ldots, N$.

3.1 Continuity of curvature magnitudes

The scheme described here is due to the authors [3] and for full details we refer to that paper. As in Section 3.4 of [2] we choose the curvature magnitude κ_i at I_i to be the curvature of the circle passing through I_{i-1}, I_i, I_{i+1}, and we choose the tangent direction T_i at I_i to be of the form

$$T_i = a_i L_{i-1} + b_i L_i,$$

for some $a_i \geq 0$, $b_i \geq 0$. As before we want $b_i = 0$ when I_{i-2}, I_{i-1}, I_i are collinear and $a_i = 0$ when I_i, I_{i+1}, I_{i+2} are collinear. The linearity conditions in Section 2.2 require that when $N_i.N_{i+1} = 0$ then r is linear on $[i, i+1]$ and so $a_i = 0$. Similarly we require $b_i = 0$ when $N_{i-1}.N_i = 0$. All these are satisfied by choosing

$$a_i = \frac{|N_i.N_{i+1}|}{|N_{i+1}|} \kappa_{i+1} |L_i|^2,$$
$$b_i = \frac{|N_i.N_{i-1}|}{|N_{i-1}|} \kappa_{i-1} |L_{i-1}|^2. \tag{1}$$

We note that when $N_{i+1} = 0$, then $\kappa_{i+1} = 0$ and so $a_i = 0$. Similarly when $N_{i-1} = 0$, then $b_i = 0$. When $N_i = 0$, then I_{i-1}, I_i, I_{i+1} are collinear and r is linear on $[i-1, i+1]$.

Except for the cases when r is linear on $[i, i+1]$, we define the interpolating curve between I_i and I_{i+1} by

$$r(t) = \frac{I_i \alpha (1-s)^3 + Bs(1-s)^2 + Cs^2(1-s) + I_{i+1} \beta s^3}{\alpha (1-s)^3 + s(1-s)^2 + s^2(1-s) + \beta s^3}, \quad s = t-i, \quad i \leq t \leq i+1, \tag{2}$$

where $\alpha, \beta > 0$ and B, C in R^3 depend on i. As this curve has tangent directions T_i, T_{i+1} at I_i, I_{i+1} respectively, we must have

$$B = I_i + xT_i, \quad C = I_{i+1} - yT_{i+1}, \tag{3}$$

for some $x, y > 0$. Having chosen B and C, α and β are uniquely determined by requiring the curvatures at I_i and I_{i+1} to have magnitudes κ_i and κ_{i+1} respectively.

The following definitions are suggested for x and y, where ψ denotes the angle between N_i and N_{i+1}, $0 \leq \psi \leq \pi$, a denotes the angle between L_i and T_i, and b denotes the angle between L_i and T_{i+1}, $0 < a, b < \pi$.

(a) $N_i.N_{i+1} > 0$, i.e. $0 \leq \psi < \frac{\pi}{2}$.

$$x = \frac{2|L_i| \sin b}{2\lambda \sin b \cos \psi + (1-\lambda)|L_i||\kappa_{i+1}| \cos \psi + 2|\sin a \cos b|/\cos \psi + 2|\cos a \sin b|},$$

$$y = \frac{2|L_i| \sin a}{2\mu \sin a \cos \psi + (1-\mu)|L_i||\kappa_i| \cos \psi + 2|\sin a \cos b| + 2|\cos a \sin b|/\cos \psi},$$

for some parameters λ, μ with $0 < \lambda, \mu < 1$.

When $\psi = 0$, this reduces to the formula used for the scheme in Section 3.4 of [2], which ensures that circular arcs are reproduced. The above choice also ensures that the convexity conditions are satisfied: for further details, see [3].

(b) $N_i.N_{i+1} < 0$, i.e. $\frac{\pi}{2} < \psi \leq \pi$.

Here we take the same formula for x and y as in (a) except that $\cos\psi$ is replaced by $|\cos\psi|$ and terms involving $\cos\psi$ are multiplied by a parameter $k \geq 1$. It can be shown that the inflection condition is satisfied, see [3].

We note that as $\psi \to \frac{\pi}{2}$, the above formula show that $x, y \to 0$ and so the curve approaches the linear segment $I_i I_{i+1}$. This is consistent with the linearity condition which requires that r is linear between I_i and I_{i+1} when $N_i.N_{i+1} = 0$.

Finally we remark that the rational cubic curve given by (1) is always a coil, which is ascending or descending as its Bézier polygon $I_i BC I_{i+1}$ is ascending or descending. This can be seen from the deCasteljau subdivision algorithm, which shows that the curve can be gained by successively cutting corners of its Bézier polygon. This ensures that the scheme satisfies the torsion condition, and indeed the coil condition for $j = i + 3$.

As in the scheme in Section 3.4 of [2], the curve can be modified by the user by varying, for example, the tangent direction T_i and the curvature κ_i.

3.2 G^2 continuity

The above scheme ensures continuity of the tangent direction and the magnitude of the curvature. However the binormal to (2) at I_i is in the direction of $I_i B \times BC$, i.e. $T_i \times BC$. Similarly the binormal to (1) at I_{i+1} is in the direction of $BC \times T_{i+1}$. Thus there is no guarantee that the binormal direction is continuous at the data points. In this section we remedy this with a scheme which produces G^2 interpolating curves. This scheme is due to the authors [4] and full details are in that paper.

At each data point I_i we specify the binormal direction M_i. For $2 \leq i \leq N-2$, this is given by

$$M_i = N_i - l_i(L_{i-1} \times L_{i+1}) - r_i(L_{i-2} \times L_i),$$

for some $l_i, r_i > 0$. (For the remaining data points this formula must be modified.) The choice of l_i, r_i will be discussed later. We note that if $I_{i-2}, \ldots, I_{i+2}$ lie in a plane, then M_i is in the direction of N_i, the normal to that plane.

The curvature magnitudes κ_i at I_i are chosen as before. The tangent direction T_i must lie in the osculating plane, i.e. it must be orthogonal to M_i. We choose it so that its projection onto the plane of I_{i-1}, I_i, I_{i+1} is as before, i.e.

$$T_i = a_i L_{i-1} + b_i L_i - k_i N_i,$$

where a_i, b_i are given by (1), and

$$k_i = \frac{a_i r_i [L_{i-2}, L_{i-1}, L_i] + b_i l_i [L_{i-1}, L_i, L_{i+1}]}{|N_i|^2 - l_i [N_i, L_{i-1}, L_{i+1}] - r_i [N_i, L_{i-2}, L_i]}. \tag{4}$$

Between I_i and I_{i+1}, the curve is given as before by (2). The binormal to this curve at I_i is in the direction of $I_i B \times BC$ and so we require that the plane containing I_i, B, C is orthogonal to M_i. Similarly we need the plane through B, C, I_{i+1} to be orthogonal to M_{i+1}. Hence B and C lie on the line of intersection of the plane through I_i orthogonal to M_i and the plane through I_{i+1} orthogonal to M_{i+1}. Hence B and C are determined uniquely as the points of intersection of this line and the tangent lines through I_i, I_{i+1} respectively. As before the weights α and β are determined by the curvature magnitudes κ_i and κ_{i+1}. So it remains only to choose the numbers l_i, r_i, and this must be done so that the shape preserving conditions are satisfied.

The torsion condition requires that the sign of the torsion of r on $[i, i+1]$ has the same sign as that of $[L_{i-1}, L_i, L_{i+1}]$. Now the torsion of the curve (2) has the same sign as that of its Bézier polygon, i.e. the same as the sign of $[B - I_i, C - B, I_{i+1} - C] = [B - I_i, L_i, I_{i+1} - C]$ which, after some calculation, can be seen to be the same as the sign of

$$(a_i b_{i+1} - k_i k_{i+1} |L_i|^2)[L_{i-1}, L_i, L_{i+1}] - (a_i k_{i+1} + b_{i+1} k_i) N_i . N_{i+1}.$$

Thus the torsion condition (and indeed the coil condition with $j = i + 3$) will hold provided that the parameters k_i, k_{i+1} are small enough. ¿From (4) we see that this will hold if the parameters $l_i, l_{i+1}, r_i, r_{i+1}$ are small enough.

It can be seen that the convexity and inflection conditions on $[i, i+1]$ are satisfies provided

$$M_i . N_i > 0, \quad M_{i+1} . N_{i+1} > 0$$

and

$N_i . N_{i+1} > 0$ (resp. < 0) implies that $M_i . N_{i+1}, M_{i+1} . N_i > 0$ (resp. < 0).

These conditions are linear in $l_i, r_i, l_{i+1}, r_{i+1}$. Now $M_i \to N_i$ as $l_i, r_i \to 0$, and so these conditions will be satisfied provided that the parameters $l_i, l_{i+1}, r_i, r_{i+1}$ are small enough.

The approach of this interpolation scheme is to first choose 'optimal' values l_i^*, r_i^* for the parameters l_i, r_i, which are chosen so that the values x, y in (3) are approximately those given by the method of Section 3.1. These are then reduced, if necessary, until the shape preserving conditions are satisfied. An alternative approach is now under investigation with the help of Miss L. Sampoli. In this approach the above non-linear torsion conditions are replaced by sufficient linear conditions and then

$$\sum (l_i - l_i^*)^2 + \sum (r_i - r_i^*)^2$$

is minimised subject to the linearised shape preserving conditions. The resulting curve can then be interrogated and, if necessary, the values l_i^*, r_i^* modified and the process repeated. This iterative procedure is performed automatically by the algorithm. However it is also possible, as in Section 3.1, for the user to modify the curve by varying certain parameters.

Finally we recall the difficulties mentioned at the end of Section 2 concerning G^2 continuity and coplanar data. In order to preserve G^2 continuity, the scheme relaxes the torsion condition on $[i, i+1]$ if $I_{i-2}, \ldots, I_{i+1}$ but not $I_{i-2}, \ldots, I_{i+2}$ are coplanar, or if $I_i, \ldots, I_{i+3}$ but not $I_{i-1}, \ldots, I_{i+3}$ are coplanar. In such cases the curve on $[i, i+1]$ is represented by two rational cubic segments.

References

[1] Goodman, T. N. T.: *Inflections on Curves in Two and Three Dimensions.* Computer Aided Geometric Design **8** (1991), 37–51.

[2] Goodman, T. N. T.: *Shape Preserving Interpolation by Planar Curves*. In this volume.

[3] Goodman, T. N. T., Ong, B. H.: *Shape Preserving Interpolation by Space Curves*. University of Dundee Report AA/963 (1996).

[4] Goodman, T. N. T., Ong, B. H.: *Shape Preserving Interpolation by G^2 Space Curves*. University of Dundee Report AA/964 (1996).

[5] Kaklis, P. D., Sapidis, N. S.: *Convexity-Preserving Interpolatory Parametric Splines of Non-Uniform Degree*. Computer Aided Geometric Design **12** (1995), 1–26.

[6] Kaklis, P. D., Karavelas, M. I.: *Shape-Preserving Interpolation in R^3*. Submitted to Computer Aided Geometric Design.

[7] Labenski, C., Piper, B.: *Coils*. To appear in Computer Aided Geometric Design.

Authors

Prof. Dr. T. Goodman
University of Dundee
Dept. of Mathemaitics & Computer Science
DUNDEE DD1 4HN
Scotland (UK)
E–mail: tgoodman@mcs.dundee.ac.uk

Boon–Hua Ong
Universiti Sains Malaysia

A coparative study of two curve fairing methods in Tribon Initial Design

Andrew Ives-Smith
KCS Ltd., Armstrong Technology Centre,
Davy Bank
Wallsend
TYNE and WEAR NE28 6UY
UK

Abstract

A comparative study of two methods for the automatic fairing of B-Splines curves is made. The methods discussed are the Eck/Hadenfeld algorithm and the Pigounakis/Sapidis/Kaklis method. By using existing datastores from the Tribon Initial Design module LINES test were carried out to evaluate the effectiveness of the two algorithms. The work was carried out as part of the European Union funded project FAIRSHAPE (Automatic FAIRing and SHAPE-Preserving Methodologies for CAD/CAM.

1. Introduction

One of KCS' roles as an industrial partner in the European Union funded project FAIRSHAPE (Automatic FAIRing and SHAPE-Preserving Methodologies for CAD/CAM, KCS is evaluation of state-of-the-art fairing and shape-preserving methods for curves and surfaces.

The fairing of curves and surfaces plays an important role in ship design. For the purposes of evaluating the effectiveness of the fairing algorithms, datastores from the Tribon Initial Design module LINES were used. LINES is a hullform definition system which enables Naval Architects to define and fair a hullform by means of a series of orthogonal and space curves. Input to the system is generally via an offset file, digitisation of a preliminary lines plan, or from other programs via suitable interfaces such as IGES and DXF. Once inserted, the hullform and its appendages can be developed by the progressive refinement of frames, waterlines, buttocks, 3D curves, angle curves and boundary curves, which between them provide a complete three-dimensional wire frame definition of the hullform.

Fairness can be checked by viewing these curves drawn out on scaled plots. However the most usual method of evaluating the fairness is visually is displaying the curvature as porcupines. Methods which would assist the loftsman in the fairing process would reduce the time taken to complete the design of a ship and also produce more fair curves than could be achieved by hand. Of increasing importance is the role of three dimensional curves. The current practice is to project the 3D curve in to two mutual orthogonal planes, fair the curves in their constituent planes and then merge the two faired curves back together. Any method which would allow users to automatically fair the 3D curve would be beneficial. In the literature there exists many different viewpoints as to what constituents a fair curve. The algorithms reviewed in this paper use different methods.

Section 2 defines a set of fairness measures which have been evaluated in order to compare the two fairing methods. In section 3 the two methods Eck and Hadenfeld (EH) and the Pigounakis, Sapidis, Kaklis method (PSK) are presented together. Section 4 details how the algorithms were implemented. Section 5 contains examples of the applying the methods to curves contained in KCS datastores and Section 6 presents the conclusions of the author.

2. Fairness Measures

In order to obtain information regarding the comparative performance of the fairing algorithms in the study, several measures of fairness will be introduced. These measures were agreed upon at the 2nd Internal workshop of Fairshape partners in Berlin, August 1995.

Definition for B-Splines

Let $D = \{\mathbf{d}_i, i = 0, 1, \ldots M\}$ be the polygon points in 2D or 3D space and

$U = \{u_0, u_1, \ldots, u_N\}$ be a montonic increasing sequence of real numbers

then the B-spline curve $\mathbf{Q}(u)$ of degree n is given by

$$\mathbf{Q}(u) = \sum_{j=0}^{M} \mathbf{d}_j N_j^n(u) \,,\, u \in [u_n, u_{M+1}]$$

Let $\kappa(u)$ be the curvature, we shall also introduce a quantity which measures the $\kappa' = \dfrac{d\kappa}{ds}$ discontinuity, where s is the arc length of the curve.

let $z_j = \left| \dfrac{d\kappa}{ds}(u_j^+) - \dfrac{d\kappa}{ds}(u_j^-) \right|$.

If $\tau(u)$ be the torsion of the curve, then a corresponding measure of τ discontinuity is defined as

$l_j = \left| \tau(u_j^+) - \tau(u_j^-) \right|$.

1) zed - The absolute maximum discontinuity of the derivative of curvature at a knot

   $z_j = \left| \dfrac{d\kappa}{ds}(u_j^+) - \dfrac{d\kappa}{ds}(u_j^-) \right|$

2) zeta - The sum of the absolute discontinuities of the derivative of curvature $\varsigma_Q = \sum_{4}^{M} z_j$

3) Kmax - Absolute maximum value of curvature $|\kappa(u)|$.
4) CMS - Number of changes in Monotonsity in the curvature plot.
5) E2 - Integral of the squared second derivative of the curve $\int_{u_n}^{u_{M+1}} \left\| \dfrac{d^2}{du^2} \mathbf{Q}(u) \right\|^2 du$
6) E3 - Integral of the squared third derivative of the curve $\int_{u_n}^{U_{M+1}} \left\| \dfrac{d^3}{du^3} \mathbf{Q}(u) \right\|^2 du$
7) Ek - Integral of the squared curvature of the curve w.r.t. arc length. $\int \kappa^2(s) ds$
8) Ek` - Integral of the squared derivative of curvature. $\int (\kappa'(s))^2 ds$
9) IP - Number of inflection points in 2D curves.
10) TD - Total deviation of the nodal points $\sum \left\| \mathbf{Q}_1(u_j) - \mathbf{Q}_2(u_j) \right\|$
11) MD - Maximum deviation of the nodal points.

12) EL - Absolute maximum discontinuity in the torsion plot $l_j = |\tau(u_j^+) - \tau(u_j^-)|$

13) LAMBDA - Sum of the absolute discontinuities in the torsion plot. $\lambda_Q = \sum_{4}^{M} l_j$

14) TMAX - Absolute maximum value of torsion
15) TSC - Number of torsion sign changes
16) Et - Integral of the squared torsion of the curve w.r.t. arc length.
17) It - Number of iterations of the fairing method

3. The Algorithms

3.1 Eck and Hadenfeld Method

(Eck and Hadenfeld, 1995) developed an algorithm that incorporates the local fairing of a B-spline using the minimisation of energy approach. The Eck/Hadenfeld is based upon the assumption that a curve is fair *if it minimises the squared curvature with respect to arc length*. In order to create a solvable problem, it is linearised by assuming that the parameter of the curve t is approximately the arc length. This leads to the solving of

$$E_l(d_r) = \int_{u_n}^{u_M} \left(\frac{d^l}{dt^l} \mathbf{Q}(t)\right)^2 dt$$

where l is either 2 or 3. The above may be generalised as:

$$E_l(d_r) = A \int_{u_n}^{u_M} \left(\frac{d^2}{dt^2} \mathbf{Q}(t)\right)^2 dt + B \int_{u_n}^{u_M} \left(\frac{d^3}{dt^3} \mathbf{Q}(t)\right)^2 dt \quad A \geq 0, B \geq 0, A+B=1$$

In the special cases of A=1,B=0 we obtain the minimisation of the second derivative and A=1,B=0 the minimisation of the third derivative. It may be shown that the above leads to a solution for the polygon point to be moved

$$\mathbf{d}_r = \sum_{\substack{j=r-n \\ j \neq r}}^{r+n} \alpha_j \mathbf{d}_j$$

where α_j are constants dependent upon the knot set and the integral to be minimised.

The above results in a minimisation of the fairness criteria imposed by the integral to be minimised. (Eck and Hadenfeld, 1995) also add a distance control into their algorithm. This is of significant benefit when incorporated into a fairing method, because typically the user requires the faired curve to be less than a specified distance from the original.

3.2 Pigounakis, Sapidis, Kaklis Algorithm

In (Farin and Sapidis 1990) an automatic method was presented for the fairing of planar curves. The method (Conv) is based on the criterion that;

A curve is fair if the curvature plot is continuous, has the appropriate sign (if stated) and is as close as possible to a piecewise monotone function with as few as possible monotone pieces. In the method the worst discontinuity in the derivative of curvature $z_j = \left|\frac{d\kappa}{ds}(u_j^+) - \frac{d\kappa}{ds}(u_j^-)\right|$ is found, the knot u_j is then removed then re-inserted. This action has the effect of making the curve C^3 at the stated knot. The process is repeated until the sum of the discontinuities at a given step of the algorithm is greater than the corresponding sum of the discontinuities from the previous step.

(Pigounakis, Sapidis and Kaklis, 1995) developed the ideas further so that the method may be applied to 3D curves. They introduce a new fairing criteria namely,

A C^2 curve is characterised as fair if
(a) the curvature plot is comprised of as few as possible monotonic segments,
(b) its torsion plot is as close to possible of being continuous also with the fewest number of monotonic pieces,
(c) sign changes in the torsion plot are as few as possible,
(d) The value of torsion, at each point of the curve, is as small as possible.

In addition to the discontinuity in the curvature or k' measure, a fairness indicator $l_j = \left| \tau(u_j^+) - \tau(u_j^-) \right|$ is introduced together with the measure $\lambda_Q = \sum_4^M l_j$ additionally a further measure corresponding the maximum torsion is also introduced.

4. Implementation

Code was written to implement the EH, Conv and PSK algorithms. These algorithms were linked to access routines so that curves could be obtained from a KCS Lines datastore. This enabled many curves to be faired. Both methods under review require the curve to be C^2 continuous at all knot locations. However it is usual, for example, for the Flat of Side and Flat of Bottom curve, to be joined to the rest of a section curve with C^1 continuity. Consequently the first part of the algorithm splits a curve into pieces which are C^2 continuous each piece is then processed by the appropriate algorithm.

4.1 Eck and Hadenfeld Algorithm

1) For every polygon point calculate the fairness indicator
2) Produce a ranking of the fairness indicators in descending order
3) Consider the first L items in the list where L = Int(n/2) where n is the number of polygon points.
4) Move each polygon point in turn starting with the polygon point associated with the largest fairing indicator.
5) If $\left| E(Q_{NEW}) - E(Q_{OLD}) \right| < \delta$ where δ is some small tolerance and $E(Q)$ is the fairness measure being minimised, then stop otherwise go to 1.

4.2 Pigounakis, Sapidis, Kaklis Algorithm

1) For each internal knot find the following quantities
2) $z_j = \left| \dfrac{d\kappa}{ds}(u_j^+) - \dfrac{d\kappa}{ds}(u_j^-) \right|$ and $\zeta_Q = \sum_4^M z_j$
3) $l_j = \left| \tau(u_j^+) - \tau(u_j^-) \right|$ and $\lambda_Q = \sum_4^M l_j$
4) Calculate the torsion $\tau(u)$ of $\mathbf{Q}(u)$ at equidistant points in each span. If $|\tau(v)|$ is the maximum value of torsion then, if u_r is a knot set u_r equal to the knot, otherwise if $u_i < v < u_{i+1}$ set u_r when $l_i > l_{i+1}$ or u_{r+1} when $l_i < l_{i+1}$.

5) In order to carry out a comparison between the three Fairness Indicators they are normalised so that u_h corresponds to the Normalised Fairness Indicator $F_h = \frac{z_h}{\zeta_Q}$, u_m $F_m = \frac{l_m}{\lambda_Q}$ and u_r,

$F_r = \frac{\tau_Q}{\tau_{ref}}$ where τ_{ref} is the maximum absolute torsion of the original curve.

The knot selected for removal is the knot associated with the largest Normal Fairness Indicator. The knot is then re-inserted into the spline.

6) Steps 1 to 5 are repeated until $\|\mathbf{Q}^i(u_j) - \mathbf{Q}^{i+1}(u_j)\| > \delta$ for some $u \in [u_4, \ldots, u_M]$, where $\mathbf{Q}^i(u)$ is the spline at step I of the algorithm.

Note: That the termination criteria for the PSK algorithm is different to that in (Pigounakis, Sapidis and Kaklis, 1995) this is because the only meanifull criteria is that the nodal values are within the user specified tolerance. In (Pigounakis et al) the termination criteria is $\sum_{j=4}^{m} \|\mathbf{Q}^i(u_j) - \mathbf{Q}^{i+1}(u_j)\| > \gamma$

where γ is some user of system defined tolerance. However this is inappropriate for users as usually they require the deviation of any point on the curve to be less than a given tolerance. The author in his implementation has also modified the PSK algorithm so that, when the termination criteria has been met, the previous step is undone and the knot excluded. The algorithm then searches to find other areas of the spline which, due to the local nature of B-spline curves, can be faired without violating the termination criteria.

5. Results

Numerous tests were undertaken, however only two cases are presented here due to space restrictions. In Example 1 the fairing of a 3D curve is shown. The input to the system was a tolerance of 0.02m. In all the figures the solid line is the original curve and the dashed line is the faired curve.

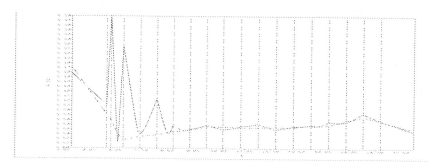

Figure 1 : Curvature Psk algorithm

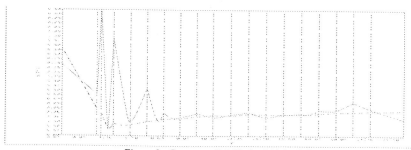

Figure 2 : Curvature Eck/Hadenfeld A=1, B=0

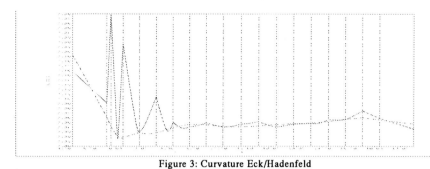

Figure 3: Curvature Eck/Hadenfeld

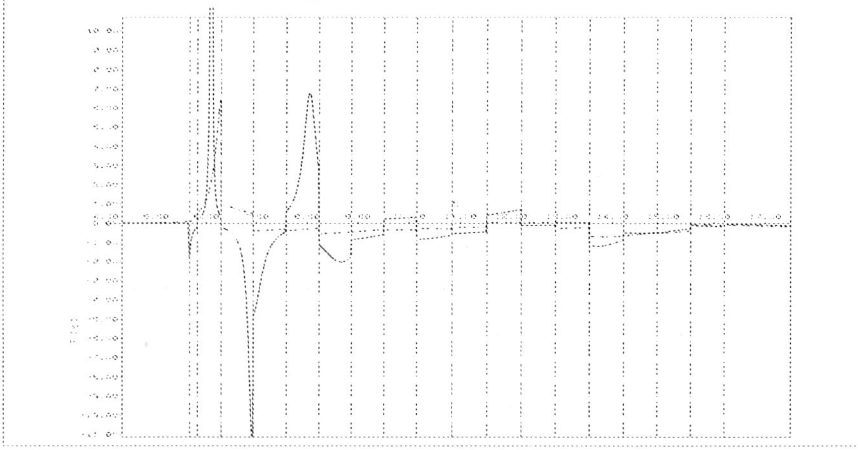

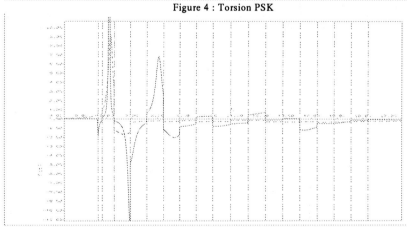

Figure 4 : Torsion PSK

Figure 5 : Torsion Eck/Hadenfeld A= 1,B=0

A coparative study of two curve fairing methods in Tribon Initial Design

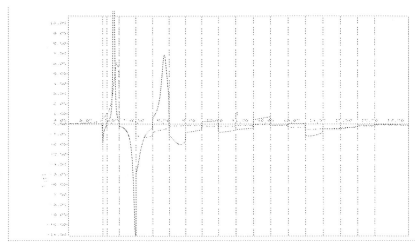

Figure 6 : Torsion Eck/Hadenfeld A= 0,B=1

Table 1: Summary for 3D curve

FAIRNESS MEASURES	Original	PSK	Eck/Hadenfeld a=1 b=0	Eck/Hadenfeld a=0 b=1
zed	2.224	0.041	0.142	0.033
zeta $\zeta_Q = \sum_{4}^{M} z_j$	5.748	0.164	0.341	0.193
Kmax- Absolute maximum value of curvature $|\kappa(u)|$.	0.281	0.171	0.193	0.193
CMS -Number of changes in Monotonsity in the curvature plot.	18	8	13	6
E2 - $\int_{u_n}^{u_{M+1}} \left\| \frac{d^2}{du^2} Q(u) \right\|^2 du$	0.077	0.047	0.046	0.047
E3 - $\int_{u_n}^{U_{M+1}} \left\| \frac{d^3}{du^3} Q(u) \right\|^2 du$	0.879	0.025	0.028	0.024
Ek - $\int \kappa^2(s) ds$	0.077	0.047	0.048	0.049
Ek' - $\int (\kappa'(s))^2 ds$	0.788	0.013	0.019	0.018
TD Total deviation of the nodal points $\sum \left\| Q_1(u_j) - Q_2(u_j) \right\|$		0.06381	0.115934	0.093802
MD Maximum deviation of the nodal points.		0.019416	0.014062	0.013106
EL - $l_j = \left| \tau(u_j^+) - \tau(u_j^-) \right|$	5.783	5.632	4.301	4.195
LAMBDA $\lambda_Q = \sum_{4}^{M} l_j$	20.135	8.881	6.891	7.536
TMAX - Maximum Torsion	75.832	6.632	9.232	6.121
TSC - Number of torsion sign changes	11	3	1	1
Et - Integral of the squared torsion of the curve w.r.t. arc length. $\int |\tau(s)|^2 ds$	148.279	6.271	16.048	9.648
It - Number of iterations of the fairing method		31		

As is demonstrated by the graphs and the figures in Table 1 the PSK method gives the smallest value of torsion. As expected the Eck/Hadenfeld algorithms produce the smallest values of E2 and E3. However the above demonstrates that because the PSK algorithm does not limit the polygon point movement, there is at least one knot in the curve that cannot be moved. If it was allowed to then the termination criteria would be breached. One method that could possibly help to improve the performance of the algorithm would be the use of error control knots, first mentioned in (Farin and Sapidis 1989).

In a second example, a section curve is considered. To simulate 'fine' fairing a tolerance of 0.005m was used. In all the figures the solid line is the original curve and the dashed line is the faired curve. In figures 7-10 the effect of using the different methods may be seen.

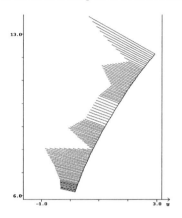

Figure 7 : Original Curve

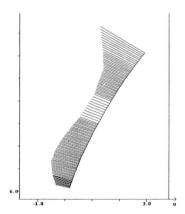

Figure 8 : Conv

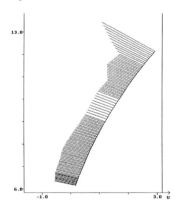

Figure 9 : EH A=1,B=0

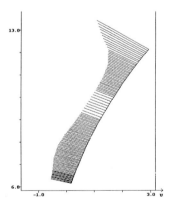

Figure 10: EH A=0,B=1

A coparative study of two curve fairing methods in Tribon Initial Design

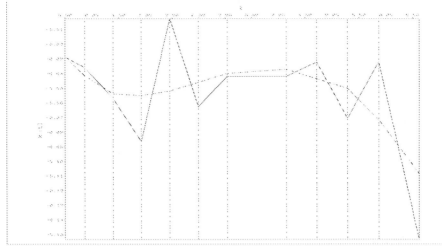

Figure 11:Curvature for Conv

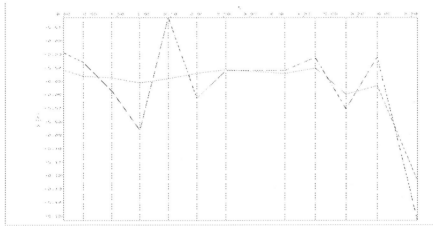

Figure 12 : Curvature for Eck/Hadenfeld A=1,B=0

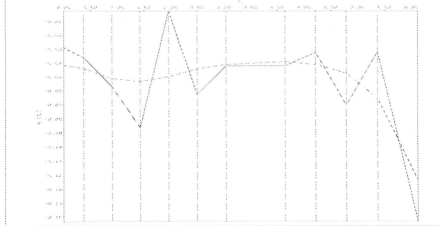

Figure 13: Curvature for Eck/Hadenfeld A=0,B=1

FAIRNESS MEASURES		PSK	Eck/Hadenfeld a=1 b=0	Eck/Hadenfeld a=0 b=1
zed		0.0268	0.025	0.104	0.046
zeta $\zeta_Q = \sum_{4}^{M} z_j$		1.164	0.105	0.241	0.118
Kmax- Absolute maximum value of curvature $|\kappa(u)|$.		0.153	0.108	0.124	0.124
CMS -Number of changes in Monotonsity in the curvature plot.		8	2	6	2
E2 - $\int_{u_n}^{u_{M+1}} \left\| \frac{d^2}{du^2} \mathbf{Q}(u) \right\|^2 du$		0.021	0.019	0.019	0.020
E3 - $\int_{u_n}^{u_{M+1}} \left\| \frac{d^3}{du^3} \mathbf{Q}(u) \right\|^2 du$		0.048	0.004	0.008	0.003
Ek - $\int \kappa^2(s) ds$		0.021	0.019	0.019	0.020
Ek' - $\int (\kappa'(s))^2 ds$		0.048	0.004	0.008	0.003
IP - Number of Inflection points in 2D curves		0	0	0	0
TD Total deviation of the nodal points $\sum \|\mathbf{Q}_1(u_j) - \mathbf{Q}_2(u_j)\|$			0.01917756	0.028082	0.0198109
MD Maximum deviation of the nodal points.			0.00483	0.0045823	0.0034489
It - Number of iterations of the fairing method			9		

Table 2

In this example the value of zeta is the smallest for the Conv algorithm. The Eck/Hadenfeld algorithms produce the lowest values in E2 and E3 respectively.

Figures 7-9 all show significant improvements in the curvature graphs with Conv being the best.

6. Conclusions

This paper has only discussed two examples from the datastores analysed. Both algorithms generally provide good results. The advantages of the PSK algorithm is that it is the only algorithm to fair the torsion curve. The impact of this in the LINES system needs to be investigated however indications views are that it is a desirable feature. The Eck/Hadenfeld algorithm incorporates a user specified tolerance within the fairing algorithm. This enables the method to fair curves in which a *fine tolerance* is used which are not sometimes currently possible with the PSK algorithm. As previously mentioned the author has discussed with Mr. Pigounakis the use of error control knots as referenced in (Farin and Sapidis 1989).

Acknowledgements

The author would like to thank Kostis Pigounakis of N.T.U.A for providing an implementation of his PSK algorithm and also for his comments concerning the PSK algorithm.

REFERENCES

1. Pigounakis K. G., Sapidis N.S., Kaklis P.D. : Fairing Spatial B-Spline Curves.
2. Eck M., Hadenfeld J. : Local Energy Fairing of B-Splines Curves. In G. Farin, H. Hagen, and H. Noltemeier (eds.): Computing, Supplementum **10**, Springer (1995), 203-212
3. Sapidis N.S., Farin G., Automatic Fairing Algorithm for B-Spline Curves, Computer Aided Design, Vol. 22, pp. 121-129, 1990Design, Vol. 22, pp. 121-129, 1990
4. Lines Techinical Brief, KCS

Fairing Curves and Surfaces

Fairing of B–Spline Curves and Surfaces

Jan Hadenfeld

Technische Hochschule Darmstadt

Abstract: We want to give an overview on our methods for fairing B–spline curves [7] and surfaces [16]. The extension to Bézier–splines and rational curves and surfaces will be described as well. These methods are fairing a curve or surface in a local scheme. Changing only one control point in every step and using a quadratic fairness functional the main advantage is that we can determine an exact solution of this new control point in every step and that we simply can fulfill a given distance tolerance.

1   Introduction

If a designer constructs a B–spline curve or surface with the help of his CAD system or any other application the result might be *non-smooth*. To avoid this affect, which can originate from e.g. digitizing errors of data points, two different principles are generally used. The first one incorporates a suitable fairness criterion into the interpolation or approximation process; for more informations see [4, 17]. The second one is to separate the construction and the fairing process which again can be subdivided into two principles: global methods as described in [20, 21, 22] and local, iterative methods like [10, 12, 13, 26, 27, 29]. Our method we are going to describe in the next sections also belongs to the second principle.

Although the fairness of a curve or surface can only be described in a subjective manner the following two definitions are very common in literature (see also [28]): a (planar) curve is smooth if it minimizes the *strain energy* of a thin elastic beam described by

$$E = \int \kappa^2(s)\, ds \quad , \tag{1}$$

where κ represents the curvature of the curve. A surface is smooth if it minimizes the potential energy of a thin elastic plate (*thin-plate energy*) given by

$$\Pi_P = \iint_S a\left(\kappa_1^2 + \kappa_2^2\right) + 2(1-b)\,\kappa_1\kappa_2 \, dS \quad , \tag{2}$$

where κ_1 and κ_2 are the principle curvatures of the plate and the factors a, b are constants depending on the material (see [3]).

But both fairing criteria lead to a non-linear problem which has to be solved with the help of numerical tools whereas explicit solutions cannot be given in general. Following this, both criteria are linearized in most fairing algorithms. Concerning curves we assume that the parameter t of the curve represents the arc length. So, the simpler integral

$$E_2 = \int \left(\mathbf{x}''(t)\right)^2 \, dt \qquad (3)$$

is minimized instead. Furthermore, we do use the third derivative also described in [17, 24]. If the curve is parameterized with respect to the arc length it can be shown that this integral is equivalent to the integral over $(\dot{\kappa})^2 + \kappa^2(\kappa^2 + \tau^2)$.
In case of given surfaces the functional

$$\Pi = \iint_A \mathbf{X}_{uu}^2 + 2\,\mathbf{X}_{uv}^2 + \mathbf{X}_{vv}^2 \, du\, dv \qquad (4)$$

is a good approximation of (2) if the parameterization of the surface is nearly isometric (and by choosing $a = 1$ and $b = 0$). More details about this linearization can be found in [14]. But for general curves or surfaces, (3) and (4) are far away from being a good approximation and, therefore, could produce strange results. In [23] some effects are pointed out for curves. Another functional suggested in [15] is

$$G = \iint_A (\text{grad div grad } \mathbf{X})^2 \, du\, dv \qquad (5)$$

or the following from [1]

$$H = \iint_A \mathbf{X}_{uuu}^2 + \mathbf{X}_{vvv}^2 \, du\, dv \quad . \qquad (6)$$

In this paper a B–spline curve $\mathbf{x}(t)$ of general order k with the knot sequence $T = (t_j)_{j=0}^{n+k}$, which is always described by

$$\mathbf{x}(t) = \sum_{i=0}^{n} \mathbf{d}_i \cdot N_{i,k}(t) \quad , \quad t \in [t_{k-1}, t_{n+1}] \qquad (7)$$

(see [11, 19] for details), or a B–spline surface $\mathbf{X}(u,v)$ of order (k,l) with the knot sequences $U = \{u_i\}_{i=0}^{m+k}$ and $V = \{v_j\}_{j=0}^{n+l}$, which is described by

$$\mathbf{X}(u,v) = \sum_{i=0}^{m}\sum_{j=0}^{n} \mathbf{d}_{ij} \cdot N_{i,k}(u)\, N_{j,l}(v) \quad , \quad (u,v) \in [u_{k-1}, u_{m+1}] \times [v_{l-1}, v_{n+1}] \quad , \qquad (8)$$

should be faired in the following way: by modifying only one control point at a time and keeping all others fixed, we want to minimize the (linearized) fairness functional. This idea had been successfully used before in [8] for fairing polylines.

To smooth a whole curve or surface the following points have to be done:

1. Find the control point which has to be changed.
2. Change the control point to a location in such a way that the new curve or surface minimizes the fairness criterion and fulfills a given distance tolerance.
3. Follow step 1 and 2 so often as a suitable criterion to stop is fulfilled.

In the first part of this contribution we want to describe our method for fairing B–spline curves (c.f. [7]) and Bézier–spline curves. A possible solution for fairing rational B–spline curves will also be discussed. The second part contains the extension of this method to surfaces (c.f. [16]). In both parts the benchmark curves and surfaces of the workshop in Lambrecht and some other examples are treated. The results are visualized in some pictures.

2 Fairing of B–Spline Curves

2.1 The New Control Point

First of all we introduce the following notations to make a distinction of the different stages of the curve:

1. The given curve which is to be faired:

$$\mathbf{x}(t) = \sum_{i=0}^{n} \mathbf{d}_i \cdot N_{i,k}(t) \quad , \quad t \in [t_{k-1}, t_{n+1}] \quad . \tag{9}$$

2. The B–spline curve which has already been faired by a certain number of iterations:

$$\bar{\mathbf{x}}(t) = \sum_{i=0}^{n} \bar{\mathbf{d}}_i \cdot N_{i,k}(t) \quad , \quad t \in [t_{k-1}, t_{n+1}] \quad . \tag{10}$$

3. The new curve in the next iteration step:

$$\tilde{\mathbf{x}}(t) = \sum_{\substack{i=0 \\ i \neq r}}^{n} \bar{\mathbf{d}}_i \cdot N_{i,k}(t) + \tilde{\mathbf{d}}_r \cdot N_{r,k}(t) \quad , \quad t \in [t_{k-1}, t_{n+1}] \quad . \tag{11}$$

Here we restrict the index r by $\alpha \leq r \leq \beta$ to achieve a wanted continuity between the original and the faired B–spline curve. If the curve is also a part of a set of curves and has any continuity with respect to its neighbours it may be wanted to fix the boundaries.

Our task now is to find a new location for the control point $\tilde{\mathbf{d}}_r$ in (11) in such a way that the curve $\tilde{\mathbf{x}}(t)$ minimizes the fairness functional.

This local minimization problem

$$E_l(\tilde{\mathbf{d}}_r) = \int_{t_{k-1}}^{t_{n+1}} \left(\frac{d^l}{dt^l}\tilde{\mathbf{x}}(t)\right)^2 dt = \int_{t_{k-1}}^{t_{n+1}} \left(\tilde{\mathbf{x}}^{(l)}(t)\right)^2 dt \tag{12}$$

is a quadratic form in $\tilde{\mathbf{d}}_r$. It has a unique minimum and can be solved explicitly. Here we introduce the value l, where $l = 2$ or $l = 3$ are appropriate choices.

Inserting the curve (11) into (12), the unique minimum $\tilde{\mathbf{d}}_r$ is determined by

$$\frac{\partial E_l(\tilde{\mathbf{d}}_r)}{\partial \tilde{\mathbf{d}}_r} \stackrel{!}{=} 0 \quad . \tag{13}$$

This equation can be solved explicitly for the control point $\tilde{\mathbf{d}}_r$ and we obtain

$$\tilde{\mathbf{d}}_r = \sum_{\substack{i=i_0 \\ i \neq r}}^{i_1} \gamma_i \cdot \bar{\mathbf{d}}_i \tag{14}$$

with the weighting factors γ_i of the form

$$\gamma_i = -\frac{\int_a^b N_{i,k}^{(l)}(t)\, N_{r,k}^{(l)}(t)\, dt}{\int_a^b \left(N_{r,k}^{(l)}(t)\right)^2 dt} \tag{15}$$

and the following abbreviations for the limits of the integrals and the sum:

$$a = \max\{t_r, t_{k-1}\} \quad \text{and} \quad b = \min\{t_{r+k}, t_{n+1}\} \ ,$$
$$i_0 = \max\{0, r-k+1\} \quad \text{and} \quad i_1 = \min\{r+k-1, n\} \ .$$

To control the calculation of the integrals we can use the property that the control point $\tilde{\mathbf{d}}_r$ is an affine combination of the neighbouring control points because of

$$\sum_{\substack{i=i_0 \\ i \neq r}}^{i_1} \gamma_i = 1 \ . \tag{16}$$

The integrals could be calculated exactly with the help of Gaussian quadrature. Further information can be found in [31].

2.2 The Ranking–List

The best location has now be found by changing a control point. Solving the next problem is to determine the faired control point. That means we have to find the index r of the control point.

We want to change the control point where the largest improvement of the fairness functional is given. We call this improvement

$$z_r = E_l(\bar{\mathbf{d}}_r) - E_l(\tilde{\mathbf{d}}_r) \geq 0 \tag{17}$$

ranking-number. Following up some calculations we obtain by inserting the control point $\tilde{\mathbf{d}}_r$ (14)

$$z_r = \left(\bar{\mathbf{d}}_r - \tilde{\mathbf{d}}_r\right)^2 \cdot \int_a^b \left(N_{r,k}^{(l)}(t)\right)^2 dt \ . \tag{18}$$

This ranking-number is a weighted function of the squared change of control point $\bar{\mathbf{d}}_r$.

The ranking-numbers have to be calculated for all involved control in order to find the largest improvement. This *ranking-list* then has to be sorted.

The whole ranking-list has to be calculated only in the first step for all control points because the control point $\tilde{\mathbf{d}}_r$ influences only the ranking-numbers z_i with the index $r-k+1 \leq i \leq r+k-1$ and these ranking-numbers only have to be recalculated in the following step.

Fairing of B–Spline Curves and Surfaces

2.3 Distance Tolerance

Up to now we did not take care of any distance tolerance. But it is often necessary to fulfill a prescribed tolerance δ between the old and the smoothed curve.
So, we have to take care of the constraint

$$\max\{|\mathbf{x}(t) - \tilde{\mathbf{x}}(t)| \ : \ t \in [t_{k-1}, t_{n+1}]\} \leq \delta \tag{19}$$

in each step. But this constraint also leads to a nonlinear-problem and resulting we do use

$$|\mathbf{d}_r - \tilde{\mathbf{d}}_r| \leq \delta \tag{20}$$

as an upper bound for (19) (c.f. [30]).

Let $\tilde{\mathbf{d}}_r^*$ be the control point which minimizes the fairness functional under the constraint (20). Two cases have to be distinguished: firstly, the new control point $\tilde{\mathbf{d}}_r$ satisfies the constraint (20). In this case nothing else has to be done (see Fig. 1 left). Secondly, the constraint is not fulfilled. In this case we are searching for a new location $\tilde{\mathbf{d}}_r^*$. Here we can use the fact that the isolines of the energy integral are in the planar case concentric circles (spheres in $\mathbb{R}^3$) with center $\tilde{\mathbf{d}}_r$. Then the new point is determined by (see Fig. 1 right)

$$\tilde{\mathbf{d}}_r^* = \mathbf{d}_r + \delta \cdot \frac{\tilde{\mathbf{d}}_r - \mathbf{d}_r}{|\tilde{\mathbf{d}}_r - \mathbf{d}_r|} \tag{21}$$

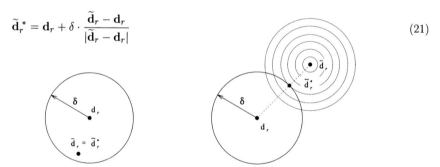

Fig. 1 The distance tolerance is fulfilled (left) or not (right).

2.4 The Algorithm

We can now build an *automatic* fairing algorithm for fairing B–spline curves with the results of the previous sections:

1. Compute the ranking-numbers z_i ; $0 \leq i \leq n$.
2. Find $z_r = \max\{z_i \ : \ 0 \leq i \leq n\}$.
3. Compute the new location $\tilde{\mathbf{d}}_r$ (resp. $\tilde{\mathbf{d}}_r^*$ satisfying the δ-distance constraint).
4. If a suitable criterion to stop is fulfilled, then exit, else goto step 1.

There are two criteria to stop:

- The ranking-numbers are less than a certain value.
- The number of iterations of the fairing algorithm is restricted.

If the number of control points is very large the calculation of the ranking-list is the most time-consuming part. A faster strategy to get a smooth curve is to change more than one control point from the beginning of the ranking-list. The list is then rebuild and the process is repeated again.

2.5 Examples

The next given three examples are benchmark curves of the workshop in Lambrecht which have to be faired. In all cases only one control point on each side is fixed and the maximum perpendicular error (max_error) is given in relation to the maximal diagonal. All calculations had been done on a HP 9000/735 workstation.

In the first example the curve is a spatial one. To get a better impression the values of the curvature are not plotted as porcupines but as circles and the direction of the normals are visualized as lines[3]. The fairing result in the case of minimizing the second derivative is not of the same quality as in the case of the third derivative. The reason could be that by minimizing the third derivative this integral is the linearization of $(\dot{\kappa})^2 + \kappa^2(\kappa^2 + \tau^2)$ and the torsion is also involved as well.

The second and third curves are nearly planar. Only the *porcupines* here are plotted to visualize the fairness of the curves.

Fig. 2 The given B-spline curve named chine3, the faired one with E_2 (max_error=0.19%, 0.19 sec.) and E_3 (max_error=0.15%, 0.19 sec.).

2.6 Extension to sets of Bézier- and B-Spline Curves

If a designer constructs a curve in a way that he splits this curve into a set of Bézier- or B-splines, one possible solution to smooth these curves is to handle each one separately and fix as much as needed control points at the connections to hold the continuity. But by using this method it is not possible to control the whole shape of the curve.

A further method is to convert the curves into one B-spline curve. This can easily be done by using a knot-vector with multiple inner knots. If the degrees of the curves are not all the same we have to do a degree elevation to the highest one. By the way, this is also the procedure if you import a VDA-FS file into your CAD system.

---

[3]Pictures made with *Surfacer* from IMAGEWARE

Fig. 3 The given B–spline curve named pr_krumm and the faired one (max_error=0.20%, 0.09 sec.).

Fig. 4 The given B–spline curve named np113 and the faired one (max_error=0.30%, 0.06 sec.).

By fairing this B–spline curve with degree-folded inner knots the result may look like the curve in figure 5.

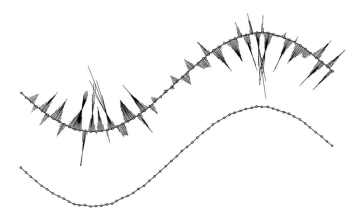

Fig. 5 The disturbed B–spline curve with multiple inner knots and the faired one.

The shown segments of the curve are smooth (they are nearly lines) whereas the whole curve is not smooth as the continuity of the segments is not optimized. To avoid this, two possible extensions will be described:

1. Remove as much as possible inner knots to get a minimum continuity of C^1 (or higher).
2. Use an extended fairness functional which is a mixture of energy and continuity.

The usual way to construct a set of splines is to obtain a continuity at the connecting points of at least C^1. If the curve has a continuity of C^1 at the degree-folded knot t_r, $t_r \neq t_{r-1}$,

$$\frac{k-1}{t_r - t_{r-1}}(\mathbf{d}_{r-1} - \mathbf{d}_{r-2}) = \frac{k-1}{t_{r+k-1} - t_{r+k-2}}(\mathbf{d}_r - \mathbf{d}_{r-1}) \tag{22}$$

must hold. In comparison with e.g. theorem 8 in [6] we see that this knot can be removed exactly and the new control points are determined by

$$\bar{\mathbf{d}}_i = \begin{cases} \mathbf{d}_i, & 0 \leq i \leq r-1 \\ \mathbf{d}_{i+1}, & r \leq i \leq n-1 \end{cases} \tag{23}$$

In the cases, the curve is not overall C^1 another possible procedure is to remove as much as possible inner knots with the method described in [6] to get a C^1 or higher continuity. Here, also an upper bound for the maximum deviation can be given. Thereafter this curve can be faired with the algorithm described before.

The second extension is to optimize not only the energy functional but also the continuity at the inner knots. By doing so the extended fairness functional

$$F = (1-\lambda)\int_a^b \left(\tilde{\mathbf{x}}^{(l)}(t)\right)^2 dt + \lambda \sum_{\substack{j=k \\ t_j \neq t_{j-1} \\ \texttt{mult}=k-1}}^n \left(\frac{\partial \tilde{\mathbf{x}}(t_j^-)}{\partial t} - \frac{\partial \tilde{\mathbf{x}}(t_j^+)}{\partial t}\right)^2, \quad 0 \leq j < 1 \tag{24}$$

is going to be used. Here `mult` denotes the multiplicity of the knot. Because of the multiplicity of the knots, the left and right derivative in (24) can be calculated easily. With

$$\frac{\partial \tilde{\mathbf{x}}(t_j^-)}{\partial t} = \frac{k-1}{t_j - t_{j-1}}(\tilde{\mathbf{d}}_{j-1} - \tilde{\mathbf{d}}_{j-2}), \tag{25}$$

$$\frac{\partial \tilde{\mathbf{x}}(t_j^+)}{\partial t} = \frac{k-1}{t_{j+k-1} - t_{j+k-2}}(\tilde{\mathbf{d}}_j - \tilde{\mathbf{d}}_{j-1}), \tag{26}$$

we obtain

$$\left(\frac{\partial \tilde{\mathbf{x}}(t_j^-)}{\partial t} - \frac{\partial \tilde{\mathbf{x}}(t_j^+)}{\partial t}\right)^2 = \tag{27}$$

$$(k-1)^2 \left(\frac{1}{\triangle_{j+k-2}}\tilde{\mathbf{d}}_j - \left(\frac{1}{\triangle_{j+k-2}} + \frac{1}{\triangle_{j-1}}\right)\tilde{\mathbf{d}}_{j-1} + \frac{1}{\triangle_{j-1}}\tilde{\mathbf{d}}_{j-2}\right)^2 \tag{28}$$

with the abbreviation $\triangle_j = t_{j+1} - t_j$. Here we can already see that we have to distinguish three cases: firstls, the control point is an *inner* point or the multiplicity is less than $k-1$, secondly, the control point lies at the segment connection point and thirdly, the control point is a neighbour of the segment connection point.

To find the optimal control point $\tilde{\mathbf{d}}_r$ we have to solve

$$\frac{\partial F(\tilde{\mathbf{d}}_r)}{\partial \tilde{\mathbf{d}}_r} \stackrel{!}{=} 0. \tag{29}$$

After some calculations and with the following abbreviations

$$\mathbf{D} = \begin{cases} \frac{1}{\Delta_{r+k-2}^2}\tilde{\mathbf{d}}_r - \frac{1}{\Delta_{r+k-2}}\left(\frac{1}{\Delta_{r+k-2}} + \frac{1}{\Delta_{r-1}}\right)\bar{\mathbf{d}}_{r-1} + \frac{1}{\Delta_{r-1}}\frac{1}{\Delta_{r+k-2}}\bar{\mathbf{d}}_{r-2} & , t_r \neq t_{r-1} \\ \left(\frac{1}{\Delta_{r+k-1}} + \frac{1}{\Delta_r}\right)^2 \tilde{\mathbf{d}}_r - \frac{1}{\Delta_{r+k-1}}\left(\frac{1}{\Delta_{r+k-1}} + \frac{1}{\Delta_r}\right)\bar{\mathbf{d}}_{r+1} - \\ \qquad\qquad \frac{1}{\Delta_r}\left(\frac{1}{\Delta_{r+k-1}} + \frac{1}{\Delta_r}\right)\bar{\mathbf{d}}_{r-1} & , t_{r+1} \neq t_r \\ \frac{1}{\Delta_{r+1}^2}\tilde{\mathbf{d}}_r - \frac{1}{\Delta_{r+1}}\left(\frac{1}{\Delta_{r+k}} + \frac{1}{\Delta_{r+1}}\right)\bar{\mathbf{d}}_{r+1} + \frac{1}{\Delta_{r+1}}\frac{1}{\Delta_{r+k}}\bar{\mathbf{d}}_{r+2} & , t_{r+2} \neq t_{r+1} \\ 0 & , else \\ & \text{or mult} \neq k-1 \end{cases} \quad (30)$$

and $\mathbf{D} = \tilde{D}_1\,\tilde{\mathbf{d}}_r + \tilde{\mathbf{D}}_2$ we do obtain the following formula for the new control point:

$$\tilde{\mathbf{d}}_r = -\frac{(1-\lambda)\sum_{\substack{i=i_0 \\ i\neq r}}^{i_1} \bar{\mathbf{d}}_i \int_a^b N_{i,k}^{(l)}(t)\,N_{r,k}^{(l)}(t)\,dt + \lambda(k-1)^2\,\tilde{\mathbf{D}}_2}{(1-\lambda)\int_a^b (N_{r,k}^{(l)}(t))^2\,dt + \lambda(k-1)^2\,\tilde{D}_1}. \quad (31)$$

Another spelling for (31) is

$$\tilde{\mathbf{d}}_r = (1-\alpha)\,\tilde{\mathbf{d}}_r^E + \alpha\,\tilde{\mathbf{d}}_r^C \quad (32)$$

with

$$\alpha = \frac{\lambda(k-1)^2\,\tilde{D}_1}{(1-\lambda)\int_a^b (N_{r,k}^{(l)}(t))^2 + \lambda(k-1)^2\,\tilde{D}_1} \quad (33)$$

where $\tilde{\mathbf{d}}_r^E$ and $\tilde{\mathbf{d}}_r^C$ are the new control points obtained by minimizing only the energy functional ($\lambda = 0$) or only the continuity ($\lambda = 1$). The incorporation of the distance tolerance (20) and the new control point is the same as described in section 2.3.

In the special case of a cubic B–spline curve with an uniform knot vector, the new control points are determined by (for $l = 2$):

$$\tilde{\mathbf{d}}_r^E = \begin{cases} -\frac{1}{4}\bar{\mathbf{d}}_{r-3} + \frac{3}{4}\bar{\mathbf{d}}_{r-1} + \frac{3}{4}\bar{\mathbf{d}}_{r+1} - \frac{1}{4}\bar{\mathbf{d}}_{r+3} & , \text{segment connections} \\ \frac{1}{2}(\bar{\mathbf{d}}_{r+1} + \bar{\mathbf{d}}_{r-1}) & , \text{inner points} \end{cases}$$

and

$$\tilde{\mathbf{d}}_r^C = \begin{cases} 2\,\bar{\mathbf{d}}_{r-1} - \bar{\mathbf{d}}_{r-2} & , t_r \neq t_{r-1} \\ \frac{1}{2}(\bar{\mathbf{d}}_{r+1} + \bar{\mathbf{d}}_{r-1}) & , t_{r+1} \neq t_r \\ 2\,\bar{\mathbf{d}}_{r+1} - \bar{\mathbf{d}}_{r+2} & , t_{r+2} \neq t_{r+1} \end{cases}$$

2.7 Example

In the following example you can see the given disturbed curve from figure 5 and the results of the two methods. Removing the inner knots in order to get a C^1 continuity leads to a smooth curve in a very short time. Using the new functional, the result in not of the same quality as in the case before and also the time is much higher, although the resulting curve is much better than the given one.

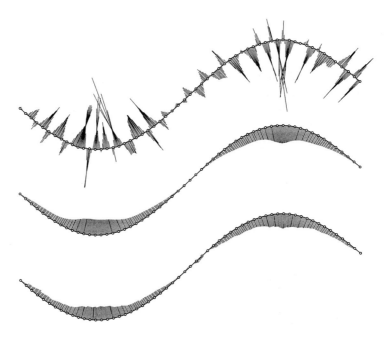

Fig. 6 Fairing the curve from figure 5(top) by removing knots (max_error=0.16%, 0.58 sec.) and with the new functional (max_error=0.66%, 2.3 sec.).

2.8 Fairing of Rational B–Spline Curves

Although the use of rational B–splines is not very common in CAD systems most of them can handle NURBS. One reason is that most of the algorithms for integral B–splines like the de Boor or the knot insertion algorithm can simply extend to rational B–splines but it is difficult for the designer to handle control points *and* weights.

A rational B–spline curve is defined by

$$\mathbf{x}(t) = \frac{\sum_{i=0}^{n} w_i \, \mathbf{d}_i \cdot N_{i,k}(t)}{\sum_{i=0}^{n} w_i \cdot N_{i,k}(t)} \quad , \quad t \in [t_{k-1}, t_{n+1}] \quad , \tag{34}$$

whereas the w_i are the so-called weights. We assume $w_i > 0$. Another spelling for (34) is

$$\mathbf{x}(t) = \sum_{i=0}^{n} \mathbf{d}_i \cdot R_{i,k}(t) \quad , \quad t \in [t_{k-1}, t_{n+1}] \tag{35}$$

with the rational basis functions

$$R_{i,k}(t) = \frac{w_i \cdot N_{i,k}(t)}{\sum_{j=0}^{n} w_j \cdot N_{j,k}(t)} \quad . \tag{36}$$

More details about NURBS can be found e.g in [9, 25].

In [18] a fairing method for NURBS is presented. In an optimization process the authors calculate a new set of weights to get a smoother curve. In contrast to their method we fixed the weights to get a linear problem.

Doing the same steps as for integral curves the new control point $\tilde{\mathbf{d}}_r$ is determined by

$$w_r \, \tilde{\mathbf{d}}_r = \sum_{\substack{i=i_0 \\ i \neq r}}^{i_1} \gamma_i \cdot w_i \, \mathbf{d}_i \tag{37}$$

with the weighting factors γ_i of the form

$$\gamma_i = -\frac{\int_a^b R_{i,k}^{(l)}(t) \, R_{r,k}^{(l)}(t) \, dt}{\int_a^b \left(R_{r,k}^{(l)}(t)\right)^2 \, dt} \quad . \tag{38}$$

The new control point is also an affine combination of the involved one but the integrals can not be calculated exactly as in the non-rational case. We use the Romberg quadrature to integrate the products of the rational basis function; see [2].

The ranking-number

$$z_r = \left(w_r \, \bar{\mathbf{d}}_r - w_r \, \tilde{\mathbf{d}}_r\right)^2 \cdot \int_a^b \left(R_{r,k}^{(l)}(t)\right)^2 \, dt \tag{39}$$

looks also very similar to the non-rational case.

If we want to fair with a distance tolerance we can use the upper bound

$$\max\{|\mathbf{x}(t) - \tilde{\mathbf{x}}(t)|\} \leq \frac{\max\{w_j\}}{\min\{w_j\}} \cdot \max\{|\mathbf{d}_i - \tilde{\mathbf{d}}_i|\} \leq \delta \quad , \tag{40}$$

which is a (roughly) upper bound for (19) in the rational case.

2.9 Examples

For the example shown in figure 7 the curve from figure 5 was handled as a rational curve with $w_i = 1$. Then the control points and the weights of the curve were disturbed with the help of random numbers. Because of the numerical quadrature the CPU-time is much larger than in the non-rational case.

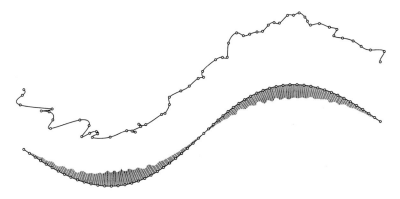

Fig. 7 The disturbed rational B–spline curve and the faired one (`max_error`=0.82%, 0.43 sec.).

Fig. 8 The disturbed rational B–spline curve and the faired one (`max_error`=0.64%, 8.523 sec.).

3 Fairing of B–Spline Surfaces

The fairing of B–spline surfaces is very similar to the method for fairing B–spline curves described in section 2. Therefore not all steps are described in detail.

As for curves we want to do the following: a given B–spline surface $\mathbf{X}$ has already been faired to a surface $\bar{\mathbf{X}}$. In the following step we want to change only one control point in order to get the smoother surface

$$\widetilde{\mathbf{X}}(u,v) = \sum_{\substack{i=0 \\ (i,j)\neq(r_1,r_2)}}^{m} \sum_{j=0}^{n} \bar{\mathbf{d}}_{ij} \cdot N_{i,k}(u)\, N_{j,l}(v) + \widetilde{\mathbf{d}}_{r_1 r_2} \cdot N_{r_1,k}(u)\, N_{r_2,l}(v) \quad . \tag{41}$$

Fairing of B–Spline Curves and Surfaces

This new surface should minimize a given fairness functional like (4), (5) or (6). An existing unique minimum is determined by the control point

$$\tilde{\mathbf{d}}_{r_1 r_2} = \sum_{\substack{i=i_0 \\ (i,j)\neq(r_1,r_2)}}^{i_1} \sum_{j=j_0}^{j_1} \gamma_{ij} \cdot \bar{\mathbf{d}}_{ij} \qquad (42)$$

with weighting factors γ_{ij} which depend on the fairness functional. In the case of the thin-plate energy (4), the following results are given:

$$\gamma_{ij} = -\frac{U_{i,r_1}^{2,2} V_{j,r_2}^{0,0} + 2 U_{i,r_1}^{1,1} V_{j,r_2}^{1,1} + U_{i,r_1}^{0,0} V_{j,r_2}^{2,2}}{U_{r_1,r_1}^{2,2} V_{r_2,r_2}^{0,0} + 2 U_{r_1,r_1}^{1,1} V_{r_2,r_2}^{1,1} + U_{r_1,r_1}^{0,0} V_{r_2,r_2}^{2,2}} \qquad (43)$$

with the abbreviations for the integrals

$$U_{i,j}^{r,s} = \int_{u_0}^{u_1} N_{i,k}^{(r)}(u) \, N_{j,k}^{(s)}(u) \, du \; ,$$

$$V_{i,j}^{r,s} = \int_{v_0}^{v_1} N_{i,l}^{(r)}(v) \, N_{j,l}^{(s)}(v) \, dv$$

and for the limits of the integrals and sums

$$i_0 = \max\{0, r_1 - k + 1\} \quad ; \quad i_1 = \min\{r_1 + k - 1, m\}$$
$$j_0 = \max\{0, r_2 - l + 1\} \quad ; \quad j_1 = \min\{r_2 + l - 1, n\}$$
$$u_0 = \max\{u_{r_1}, u_{k-1}\} \quad ; \quad u_1 = \min\{u_{r_1+k}, u_{m+1}\}$$
$$v_0 = \max\{v_{r_2}, v_{l-1}\} \quad ; \quad v_1 = \min\{v_{r_2+l}, v_{n+1}\} \; .$$

Fairing with a distance tolerance is nearly the same as in the curve case and also the fairing algorithm is the same. So, we do not go into more details in the surface case.

3.1 Examples

The following B-spline surfaces are benchmarks surfaces from *Mercedes–Benz AG*. The first one is part of a fender and the second one is part of a motor hood with a small bump in the middle. In both pictures the isophotes are used to visualize the quality of the surfaces and in both cases only the boundary curves are left fixed.

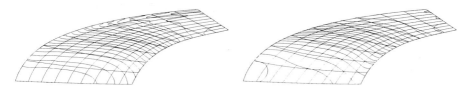

Fig. 9 Part of a fender before and after smoothing (`max_error`=0.02%, 0.43 sec.).

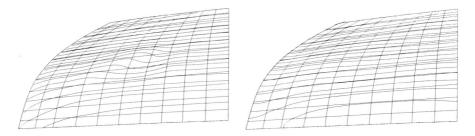

Fig. 10 Part of a motor hood before and after smoothing (max_error=0.16%, 0.76 sec.).

3.2 Extension to sets of Bézier– and B–Spline Surfaces

If one B–spline surface was a set of Bézier– or B–spline surfaces then the same problems which we had with curves do occur. Fairing without changing the method, the resulting patches are smooth but not the whole surface. Two possible solutions are given for this problem: remove as much as possible knots (see [5]) or use an extended functional.

One proposal for this new functional is

$$F = (1 - \lambda) \iint_A \mathbf{X}_{uu}^2 + 2\,\mathbf{X}_{uv}^2 + \mathbf{X}_{vv}^2 \, du \, dv$$

$$+ \lambda \sum_{\substack{i=k \\ u_i \neq u_{i-1} \\ \mathtt{mult}=k-1}}^{m} \sum_{j=l-1}^{n+1} \left(\frac{\partial \widetilde{\mathbf{X}}(u_i^-, v_j)}{\partial u} - \frac{\partial \widetilde{\mathbf{X}}(u_i^+, v_j)}{\partial u} \right)^2 \qquad (44)$$

$$+ \lambda \sum_{\substack{i=k-1 \\ v_j \neq v_{j-1} \\ \mathtt{mult}=l-1}}^{m+1} \sum_{j=l}^{n} \left(\frac{\partial \widetilde{\mathbf{X}}(u_i, v_j^-)}{\partial v} - \frac{\partial \widetilde{\mathbf{X}}(u_i, v_j^+)}{\partial v} \right)^2 ,$$

which is also a mixture of energy and continuity minimization.

As for curves, the solution of minimizing this functional can be splitted into an energy and continuity one, which should not be given here.

3.3 Example

Without going into details of the solution (its similar to curves), one example is given. It is constructed in such a way, that multiple knots are inserted in the given surface from figure 10 and the surface is disturbed with help of random numbers.

3.4 Conclusion

With this contribution we have given an overview of our methods for fairing B–spline curves and surfaces. The main idea was to minimize a quadratic fairness functional in a local scheme.

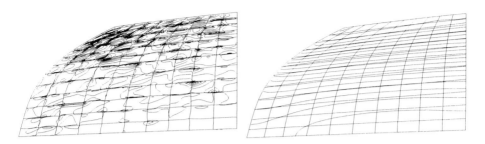

Fig. 11 The isophotes of the disturbed Bézier–spline surface and after fairing with the extended functional (max_error=0.6247%, 305 sec.).

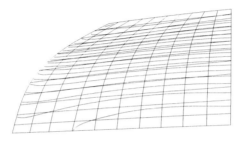

Fig. 12 The isophotes of the faired Bézier–spline surface with previous knot removal (max_error=0.66%, 8.34 sec.).

Changing only one control point in each step, solutions are given to smooth B–splines of general order and with general knot-vectors fulfilling a given distance tolerance.

Two strategies have been described for fairing sets of Bézier– and B–spline curves or surfaces. We came to the conclusion that if the curves or surfaces contain a large number of control points the better method would be to remove inner knots in order to get a C^1 continuity and then fairing with our *normal* method.

One possible extension to rational B–splines by using this method has been described as well.

3.5 Acknowledgment

Parts of the surface development were supported by the German *BMBF*.

References

[1] G. Brunnett, H. Hagen, and P. Sntarelli. Variational design of curves and surfaces. *Surveys on Mathematics for Industry*, (3):1–27, 1993.

[2] R. Bulirsch and J. Stoer. Numerical Quadratur by Extrapolation. *Numerische Mathematik*, (9):271–278, 1967.

[3] R. Courant and D. Hilbert. *Methoden der Mathematischen Physik, Zweite Auflage.* Springer, 1931.

[4] U. Dietz. Erzeugung glatter Flächen aus Meßpunkten. Preprint 1717, Technische Hochschule Darmstadt, Fachbereich Mathematik, 1995.

[5] M. Eck and J. Hadenfeld. A Stepwise Algorithm for Converting B-Splines. In P. J. Laurent, A. L. Méhauté, and L. L. Schumaker (eds.), *Curves and Surfaces in Geometric Design*, 131–138. A K Peters, Wellesley, MA, 1994.

[6] M. Eck and J. Hadenfeld. Knot removal for B-spline curves. *Computer Aided Geometric Design*, (12):259–282, 1995.

[7] M. Eck and J. Hadenfeld. Local Energy Fairing of B-Spline Curves. In G. Farin, H. Hagen, and H. Noltemeier (eds.), *Computing, Supplementum 10*, 129–147. Springer, 1995.

[8] M. Eck and R. Jaspert. Automatic Fairing of Point Sets. In Sapidis [28], 45–60.

[9] G. Farin (ed.). *NURBS for Curve and Surface Design*. SIAM, 1991.

[10] G. Farin. Degree Reduction Fairing of Cubic B-Spline Curves. In R. E. Barnhill (ed.), *Geometry Processing for Design and Manufacturing*, 87–99. SIAM, 1992.

[11] G. Farin. *Curves and Surfaces for Computer Aided Geometric Design. A Practical Guide.* Academic Press, 3rd edition, 1993.

[12] G. Farin, G. Rein, N. S. Sapidis, and A. J. Worsey. Fairing cubic B-spline curves. *Computer Aided Geometric Design*, (4):91–103, 1987.

[13] G. Farin and N. S. Sapidis. Curvature and the Fairness of Curves and Surfaces. *IEEE Computer Graphics & Applications*, 52–57, 1989.

[14] G. Greiner. Surface Construction Based on Variational Principles. In P. J. Laurent, A. L. Méhauté, and L. L. Schumaker (eds.), *Wavelets, Images and Surface Fitting*, 143–154. A K Peters, Wellesley, 1994.

[15] G. Greiner. Variational Design and Fairing of Spline Surfaces. *Computer Graphics Forum*, 13(3):143–154, 1994.

[16] J. Hadenfeld. Local Energy Fairing of B-Spline Surfaces. In M. Dæhlen, T. Lyche, and L. L. Schumaker (eds.), *Mathematical Methods for Curves and Surfaces*, 203–212. Vanderbilt University Press, 1995.

[17] H. Hagen and G. Schulze. Extremalprinzipien im Kurven- und Flächendesign. In J. L. E. ao, J. Hoschek, and J. Rix (eds.), *Geometrische Verfahren der Graphischen Datenverarbeitung*, 46–60. Springer, 1990.

[18] W. Hohenberger and T. Reuding. Smoothing rational B-spline curves using the weights in an optimization procedure. *Computer Aided Geometric Design*, (12):837–848, 1995.

[19] J. Hoschek. *Fundamentals of Computer Aided Geometric Design*. A K Peters, Wellesley, MA, 1993.

[20] M. Kallay. Contrained optimization in surface design. In B. Falcidieno and T. L. Kunii (eds.), *Modeling in Computer Graphics*, 85–93. Springer, 1993.

[21] J. A. P. Kjellander. Smoothing of bicubic parametric surfaces. *Computer-Aided Design*, 15(5):288–293, 1983.

[22] J. A. P. Kjellander. Smoothing of cubic parametric splines. *Computer-Aided Design*, 15(3):175–179, 1983.

[23] E. T. Y. Lee. Energy, fairness, and a counterexample. *Computer-Aided Design*, 22(1):37–40, 1990.

[24] H. Meier. Der differentialgeometrische Entwurf und die analytische Darstellung krümmungsstetiger Schiffsoberflächen. *VDI-Reihe 20*, (5), 1987.

[25] L. Piegl. On NURBS: A Survey. *IEEE Computer Graphics & Applications*, 55–71, 1991.

[26] N. Sapidis. Algorithms For Locally Fairing B-spline Curves. Master's thesis, Department of Mathematics, The University of Utah, 1987.

[27] N. S. Sapidis. Toward Automatic Shape Improvement of Curves and Surfaces for Computer Graphics and CAD/CAM Applications. In G. W. Zobrist and C. Sabharwal (eds.), *Progress in Computer Graphics*, 216–253. Ablex Publishing Corporation, 1992.

[28] N. S. Sapidis (ed.). *Shape Qualitiy in Geometric Modeling and Computer-Aided Design*. SIAM, 1994.

[29] N. S. Sapidis and G. Farin. Automatic fairing algorithm for B-spline curves. *Computer-Aided Design*, 22(2):121–129, 1990.

[30] R. Schaback. Error estimates for approximations from control nets. *Computer Aided Geometric Design*, (10):57–66, 1993.

[31] A. H. Vermeulen, R. H. Bartels, and G. R. Heppler. Integrating Products of B-Splines. *SIAM Journal of Scientific and Statistical Computing*, 13(4):1025–1038, 1992.

Author

Jan Hadenfeld
Technische Hochschule Darmstadt
Fachbereich Mathematik
Schloßgartenstraße 7
D-64289 Darmstadt
E–mail: hadenfeld@mathematik.th-darmstadt.de

Declarative Modeling of fair shapes: An additional approach to curves and surfaces computations

Marc Daniel

Institut de Recherche en Informatique, Ecole Centrale de Nantes

Abstract: Present day modeling systems constrain the designer to describe the studied object by means of lists of coordinates, values or geometric primitives which are often complex and tedious to create. Declarative modeling allows us to create shapes by only providing a set a geometric, topological or physical properties. The computer explores the universe of potential shapes, selecting those corresponding to the given definition. Using suitable tools, the designer chooses one or several shapes. A first example of curve modeling using declarative mechanisms is detailed.

1 Introduction

Current application packages frequently make extensive use of geometric modelers and image synthesis techniques. There exist powerful algorithms for visualizing highly realistic scenes. All available geometric modelers make it possible to construct complex shapes. Nevertheless, these modelers constrain the designer to describe the studied object by means of lists of coordinates, values or geometric primitives, which are often complex and tedious to create. Their role is limited to converting the data defining the already designed object into an internal numerical model. We call this way of working imperative modeling.

In our opinion, research must now focus on the development of modelers allowing us to describe objects using more abstract notions, based on properties and constraints. This is why we have introduced the concept of declarative modeling [1, 2]. The goal is to permit the creation of objects by only providing a set of abstract specifications, generally based on geometric, topological or physical properties. The role of the computer is to explore the universe of potential shapes to select those corresponding to the given definition. Using suitable tools, the designer chooses one or several shapes meeting his/her requirements.

The general notions exposed above have been tested through numerous studies which have led to an improved analysis of the mechanisms and the underlying difficulties of such an approach [3]. Promising results have been obtained, essentially in the field of polyhedra, imaginary landscapes or architecture [4, 5].

Our objective is now to apply the knowledge we gained in other fields to curve and surface modeling. We should note that programming under constraints and parameterization [6, 7, 8] are approaches partly pertaining to the declarative modeling process and their associated techniques must be included in a declarative modeler. The results stated in this book lead to mathematical techniques of higher level than manipulating control points directly and are also very important for declarative modeling. In the next paragraph, we present the different steps of a declarative design process. A first implementation for B-spline curve modeling (detailed in [9]) is proposed in paragraph 3. This implementation corresponds to an initial mock-up which only proves the feasibility and only allows us to illustrate the concept of declarative modeling. We are currently working on an improved version.

2 Principles of declarative modeling

Geometric declarative modeling consists in expressing the desired properties of the object, the modeler being in charge of computing all the numerical values leading to the definition of the object, for example the control points for curves and surfaces. The properties to be satisfied for an object are geometric or correspond to functional constraints. In order to describe, construct, and study the objects pertaining to the universes of shapes, one must have available the tools for description, generation and understanding. An overview of declarative modelers and its associated bibliography is proposed in [3].

2.1 Description tools

The basic idea consists in using a set of necessary and sufficient conditions to describe a set of objects completely. The main difficulty is to determine whether a precise vocabulary is associated to the given field. Specific vocabularies are shared by different applications (absolute or relative location (object, observer), space dividing (in 2 or 3 dimensions), ...). Descriptions can be entered in literal form (with keywords, or sentences in pseudo-natural language) with drawings or graphical inputs. This means that the user must choose the most convenient techniques for entering the description.

We can notice that stating properties does not exclude the need for very accurate modifications. Moving control points directly, and thus their coordinates seems unavoidable. Experience tells us that it may be more difficult to apply an accurate modification by giving a set of properties than operate directly on point coordinates (using a more or less automatic process). Declarative modeling must then be considered as a powerful tool for rough draft realization, obtained from the given properties. These drafts can evidently be the inputs of classical modelers which then appear as complementary to declarative modelers, the user getting rid of the most tedious part of the design.

2.2 Generation techniques

A problem solved through a declarative approach can have no solution, one solution, several or even an infinite number of solutions. The main problem is to transform the formal model into the geometric model. Algorithms specific to the given object, random sampling (which requires a parameterization of the objects) or methods allowing an accurate control on the produced solutions can be applied. There exist two different approaches: a generation in exploration mode which consists in computing all the solutions or a set of them, and a generation in sampling mode which provides one solution which can be modified by the designer. The latter can call for another one.

It seems that curve and surface modeling can be better studied with the sampling mode. This does not mean that several solutions cannot be obtained, which is necessary for a creative design. The declarative modeler must propose, on request, new solutions taking into account initial properties and additional constraints that may be added by the designer.

2.3 Understanding tools

Declarative modeling requires mechanisms for a quick understanding of created objects. This leads us to develop visualization techniques making the most of the known properties of the system. Different visualizations must be available (wire frame, removal of hidden parts, transparencies). All the object components, a part of them, or even a skeleton, will be shown. These different modes can simultaneously appear on the same image.

The possiblity of exploring visually a big number of potential solutions warrants the success of systems for the exploration of universes. Mechanisms for browsing through sets of solutions are absolutely necessary.

What is more, techniques to select a good view point automatically have also been studied. The underlying idea is that algorithms taking into account given and deduced properties would allow the modeler to automatically select a view point emphasizing such and such a property. Selecting a good view point is based on the notion of observation zones, themselves linked to the property to be emphasized.

2.4 Contributions of declarative modeling

There are numerous avantages to be derived from the declarative approach. First, of all, it frees the designer from a mathematical technique for modeling by having him describe the results he/she wishes to obtain in a "simple" vocabulary. The description is more or less complete, the software finds automatically the missing information by using default values or exploring the set of solutions. The vocabulary for description can be specific to a given field which provides the opportunity of using the system even to a non-specialist of geometric modeling. Numerous tests can also be easily carried out. Moreover, unexpected but original and interesting solutions can also be computed. The designer's pencil strokes can be translated into properties and modelled with our approach.

The advantages of declarative modeling are not to be found only at the stage of the initial creation of the objet but can also be felt during all the design process. As a matter of fact, it is possible to benefit from the modeler retaining all the knowledge about object properties. We have already introduced the notion of good view point. The automatic lighting of surfaces in order to detect drawbacks could be another application: we can imagine that all the techniques based on light rays automatically choose view points according to the properties and therefore the geometry of the object being studied. Problems of intersection are another example. In fact, it is well-known that the knowledge of properties on the objects often enables the simplification of the intersection algorithms.

3 An example: the declarative modeling of curves

3.1 Search for vocabulary

The study of vocabulary can be summarized as follows: how to describe a curve without giving a list of coordinates? Various past experiences proved that was possible. This does not mean that a set of words and/or an associated syntax can be easily deduced. The relevant vocabulary can be divided into three categories:

* mathematical vocabulary which is universal and undisputable. Shades of meaning cannot be introduced. Concave, convex, inflection, curvature, cups, ... belong to this category.

* qualitative vocabulary which allows shades of meaning, but is subjective and sometimes differently understood. Words such as flat, round, slender, ... belong to this category. The set of words best interpreted by a maximum number of persons must be defined.

* Quantifiers such as as too much, little, much, more, less very, ... enrich the description and allow modifications to occur.

On the other hand, a set of terms allows us to describe the functional constraints to be respected (reaching, beginning in, ending in, with such a length, surrounding such an area, ...). Introducing the constraints of a given application could be particularly valuable for the user.

The study of the description must be improved, but the principles of declarative modeling can be applied using a minimum vocabulary. Managing a dictionary of synonyms and antonyms is necessary in order to allow each designer to use his own words. Moreover, simply juxtaposing words is insufficient to describe a curve. Building sentences according to grammar rules is actually studied.

We consider here a formal description. We should recall (section 2.1) that entering the description can be achieved using different techniques. For example, it is easier for the user to state that a curve has no inflection rather than trying to manage the position of control points in order to satisfy this property. Let us consider a surface of revolution as a second example. It may be more convenient to enter a drawing for the profile curve, but it is always easier to state that the surface is a surface of revolution rather than to manually generate its control net.

3.2 Modeling technique - parameters

We chose to study the most frequently encountered curves: the B-splines (described for example in [10]. Uniform B-splines of order 4 (cubics) have been selected. The knot vector satisfies the classical multiplicities of extreme knots. It is thus completlely defined when the number of control points (and the order) is chosen. The relationships between a curve and the location of its control points are well mastered for cubics [11, 12]. These curves have many practical advantages (local control, computing time, ...). For creating a curve, we finally have to define the number of control points and their coordinates.

Note 1 *In what follows, we assume that a frame (O,x,y) is associated with our workspace so that it is possible to consider a curve segment (defined by an equation like $y=f(x)$) convex or concave.*

Declarative Modeling of fair shapes

3.3 Creation techniques

3.3.1 Bounding box

The first step is to specify the variation domain of control point coordinates. This is defined by a box bounding all the points. The techniques associated with the different cases cannot be described here. We only suggest a general sketch. The box is split into 9 regions TL, T, TR, L,C, R, BL, B, BR (see figure 1). The two extreme points of the curve are put in these regions depending on the required properties. Their coordinates are implicitely provided if conditions are requested (beginning in, ending in, ...). If necessary, default values are chosen. The box dimensions follow from the extreme point coordinates and their location in the box.

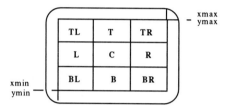

Fig. 1 Dividing the bounding box into 9 regions

3.3.2 Number of control points

The higher the number of control points, the more simple the local control of a curve is. Four points are requested to specify a cubic segment. The convexity of the associated control polygon implies the convexity (or concavity) of the corresponding B-spline curve segment. We choose to search within the curve description the number of segments whose curvature sign does not change. We call these segments: simple segments. Each of these segments will be associated to a group of 4 control points. The location of these points depends on the convexity or concavity. Figure 2 illustrates this process for a curve which at first is concave then convex. Nothing is said at the moment about the connection of these two segments. If nc is the inflection number, we obtain (nc+1) simple segments and 4(nc+1) points.

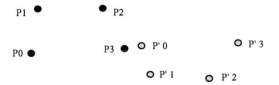

Fig. 2 Two groups of control points

The control polygon of the B-spline is obtained by considering the set of previous points and by merging the extreme points of the simple segments. The chosen number of control points is (4+ 3nc). This construction does not correspond to the definition of (nc+1) elementary cubic segments, but to that of (3nc+1) elementary cubic segments. It can be demonstrated that the number nc of inflections is automatically respected. Therefore, the creation technique relies on the manipulation and the positioning of groups of four points, two of them being called extreme ones flanking two others called middle ones. These points are not passing points. They only control indirectly the intermediate segments. However, the convex hull property and the local modeling of the B-splines allows us to master the global position of the curve by manipulating these points. This solution needs many points to be defined. If necessary, data reduction could be considered.

3.3.3 Location of control points

The proposed technique deals separately with the abscissa and the ordinates of the points. The abscissa of the extreme control points of simple segments are equally distributed between the two extreme abscissa. The variation of abscissa is therefore monotonous along the curve, this excluding shapes like loops. Then, for each simple segment, we concentrate on the abscissa of the two middle control points B and C located between two extreme points A and D which are already fixed. The interval of abscissa between A and D is divided into 6 areas of equal width. Point B can be located in one of the three areas closest to point A, point C being located in one of the three areas closest to D. The arguments given above are valid for a Bézier cubic and its four control points. It remains valid, although an approximate one, for our process. The relative positions of the abscissa of points A and B on the one hand and C and D on the other hand influence the more or less tight look of the curve segment tips. Therefore, the containing area of the abscissa of points B and C is chosen with respect to the desired properties concerning tensions at extremities, as shown on figure 3. The abscissa of each point is randomly computed within the interval defining each area.

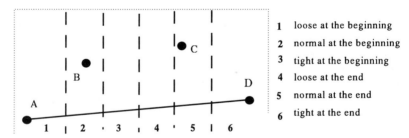

Fig. 3 Areas for abscissa location

The proposed technique for computing the ordinates considers the horizontal line passing through the known point A. The ordinates of points B and C are computed as previously, with respect to the desired convexity. Location of point D is then computed from the ordinate of point C and from the more or less bulging look of the next segment, by reproducing a division process into 6 areas between the horizontal line passing through point C and the limit of the bounding box (see figure 4).

3.4 Modification

Modifications must be possible but, the solutions do not necessarily correspond to the user's wishes. Modifications associated to simple segments or over the whole curve have been im-

Declarative Modeling of fair shapes

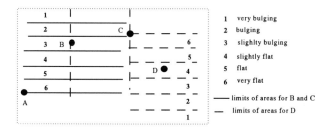

Fig. 4 Technique for ordinate location

plemented. For a simple segment, it is possible to modify the more or less tight look at each end and/or the more or less bulging look. In the first case, quantifiers more or less allow us to increment or to decrement by one unit the number of the sampling area of the abscissa. In the second case, the location areas of the ordinates are modified, and expressions such as more and more and less and less allow us to make a 2 area change of the ordinates. For each allowed modification, new values are randomly computed. Modifying the tension of the curve (strain energy) is provided. A curve can be too tight or too loose, globally or locally. Therefore, one must decrease or increase the strain energy. This is possible at the beginning, at the end, in the middle or all along the curve. We used an optimising process, with proximity constraints, in order to keep the look of the curve.

3.5 Results

We should recall that default values are used when the desired properties do not permit us to obtain a complete definition. Figure 5 illustrates a solution to a curve (left), a very flat curve (middle) and a very bulging curve (right). Figure 6 (left) is an example of a loose curve at the beginning and at the end, when the example of figure 6 (right) corresponds to a tight curve at the beginning and at the end. Figure 7 (left) is a solution to the description: concave very flat then convex very bulging. The properties less flat at the beginning and a little less bulging at the end leads to figure 7 (middle). Starting from the latter, the request of a tighter curve gives the curve on figure 7 (right).

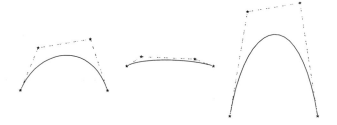

Fig. 5 A curve (left), a very flat curve (middle), a very bulging curve (right)

In conlusion, one could say that the associated software program can only be considered as a first attempt, making it possible to demonstrate that feasibility is obtainable, but also

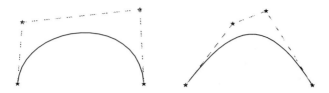

Fig. 6 Tension effect on a curve: loose (left) (tight (right)) at the beginning and at the end

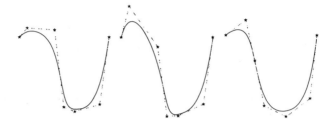

Fig. 7 Modification of a curve: initial curve (left), less flat at the beginning and little less bulging at the end (middle), globally looser (right)

pointing at the inherent difficulties. One of the main difficulties is to be able to define a set of words and a syntax, allowing the designer to describe a curve. The interface permits the user to obtain easily the properties of the different curve segments, as well as the desired modifications. He/she can control a very partial exploration of the solutions through a new series of random samples. The number of control points of the solution must be controlled carefully to avoid an unnecessarily high number of points. If this occurs, a strategy of data reduction can be applied. We must also ensure that the time used to describe the object must be less than the time required to define the object by manipulating control points. We should recall that we are more interested with a draft than a very accurate object and that although the description of the object is given once, it may be used during the entire design process (see section 2.4).

4 Conclusion

The aim of this paper is to draw attention on this new approach in curve and surface modeling and on the advantages it offers. Numerous modelers, dealing with very different universes of shapes are under development in Nantes (France), within the ExploFormes project.

We have initiated this work by studying planar curves, and what we offer here is only a first solution which proves the feasibility of the process. However, we realize this is a big challenge, and we hope the present paper gave an insight into what the next generation of geometric modelers could look like.

A new version is currently being developed. It will allow us to overcome the restrictions encountered in this first mock-up. Computer techniques will be slightly different but still correspond to the principles exposed in section 2. We also plan to study surfaces with a declarative approach.

References

[1] Lucas, M., Martin, D., Martin, P., Plémenos, D.: *Le projet ExploFormes, quelques pas vers la modélisation déclarative de formes*, Bigre n. 67 (1990), 35–49.

[2] Lucas, M.: *Equivalence Classes in Object Shape Modelling*, IFIP TCS/WG 5.10 Working Conference in Computer Graphics, Tokyo (1991), 8–12.

[3] Desmontils, E.: *Les modeleurs déclaratifs*, Research Report IRIN n. 95 (1995).

[4] Martin, D., Martin, P.: *Declarative Generation of a Family of Polyhedra*, Graphicon'93, Saint-Petersburg (1993).

[5] Chauvat, D., Martin, P., Lucas, M.: *Modélisation déclarative et contrôle spatial*, Revue Internationale de CFAO et d'Infographie **8** (1) (1994), 41–49.

[6] Hoffmann, C. M., Peters, J.: *Geometric Constraints for CAGD, in Mathematical Methods for Curves and Surfaces*, In M. Dæhlen, T. Lyche, and L. L. Schumaker (eds.): Mathematical Methods in CAGD III, (1995), 237–254.

[7] Bouma, W., Fudos, I., Hoffmann, C., Cai, J., Paige, R.: *A Geometric Constraint Solver*, Computer Aided Design **27** (1995), 487–502.

[8] Wesselink, W., Veltkamp, R. C.: *Interactive Design of Constrained Variational Curves*, Computer Aided Geometric Design **12** (1995), 533–546.

[9] Daniel, M.: *Une première approche de la modélisation déclarative de courbes*, Research Report IRIN n. 81 (1995).

[10] Farin, G.: *Curves and Surfaces for Computer Aided Geometric Design – A Practical Guide*. 3. ed. Academic Press 1993.

[11] Stone, M. C., Derose, T. D.: *A Geometric Characterization of Parametric Cubic Curves*, ACM Transactions on Graphics **8**, n. 3 (1989), 147–163.

[12] Kantorowitz, E., Schechner, Y.: *Managing the Shape of Planar Spline by their Control Points*, Computer Aided Design **25** (1993), 355–364.

Author

Prof. Dr. M. Daniel
Inst. de Rech. en Informatique de Nantes
Ecole Centrale de Nantes
1, rue de la Noe
F–44072 Nantes Cedex 03

Shape Preserving of Curves and Surfaces

Shape-preserving interpolation with variable degree polynomial splines

Paolo Costantini

Dipartimento di Matematica, Università di Siena

Abstract *The main goal of this paper is to present some results obtained in functional shape-preserving interpolation using variable degree polynomial splines, and show how these functions are emerging as a powerful tool both in tension methods and in CAGD applications.*

§ 1. Introduction

Since Schweikert's 1966 pioneering work [62] tension methods have turned out to be one of the most effective approaches for avoiding those inflections and oscillations extraneous to data which often affect the "standard" interpolating functions, and have therefore been widely studied and improved during the following years.

For a brief introduction we recall they basically consist in constructing C^k, $k \geq 1$ piecewise functions depending on a set of parameters which are selected, in a local or global way and with automatic or interactive choice, to control the shape of the interpolants, stretching their patches between data points. An analysis of the papers appeared in the literature shows how the scientific production is, like for any other aspect of interpolation or approximation, more conspicuous in the 1-D than in the 2-D setting, especially with respect to the important case of scattered data. It is also evident that the largest part of them deals with exponential or rational functions, which are in general conceived to reproduce some of the useful properties of cubic piecewise polynomials (it is well known that cubics are the main "building box" in practical interpolation/approximation/CAGD applications).

As a sample from the literature we mention [5], [48], [49], [51], [52], [53], [55], [56], [59], [60], [62], [63], [64] and [66] for 1-D exponential tension methods, [4], [20], [21], [22], [23], [24], [37], [38], [39] and [43] for 1-D rational splines and [58] for the 2-D case. For the sake of completeness we mention also T. Foley's C^1 piecewise cubics ([27], [28], [29]), whose patches can be stretched in correspondence of a proper selection of a set of weights inserted in a quadratic objective functional. Although these piecewise cubics have typical tension properties, they follow a different idea (the variational approach) and no additional parameters are used to define the polynomial patches.

Since this paper is mainly concerned with piecewise polynomials and functional interpolation, we omit detailed comparisons with parametric oriented tools, like ν-splines ([47]), β-splines ([1]), NURBS (for which the reader is referred to any textbook on CAGD) and with T.N.T. Goodman and colleagues' parametric shape-preserving rational splines (the reader is now referred to the corresponding papers in this volume, to the references quoted therein and to [31]–[36]).

With the exception of particular choices of rationals, the main drawback of these kind of tension functions is that they have no geometrical counterpart (like the control net for Bèzier polynomials); this, in turn, makes the computation of shape–constraints a difficult task – almost impossible for 2-D interpolation – and avoid their direct use in CAGD applications.

However, in the last years polynomial alternatives to rationals and exponentials have been introduced and have been successfully applied in several shape-preserving problems. Their structure and properties will be explained in detail later; here it suffices to say that the n_i-degree polynomial piece, which is a cubic for $n_i = 3$ and tends to be linear as n_i tends to infinity, has a Bèzier net as simple as the cubic one. This has allowed very easy formulations of the shape constraints, and has therefore opened the way to several 2-D extensions. The additional features of numerical stability and low computational cost (practically equivalent to the cubic case for any n_i), have made this approach a very effective one for functional shape-preserving interpolation.

At present, the interest in these variable degree polynomial splines is growing because further results in 2-D and 3-D parametric curves have been developed by P.D. Kaklis and colleagues (see the corresponding paper in this volume and the references therein) and, at the same time, applications in the areas of (shape-preserving) curve fairing and in curve and surface modeling seem possible and are under current research.

Aim of this paper is therefore to provide a short survey of some of those results (so far spread in several papers) which have practical applications in interpolation and to discuss some possible applications in CAGD. Since many of the constructive methods can, in principle, be applied to other kinds of tension functions, instead of giving heavy analytical details we intend to furnish an intuitive explanation, with the hope of providing some help for further developments.

The present paper is divided in 6 sections. The next one is devoted to introduce the basic 1-D ideas which will be extended in sections 3 and 4 to 2-D interpolation of regular and scattered data. In section 5 we will discuss some applications in curve design, and end in section 6 with final conclusions and remarks.

One of the most important features of these piecewise functions of arbitrary degree is that the polynomial pieces are taken from a four dimensional space equivalent to (that is with same characteristics of) $\mathbb{P}_3$, and, as a consequence, all the properties of piecewise cubics can in principle be reproduced. Before moving to more technical arguments, it is worthwhile to say that in sections 2–5 are discussed those applications similar to the C^1, local piecewise, cubics ones. Of course, the n_i-degree polynomial patches can be connected in a global way to form C^2 splines, equivalent to classical C^2 cubics. We omit the discussion concerning the important methods obtainable from this point of view, suggesting the reader to complete the results presented here with those reported in [41], [42], where variable degree schemes, similar to C^2 cubic splines interpolating at the knots, are developed.

We conclude this introductory section anticipating that, for reasons of space, in this paper are reported only those graphical examples which are either necessary for the comprehension of the argument or which have never appeared before. The boring task of integrating this paper with the quoted ones is therefore requested to the reader.

§ 2. Basic one–dimensional ideas

Aim of this section is to describe the structure and the properties of a particular space of polynomial functions, and discuss their applications in shape–preserving interpolation.

Let the following set of data $(x_i, f_i, f_i^{(1)})$; $i = 0, 1, \ldots, N$, with $x_0 < x_1 \ldots < x_N$, and, associated to an arbitrary interval $[x_i, x_{i+1}]$, let the integer $n_i \geq 3$ be given. Now, having set $h_i := x_{i+1} - x_i$, let $l(x; n_i)$ be the piecewise linear function connecting the following four points:

$$(x_i, f_i) \; ; \; (x_i + \frac{h_i}{n_i}, f_i + f_i^{(1)}\frac{h_i}{n_i}) \; ; \; (x_{i+1} - \frac{h_i}{n_i}, f_{i+1} - f_{i+1}^{(1)}\frac{h_i}{n_i}) \; ; \; (x_{i+1}, f_{i+1}) \; ,$$

and let $b(x; n_i)$ be its corresponding Bèzier polynomial of degree n_i:

$$b(x; n_i) := \sum_{r=0}^{n_i} \binom{n_i}{r} l\left(\frac{r}{n_i}; n_i\right) (x - x_i)^r (x_{i+1} - x)^{n_i - r} , \qquad (1.a)$$

which, in the case $n_i = 3$, reduces to a cubic polynomial.

It is immediate to check that, for any n_i:

$$b(x_i; n_i) = f_i \; , \quad b(x_{i+1}; n_i) = f_{i+1} \; ,$$

$$\frac{d}{dx} b(x; n_i)|_{x_i} = f_i^{(1)} \; , \quad \frac{d}{dx} b(x; n_i)|_{x_{i+1}} = f_{i+1}^{(1)} \; ,$$

and that, as n_i increases, $b(.; n_i)$ tends to uniformly approximate the straight line joining (x_i, f_i) and (x_{i+1}, f_{i+1}). Obviously, we are expecting a decay from the cubics' fourth order approximation to the second of linear interpolants; this can in fact be proven and we refer to [11] for several types of error estimates. Here we limit ourselves to recall the following result, because of it extensions to the bivariate schemes of the following sections: let $f_i = f(x_i)$, $f_i^{(1)} = f'(x_i)$ where $f \in C^4[x_0, x_N]$; then, for any n_i $\|f - b(\cdot; n_i)\| = O(h^2)$, where h is the maximum knot spacing.

Before going into details concerning shape–preserving properties, it is important to observe that any polynomial of the form (1) can be expressed as

$$b(\cdot; n_i) := \mathcal{H}^{n_i} f$$
$$= f_i H_i^0(\cdot) + f_i^{(1)} H_i^1(\cdot) + f_{i+1} H_{i+1}^0(\cdot) + f_{i+1}^{(1)} H_{i+1}^1(\cdot) \quad (1.b)$$

where $H_i^0 = H_i^0(\cdot; n_i)$, $H_i^1 = H_i^1(\cdot; n_i)$, $H_{i+1}^0 = H_{i+1}^0(\cdot; n_i)$, $H_{i+1}^1 = H_{i+1}^1(\cdot; n_i)$ are the Hermite basis functions given by formulas analogous to (1.a) in which the piecewise linear functions are defined setting $(f_i, f_i^{(1)}, f_{i+1}, f_{i+1}^{(1)})$ equal, respectively, to $(1, 0, 0, 0)$, $(0, h_i/n_i, 0, 0)$, $(0, 0, 0, 1)$, $(0, 0, -h_i/n_i, 0)$. We set

$$BX_{n_i} := span\{H_i^0, H_i^1, H_{i+1}^0, H_{i+1}^1\} .$$

We observe also that

$$b(x; n_i) = r(x) + e_i (x_{i+1} - x)^{n_i} + e_{i+1} (x - x_i)^{n_i} , \quad (1.c)$$

where $r(x)$ is the straight line through $(x_i + h_i/n_i, f_i + f_i^{(1)} h_i/n_i)$, $(x_{i+1} - h_i/n_i, f_{i+1} - f_{i+1}^{(1)} h_i/n_i)$ and $e_i = f_i - r(x_i)$, $e_{i+1} = f_{i+1} - r(x_{i+1})$; therefore, the computational cost in evaluating our polynomial is practically independent of the degree.

A straightforward consequence of its definition is that, for sufficiently large n_i, $b(x; n_i)$ is shape preserving, that is has the same shape induced by the data f_i, $f_i^{(1)}$, f_{i+1}, $f_{i+1}^{(1)}$. We need more precise definitions. Let $\Delta_i := (f_{i+1} - f_i)/h_i$; we will say the data are increasing (INC) (decreasing (DEC)) in $[x_i, x_{i+1}]$ if

$$f_i^{(1)} > 0, \ \Delta_i > 0, \ f_{i+1}^{(1)} > 0 \ ; \ (f_i^{(1)} < 0, \ \Delta_i < 0, \ f_{i+1}^{(1)} < 0) \quad (2)$$

and convex (CVX) (concave (CNC)) if

$$f_i < \Delta_i < f_{i+1} \quad (f_i > \Delta_i > f_{i+1}). \quad (3)$$

Suppose the data are INC; it is shown in [8] that the Bèzier polynomial defined in (1) is INC in $[x_i, x_{i+1}]$ if $(f_i^{(1)}, f_{i+1}^{(1)}) \in D_{INC}(\Delta_i, n_i)$, where $D_{INC}(\Delta_i, n_i)$ is the convex hull defined by the points $(0, 0)$, $(n_i \Delta_i, 0)$, $(\rho(n_i)\Delta_i, \rho(n_i)\Delta_i)$, $(0, n_i \Delta_i)$ and

$$\rho(n_i) := \left(\frac{n_i}{n_i - 2} \sum_{j=1}^{n_i - 2} \binom{n_i - 1}{j} \right) \Big/ \left(\frac{2}{n_i - 2} \sum_{j-1}^{n_i - 2} \binom{n_i - 1}{j} - 2 \right)$$

(see fig. 1). Since it is possible to show that $n_i/2 < \rho(n_i) \leq n_i$ and $\rho(n_i) < \rho(n_i + 1)$, it follows that $D_{INC}(\Delta_i, n_i) \subset D_{INC}(\Delta_i, n_i + 1)$ and that $D_{INC}(\Delta_i, n_i)$ tends "to cover", as n_i tends to infinity, the first quadrant of the plane. We note parenthetically that, in the case $n_i = 3$, $\rho(n_i) = 3$ and

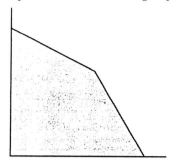

Figure 1. The increase domain $D_{INC}(\Delta_i, n_i)$.

$D_{INC}(\Delta_i, n_i)$ corresponds to the square of vertices $(0,0)$, $(3\Delta_i, 0)$, $(3\Delta_i, 3\Delta_i)$, $(0, 3\Delta_i)$, which is the well known increase domain for cubic polynomials introduced for the first time in [30].

In a similar way, let us suppose the data are CVX; the polynomial (1) is CVX if, and only if ([7]), $(f_i^{(1)}, f_{i+1}^{(1)}) \in D_{CVX}(\Delta_i, n_i)$ where

$$D_{CVX}(\Delta_i, n_i) := \{(u,v) \in \mathbb{R}^2 \text{ s.t. } u + (n_i - 1)v - n_i\Delta_i > 0 \,;$$
$$(n_i - 1)u + v - n_i\Delta_i > 0 \},$$

(see fig. 2) and from elementary computations we have again that $D_{CVX}(\Delta_i, n_i) \subset D_{CVX}(\Delta_i, n_i + 1)$ and that $D_{CVX}(\Delta_i, n_i)$ tends "to cover", as n_i tends to infinity, the set $\{(u,v) \in \mathbb{R}^2 \text{ s.t. } u < \Delta_i < v\}$. Obviously, similar properties hold for the "decrease" and "concavity" domains, $(D_{DEC}(\Delta_i, n_i), D_{CNC}(\Delta_i, n_i))$, which are defined, respectively, assuming $\Delta_i < 0$ and reversing the above inequalities. We can therefore state the following property.

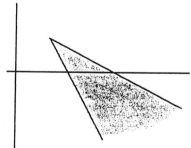

Figure 2 The concavity domain $D_{CNC}(\Delta_i, n_i)$.

Theorem 1. *Let the data be INC (DEC) and/or CVX (CNV) in $[x_i, x_{i+1}]$. There exists a "threshold" degree $\tilde{n}_i$ such that, for any $n_i \geq \tilde{n}_i$,*

$$(f_i^{(1)}, f_{i+1}^{(1)}) \in D_{INC}(\Delta_i, n_i) \quad ((f_i^{(1)}, f_{i+1}^{(1)}) \in D_{DEC}(\Delta_i, n_i))$$

and/or

$$(f_i^{(1)}, f_{i+1}^{(1)}) \in D_{CVX}(\Delta_i, n_i) \quad ((f_i^{(1)}, f_{i+1}^{(1)}) \in D_{CNC}(\Delta_i, n_i)),$$

that is the corresponding polynomial given in (1) has the same shape of the data in the same domain.

As a trivial consequence of theorem 1, we can compute a C^1, piecewise polynomial spline $s = s(x)$, which interpolates the data $(x_i, f_i, f_i^{(1)})$; $i = 0, 1, \ldots, N$ and is shape–preserving, simply connecting polynomials of the form (1) and choosing the lowest degree suggested by the theorem above (that is $n_i = \tilde{n}_i$).

So far, we have supposed to deal with Hermite-type interpolation, assuming as known the derivatives at the interpolation points, and this is not, of course, the most common situation occurring in practice. However, in this one–dimensional setting, it is not difficult to provide shape–preserving estimates for the derivatives starting from the only data (x_i, f_i) ; $i = 0, 1, \ldots, N$ (for example any average of data slopes in convex position will satisfy (3)). If we want, in addition, to obtain the lowest values for the sequence of degrees (which should produce the most "visually pleasing" graph), we can use an "abstract scheme" (see the corresponding paper in this volume) starting with an uniform sequence of low degrees (e.g. $n_i = 3$, all i) and then increasing the degrees where necessary, as suggested in [12] and [13].

We want to conclude this section observing that all the results discussed above could have been stated for C^k piecewise polynomial splines with degrees $n_i \geq 2k+1$. However, this continuity is somewhat fictitious, because $s^{(j)}(x_i) = 0$; $j = 2, \ldots, k$ and, since the corresponding "flats" at knots make them useless in practical applications, we have preferred not to waste space in further details.

§ 3. Interpolation of rectangular data

Let $(x_i, y_j, f_{i,j}^{(0,0)}, f_{i,j}^{(1,0)}, f_{i,j}^{(0,1)}, f_{i,j}^{(1,1)})$; $i = 0, 1, \ldots, N$; $j = 0, 1, \ldots, M$ be a given set of data, which, as usual, are supposed to be obtained from a sufficiently differentiable function f:

$$f_{i,j}^{(p,q)} = \frac{\partial^{p+q}}{\partial x^p \partial y^q} f(x,y)|_{(x_i, y_j)}; \quad i = 0, \ldots, N, \; j = 0, \ldots, M; \; p, q = 0, 1 , \quad (4)$$

let $h_i := x_{i+1} - x_i$, $k_j := y_{j+1} - y_j$ and let $R_{i,j} := [x_i, x_{i+1}] \times [y_j, y_{j+1}]$, $R := [x_0, x_N] \times [y_0, y_M]$. Our first goal is to extend the results of the previous section in a tensor–product structure. Let $n_0, n_1, \ldots, n_{N-1}$ and $m_0, m_1, \ldots, m_{M-1}$ be sequences of degrees, and let BY_{m_j} be the four dimensional linear space analogous to BX_{n_i} introduced in the previous section, that is

$$BY_{m_j} := span\{H_j^0, H_j^1, H_{j+1}^0, H_{j+1}^1\},$$

where $H_j^0, H_j^1, H_{j+1}^0, H_{j+1}^1$ are the cardinal Hermite basis functions defined in $[y_j, y_{j+1}]$. We start setting

$$BXY_{n_i, m_j} := BX_{n_i} \otimes BY_{m_j}$$

and, having defined $H^{p,q}_{r,s} := H^p_r H^q_s$, we take the unique $b \in BXY_{n_i,m_j}$ interpolating the data (4):

$$b = b(\cdot,\cdot;n_i,m_j) := \sum_{r=i}^{i+1}\sum_{s=j}^{j+1}\sum_{p,q=0}^{1} f^{(p,q)}_{r,s} H^{(p,q)}_{r,s}(\cdot,\cdot) . \qquad (5)$$

We will say the data are INC and/or CVX in $R_{i,j}$ along the edge $y = y_s$, $s = j, j+1$ with respect to x if (2) and/or (3) hold for the corresponding $f^{(1,0)}_{i,s}$, $(f^{(0,0)}_{i+1,s} - f^{(0,0)}_{i,s})/h_i$, $f^{(1,0)}_{i+1,s}$; we will say the data are INC and/or CVX w.r. to x in $R_{i,j}$ if they are INC and/or CVX for $s = j$ and $s = j+1$. Of course similar definition could be stated for DEC and CNC cases and w.r. to y; we remark that the mixed partials derivatives $f^{(1,1)}_{i,j}$ turn out to have no influence for our purposes because of the strict inequalities we have adopted in (2) and (3). Our objective is to construct a polynomial of the form (5) which is shape–preserving, that is has the same shape of the data in the same domain.

As in the one–dimensional case, any b can be defined and controlled by its Bèzier net, which is given by the continuous piecewise bilinear function (made up by nine patches) interpolating the 16 points

$$\left(x_i + r\frac{h_i}{n_i}, y_j + s\frac{k_j}{m_j}, f^{(0,0)}_{i,j} + \Theta(r,s)\right), \quad r,s = 0,1$$

$$\Theta(r,s) := r\frac{h_i}{n_i}f^{(1,0)}_{i,j} + s\frac{k_j}{m_j}f^{(0,1)}_{i,j} + rs\frac{h_i k_j}{n_i m_j}f^{(1,1)}_{i,j} ;$$

$$\left(x_{i+1} - r\frac{h_i}{n_i}, y_j + s\frac{k_j}{m_j}, f^{(0,0)}_{i+1,j} + \Theta(r,s)\right), \quad r,s = 0,1$$

$$\Theta(r,s) := -r\frac{h_i}{n_i}f^{(1,0)}_{i+1,j} + s\frac{k_j}{m_j}f^{(0,1)}_{i+1,j} - rs\frac{h_i k_j}{n_i m_j}f^{(1,1)}_{i+1,j} ;$$

$$\left(x_i + r\frac{h_i}{n_i}, y_{j+1} - s\frac{k_j}{m_j}, f^{(0,0)}_{i,j+1} + \Theta(r,s)\right), \quad r,s = 0,1$$

$$\Theta(r,s) := r\frac{h_i}{n_i}f^{(1,0)}_{i,j+1} - s\frac{k_j}{m_j}f^{(0,1)}_{i,j+1} - rs\frac{h_i k_j}{n_i m_j}f^{(1,1)}_{i,j+1} ;$$

$$\left(x_{i+1} - r\frac{h_i}{n_i}, y_{j+1} - s\frac{k_j}{m_j}, f^{(0,0)}_{i+1,j+1} + \Theta(r,s)\right), \quad r,s = 0,1$$

$$\Theta(r,s) := -r\frac{h_i}{n_i}f^{(1,0)}_{i+1,j+1} - s\frac{k_j}{m_j}f^{(0,1)}_{i+1,j+1} + rs\frac{h_i k_j}{n_i m_j}f^{(1,1)}_{i+1,j+1} ,$$

and it is evident that this bivariate net tends, as n_i, m_j go to infinity, to the bilinear function interpolating the four data points $f^{(0,0)}_{\cdot,\cdot}$. We have therefore the following theorem, which is analogous to, but not a trivial extension of, theorem 1 of section 1 ([14]).

Theorem 2. *Let the data be INC (DEC) and/or CVX (CNV) w.r. to x along the edge $y = y_j$ or $y = y_{j+1}$ or inside $R_{i,j}$ and/or w. r. to y along the edge $x = x_i$ or $x = x_{i+1}$ or inside $R_{i,j}$. There exist "threshold" degrees $\tilde{n}_i, \tilde{m}_j$ such that, for any $n_i \geq \tilde{n}_i$, $m_j \geq \tilde{m}_j$ the corresponding polynomial given in (5) has the same shape of the data in the same domain.*

Theorem 2 above says we can assign a certain "kind" of shape to any sub rectangle $R_{i,j}$ (for example INC and CVX along $y = y_j$, CVX along $y = y_{j+1}$ – and therefore CVX inside $R_{i,j}$ – DEC along $x = x_i$ and "nothing" along $x = x_{i+1}$) and find threshold integers s.t. any bivariate, interpolating Bèzier polynomial defined putting in (5) larger degrees, is stretched so much as to have the same shape of the bilinear.

Obviously, we obtain a C^1 piecewise polynomial function interpolating all the data in R, simply connecting together patches of the form (5). The sequences of degrees $n_0, n_1, \ldots, n_{N-1}$, $m_0, m_1, \ldots, m_{M-1}$ which satisfy the constraints, that is such that $b(\cdot, \cdot; n_i, m_j)$ is shape–preserving in $R_{i,j}$, can be computed using the following algorithm (note that n_i and m_j hold, respectively, for all the sub rectangles of the strips $x_i \leq x \leq x_{i+1}$, $y_j \leq y \leq y_{j+1}$).

Algorithm 1.

1. For $i = 0, 1, \ldots, N - 1$
   1.1 For $\mu = 0, 1, \ldots, M - 1$
     1.1.1. Compute the "threshold" degrees $\tilde{n}_{i,\mu}, \tilde{m}_{i,\mu}$ which satisfy the constraints in $R_{i,\mu}$ according to Theorem 2.
   1.2. Set $n_i := max\{n_{i,\mu}, \mu = 0, 1, \ldots, M - 1\}$.
2. For $j = 0, 1, \ldots, M - 1$
   2.1 For $\nu = 0, 1, \ldots, N - 1$
     2.1.1. Compute the "threshold" degrees $\tilde{n}_{\nu,j}, \tilde{m}_{\nu,j}$ which satisfy the constraints in $R_{\nu,j}$ according to Theorem 2.
   2.2. Set $m_j := max\{m_{\nu,j}, \nu = 0, 1, \ldots, N - 1\}$.

We observe that the correctness of algorithm 1 relies in the asymptotic behavior stated in theorem 2 and refer to [14] or to the code documented in [10] for detailed explanations on steps 1.1.1 and 2.1.1. Here we limit ourselves to say that a simple set of inequalities (with the same meaning of (2) and (3)) can be developed and algorithmically checked at a very low computational cost.

Even in this bivariate case we have supposed as given all the needed partial derivatives, and again we need formulas for their computation starting from the most common situation of having the only data $(x_i, y_j, f_{i,j}^{(0,0)})$; $i = 0, 1, \ldots, N$; $j = 0, 1, \ldots, M$. Obviously it is not difficult to use one–dimensional formulas to extract derivatives which are monotonicity and/or convexity preserving along the coordinate lines $y = y_j$ and $x = x_i$ (see [14, pp. 494–495]), but, if we want to reproduce the shape of some planar subset of data, additional, and heuristically based, modifications of partial derivatives must be introduced. Details can be found in the code of [10].

In conclusion, this tensor–product scheme is very flexible (can handle any kind of coordinate–oriented shape–constraints, and mixed Lagrange–Hermite interpolation problems) and has a very low computational cost (the code was constructed and used satisfactorily on a 12 Mh Intel 286 based machine !). On the other hand it has the drawback of not being purely local in the sense that, for evaluating the interpolant in a point $(x, y) \in R_{i,j}$ all the rectangles of the strips $x_i \leq x \leq x_{i+1}$, $y_j \leq y \leq y_{j+1}$ must be processed, and, more important, the x and y degrees are the same along the two strips. If this, as previously said, has no effect in computational time, it can on the contrary produce negative consequences on the visual aspect, because the surface may result too stretched even in "easy" sub domains.

For overcoming these problems, in [15] was presented a different approach, based on Coons patches obtained as the Boolean sum of variable degree one–dimensional Hermite operators which interpolate a network of curves and normal derivatives. These "boundary conditions" were also obtained from one–dimensional Hermite operators with variable degree; a care full choice of the tension factors induced by all these degrees was finally used to locally compute a C^1, monotonicity preserving (in the flexible sense defined above), piecewise polynomial interpolating surface.

Since the mathematical formulation of this "blending function" approach suffers of a heavy notation, we prefer not to waste space in its description, and, on the contrary, to concentrate on a simpler version of the above idea more recently introduced. This new approach has the same performances of the previous one, and, because of its neat structure, permits extensions to non-regular data, as shown in the next section.

Let the data $(x_i, y_j, f_{i,j}^{(0,0)}, f_{i,j}^{(1,0)}, f_{i,j}^{(0,1)})$; $i = 0, 1, \ldots, N$; $j = 0, 1, \ldots, M$ be given (we don't need second derivatives). The basic idea of [16] is the following: we assign x and y tension degrees to any edge of $R_{i,j}$ and compute the corresponding four interpolating boundary curves, then we form the bicubic Coons patch and, finally, increase the degrees as far as necessary to get a ahape– preserving, piecewise polmomial C^1 function.

For a formal statement of the scheme, let us consider the following integers $n_{i,s}, m_{r,j}$, $r = i, i+1$, $s = j, j+1$, respectively associated to the four edges of R_{ij} :$[x_i, x_{i+1}] \times \{y_s\}$, $\{x_r\} \times [y_j, y_{j+1}]$, $r = i, i+1$, $s = j, j+1$, which we use to define on the boundary of $R_{i,j}$ the following function $\hat{f}_{i,j}$ interpolating the data:

$$\hat{f}_{i,j}(x, y_s) := [\, f_{i,s}^{(0,0)} \quad f_{i,s}^{(1,0)} \quad f_{i+1,s}^{(0,0)} \quad f_{i+1,s}^{(1,0)} \,] \begin{bmatrix} H_i^0(x; n_{i,s}) \\ H_i^1(x; n_{i,s}) \\ H_{i+1}^0(x; n_{i,s}) \\ H_{i+1}^1(x; n_{i,s}) \end{bmatrix},$$

$$s = j, j+1, \tag{6}$$

$$\hat{f}_{i,j}(x_r, y) := [\, f_{r,j}^{(0,0)} \quad f_{r,j}^{(0,1)} \quad f_{r,j+1}^{(0,0)} \quad f_{r,j+1}^{(0,1)} \,] \begin{bmatrix} H_j^0(y; m_{r,j}) \\ H_j^1(y; m_{r,j}) \\ H_{j+1}^0(y; m_{r,j}) \\ H_{j+1}^1(y; m_{r,j}) \end{bmatrix},$$

$$r = i, i+1.$$

The surface is now defined using the well-known bicubically blended Coons patch ([2], [26]):

$$b_{i,j} := (\mathcal{P}_1 \oplus \mathcal{P}_2)\,\hat{f}_{i,j} = \mathcal{P}_1 \hat{f}_{i,j} + \mathcal{P}_2 \hat{f}_{i,j} - \mathcal{P}_1 \mathcal{P}_2 \hat{f}_{i,j}, \tag{7}$$

where:

$$(\mathcal{P}_1 \hat{f}_{i,j})(x,y) := [\, \hat{f}_{i,j}(x_i, y) \quad \hat{f}_{i,j}(x_{i+1}, y) \,] \begin{bmatrix} H_i^0(x; 3) \\ H_{i+1}^0(x; 3) \end{bmatrix},$$

$$(\mathcal{P}_2 \hat{f}_{i,j})(x,y) := [\, \hat{f}_{i,j}(x, y_j) \quad \hat{f}_{i,j}(x, y_{j+1}) \,] \begin{bmatrix} H_j^0(y; 3) \\ H_{j+1}^0(y; 3) \end{bmatrix},$$

$$(\mathcal{P}_1 \mathcal{P}_2 \hat{f}_{i,j})(x,y) := [\, H_i^0(x;3) \quad H_{i+1}^0(x;3) \,] \begin{bmatrix} f_{i,j}^{(0,0)} & f_{i,j+1}^{(0,0)} \\ f_{i+1,j}^{(0,0)} & f_{i+1,j+1}^{(0,0)} \end{bmatrix} \begin{bmatrix} H_j^0(y;3) \\ H_{j+1}^0(y;3) \end{bmatrix}$$

$$\tag{8}$$

Using the linearity of matrix multiplication, it is immediate to see that the set of functions of the form (2.5) is given by the following 12–dimensional linear space:

$$\mathrm{span}\{b_{r,s}^{(p,q)};\ r = i, i+1,\ s = j, j+1,\ 0 \le p+q \le 1\},$$

where:

$$b_{r,s}^{(0,0)}(x,y) := H_r^0(x;3) H_s^0(y; m_{r,j}) + H_r^0(x; n_{i,s}) H_s^0(y; 3) -$$
$$H_r^0(x;3) H_s^0(y;3)$$
$$b_{r,s}^{(1,0)}(x,y) := H_r^1(x; n_{i,s}) H_s^0(y;3),$$
$$b_{r,s}^{(0,1)}(x,y) := H_r^0(x;3) H_s^1(y; m_{r,j}),$$
$$\tag{9}$$

and so:

$$b_{i,j}(x,y) = \sum_{r=i}^{i+1} \sum_{s=j}^{j+1} \sum_{0 \le p+q \le 1} f_{r,s}^{(p,q)} b_{r,s}^{(p,q)}(x,y)\,. \tag{10}$$

The simple structure of (10) implies that, like for the one–dimensional and for the tensor–product cases, we have a very low computational cost for evaluating the interpolating function. From the well-known properties of Coons patches (see, e.g. [2]) we know that $b_{i,j}$ interpolates $\hat{f}_{i,j}$ on the boundary of $R_{i,j}$; at the corners we have:

$$f_{r,s}^{(p,q)} = \frac{\partial^{p+q}}{\partial x^p \partial y^q} b_{i,j}(x,y)|_{(x_r,y_s)}, \quad r=i, i+1, \quad s=j, j+1, \quad 0 \leq p+q \leq 1,$$

and, for the cross boundary derivatives, we get from (7) and (8):

$$\frac{\partial}{\partial x} b_{i,j}(x,y)|_{(x_r,y)} = \begin{bmatrix} f_{r,j}^{(1,0)} & f_{r,j+1}^{(1,0)} \end{bmatrix} \begin{bmatrix} H_j^0(y;3) \\ H_{j+1}^0(y;3) \end{bmatrix}$$
$$\frac{\partial}{\partial y} b_{i,j}(x,y)|_{(x,y_s)} = \begin{bmatrix} f_{i,s}^{(0,1)} & f_{i+1,s}^{(0,1)} \end{bmatrix} \begin{bmatrix} H_i^0(x;3) \\ H_{i+1}^0(x;3) \end{bmatrix} \quad r=i,i+1, \quad s=j,j+1.$$

Therefore, connecting patches of the form (7), (10), we get in a very easy way a C^1 piecewise polynomial function; the only drawback, inherent in any Coons bicubic blending scheme, is constituted by zero twists at knots. For any polynomial patch we have the following result.

Theorem 3. *There exist integers $\tilde{n}_{i,j}, \tilde{n}_{i,j+1}, \tilde{m}_{i,j}, \tilde{m}_{i+1,j} \geq 3$, such that for any $n_{i,j}, n_{i,j+1}, m_{i,j}, m_{i+1,j}$, with $n_{i,j} \geq \tilde{n}_{i,j}, n_{i,j+1} \geq \tilde{n}_{i,j+1}, m_{i,j} \geq \tilde{m}_{i,j}, m_{i+1,j} \geq \tilde{m}_{i+1,j}$, the corresponding $b_{i,j}$ defined in (10) is monotonicity-preserving.*

We omit the details of the proof which can be found in [16], and limit ourselves to briefly sketch the principal steps, because they have been use for other applications. The main, obvious, idea is to control the shape of the polynomial using its Bèzier net; unfortunately, although the one–dimensional basis functions $H_{\cdot,\cdot}(x)$ and $H_{\cdot,\cdot}(y)$ used in (9) have a very simple net, the bivariate $b_{i,j}(x,y)$ has degree $max\{n_{i,j}, n_{i,j+1}, 3\} \times max\{m_{i,j}, m_{i+1,j}, 3\}$ and a corresponding complicate control net. To overcome this difficulty we adopt the following strategy: first, we construct a *pseudo-control net* $L(x,y)$, following the sequence (9)–(10) but substituting to $H_r^0(x), H_r^1(x), H_s^0(y), H_s^1(y)$ their corresponding univariate Bèzier nets $l_r^0(x), l_r^1(x), l_s^0(y), l_s^1(y)$; second, we establish shape–preserving criteria for the *pseudo-net* L; third, we construct the *real* control net $\Lambda(x,y)$ repeating the process (9)–(10), but now starting with the univariate control nets $\lambda_r^0(x), \lambda_r^1(x), \lambda_s^0(y), \lambda_s^1(y)$, obtained from the corresponding $l_{\cdot}^{\cdot}(\cdot)$ elevating their degree up to $max\{n_{i,j}, n_{i,j+1}, 3\}$, $max\{m_{i,j}, m_{i+1,j}, 3\}$, and, fourth, we conclude observing that the shape of L is the same of Λ and thus of $b_{i,j}(\cdot,\cdot)$.

Now, we want to use these $b_{i,j}$ to obtain a surface $s \in C^1(R)$, with $s(x,y)|_{R_{i,j}} = b_{i,j}(x,y)$ which is interpolant and monotonicity–preserving in R. For this purpose let us denote the threshold degrees of theorem (3) with $\tilde{n}_{i,j}^{i,j}, \tilde{n}_{i,j+1}^{i,j}, \tilde{m}_{i,j}^{i,j}, \tilde{m}_{i+1,j}^{i,j}$, where the superscripts refer to the sub rectangle $R_{i,j}$ we are processing and the subscripts to the edges of $R_{i,j}$. If we take two adjacent sub rectangles $R_{i-1,j}$ and $R_{i,j}$ we would get a C^1 surface across the common edge $\{x_i\} \times [y_j, y_{j+1}]$ selecting the same degree along this edge, that is taking $\tilde{m}_{i,j}^{i-1,j} = \tilde{m}_{i,j}^{i,j}$. In such a way in fact the two pieces $b_{i-1,j}$ and $b_{i,j}$, sharing the same degree and the same interpolatory conditions along the edge, maintain the required smoothness.

For the computation of the degrees $n_{i,j}$, $n_{i,j+1}$, $m_{i,j}$, $m_{i+1,j}$, we propose the algorithm 2 below.

Algorithm 2.

1. For the five rectangles $R_{r,j}$ with $r = i - 1, i, i + 1$
   and $R_{i,s}$ with $s = j - 1, j, j + 1$
   1.1. Check the rectangle $R_{r,s}$ and compute the threshold degrees $\tilde{n}_{r,s}^{r,s}$, $\tilde{n}_{r,s+1}^{r,s}, \tilde{m}_{r,s}^{r,s}, \tilde{m}_{r+1,s}^{r,s}$ according to theorem 3.
2. Set:
$$n_{i,j}^{ij} = n_{i,j}^{i,j-1} := \max\left\{\tilde{n}_{i,j}^{i,j}, \tilde{n}_{i,j}^{i,j-1}\right\},$$
$$n_{i,j+1}^{i,j} = n_{i,j+1}^{i,j+1} := \max\left\{\tilde{n}_{i,j+1}^{i,j}, \tilde{n}_{i,j+1}^{i,j+1}\right\},$$
$$m_{i,j}^{i,j} = m_{i,j}^{i-1,j} := \max\left\{\tilde{m}_{i,j}^{i,j}, \tilde{m}_{i,j}^{i-1,j}\right\},$$
$$m_{i+1,j}^{i,j} = m_{i+1,j}^{i+1,j} := \max\left\{\tilde{m}_{i+1,j}^{i,j}, \tilde{m}_{i+1,j}^{i+1,j}\right\}.$$

We observe that, like for Algorithm 1, the correctness relies in the asymptotic form of theorem 3, and that the proposed scheme is local, because the processing of other rectangles does not change the degrees of $R_{i,j}$, since the degree along an edge does depend only on the two rectangles which share it.

We remark that the first partial derivatives can be computed from the data points using monotonicity preserving univariate strategies, as previously suggested for the tensor–product scheme.

Finally, we conclude this section observing that the second order error estimates are valid also in this two–dimensional setting.

§4. Interpolation of scattered data

We want, at this point, to move towards more interesting topics, namely interpolation of non-regular data. We will briefly sketch an extension of the blending scheme introduced in the previous section for data distributed in a general tensor–product topology, and then we will describe a method for arbitrary scattered data.

We start introducing the following set of data

$$\{(x_{i,j}, y_{i,j}, f_{i,j}) = (P_{i,j}, f_{i,j}), \ i = 0, \ldots, N, \ j = 0, \ldots, M\}, \qquad (11)$$

which is assumed to have a *tensor–product topology*, that is each "horizontal section", $\{P_{i,s}, \ i = 0, \ldots, N,\}$ and "vertical section" $\{P_{r,j}, j = 0, \ldots, M,\}$, r, s, arbitrary but fixed, belongs to a C^1 planar curve homeomorphic to a segment and horizontal and vertical sections intersect only at the data points. We are assuming, in other words, to have selected the points along privileged directions, under the suggestion of some physical or graphical criteria. Following this idea we would like to reproduce the shape of the data along the "coordinates lines" and, eventually, also the shape inside the "quadrilateral" patches. The resulting interpolating surface would therefore closely resemble the essential characteristics of the data.

For sketching the method (the reader is referred to [17] for details) we start assuming, as in the previous sections, to know the gradients $\nabla f(P_{i,j})$ at all the interpolation points or to have compute them via one-dimensional formulas. The basic idea is to look at the data (11) from a parametric point of view, constructing three surfaces $x = x(u, v)$, $y = y(u, v)$, $z = z(u, v)$ interpolating at the integer uniform grid of $[0, N] \times [0, M]$. If we are able to prove that the map $\Psi = \Psi(u, v) := [x(u, v), y(u, v)]$ is C^1 and invertible, then $f = f(x, y) := z(\Psi^{(-1)}(x, y))$ is a C^1 interpolating function, and we may try to control its shape.

We can therefore proceed as follows. We first construct a network of curves $\hat{x}_j(u), \hat{y}_j(u)$ and $\hat{x}_i(v), \hat{y}_i(v)$ which interpolate the data, that is $(x_j(i), y_j(i)) = (x_i(j), y_i(j)) = P_{i,j}$ and have no other intersection point. This can be done using, for each component, the one–dimensional scheme of section 2, because (see also the next section) large degrees force the curve to assume a piecewise linear form. (Of course, if the "coordinate lines" are known for some physical reason inherent in the problem, this part can be skipped).

Now, if we look at the x network of curves, $\hat{x}_j(u), \hat{x}_i(v)$, and to their restriction for $u \in [i, i+1]$, $v \in [j, j+1]$, we see we are dealing with four small "boundary curves" $\hat{x}_j(u), \hat{x}_{j+1}(u), \hat{x}_i(v), \hat{x}_{i+1}(v)$, which are perfectly similar to $\hat{f}$ defined in (6). At this point we may construct a bivariate function $x_{i,j} = x_{i,j}(u, v)$ using cubic Coons patches as described in (7)–(10). Obviously, the same construction will be applied on the y network.

Note that if we take the degrees $n_{i,j}, n_{i+1,j}, m_{i,j}, m_{i+1,j}$ (which, for simplicity, *are the same* for the three components of the surface) large enough, the functions $x_{i,j}(u, v)$ and $y_{i,j}(u, v)$ will result stretched enough to ensure a *one to one* map $(u, v) \mapsto (x, y)$. It is therefore obvious that we can define

a continuously differentiable, invertible map $\Psi = \Psi(u, v) := [x(u, v), y(u, v)]$ simply connecting together the patches $x_{i,j}$, $y_{i,j}$.

The third component $z = z(u, v)$ is treated as a simple bivariate function, and can be constructed to be monotonicity preserving and interpolant following with no change the blending scheme of the second section.

Summarizing, we see that the tension properties of the proposed polynomials provide again a very powerful tool; the degrees (it is worthwhile to repeat: the same for the three components) can in fact be locally chosen so that: (a) the curve network in the xy-plane forms a system of good "coordinate lines" for our data; (b) the components $x(u,v)$ and $y(u,v)$ form an invertible map, and, (c) the component $z(u,v)$ is monotonicity–preserving. The interpolating function $f = f(x,y) := z(\Psi^{(-1)}(x,y))$ is therefore monotonicity–preserving along directions "parallel to the coordinate lines".

Let us now move to "real" scattered data problems. Suppose we are given the following set of data

$$P_\eta := (x_\eta, y_\eta), \quad f_\eta := f(P_\eta), \quad \nabla f_\eta := \nabla f(P_\eta), \quad \eta = 1, \ldots, N_P, \qquad (12)$$

which are supposed have been already organized in a triangulation, where

$$T_\nu := P_{\nu_1} P_{\nu_2} P_{\nu_3}, \quad \nu = 1, \ldots, N_T,$$

is the corresponding set of non overlapping triangles which provide a partition of $D = \cup_{\nu=1}^{N_T} T_\nu$.

Our goal is to construct a surface $s(x,y)$, $s \in C^1(D)$, interpolating the data, that is,

$$s(P_\eta) = f_\eta, \quad \nabla s(P_\eta) = \nabla f_\eta.$$

with (possibly local) tension properties. In this new context (as far as we know never treated before the papers [18] and [19]) we will follow the same fundamental steps we have always adopted: we first define a suitable variable degree polynomial patch with tension properties, then we connect them to get a C^1 surface and, finally, we develop automatic criteria for the selection of the tension parameters so that to preserve some shape–constraints.

The basic idea is that cubic Clough–Tocher macro–elements are the natural extension for scattered data interpolation of the one–dimensional Hermite cubics. In fact they are made up by bivariate cubic polynomials and can be uniquely defined giving the data and derivatives (gradients) at the interpolation points whenever we assign the derivatives normal to the edges as a function (typically linear) of the gradients. We should try therefore to extend polynomials of the form (1) to a "Clough–Tocher like scheme".

We start denoting with $P_r = (x_r, y_r)$, $r = 1, 2, 3$, three non–collinear points in $\mathbb{R}^2$, and with T the triangle they form. Let also denote with e_1, e_2, e_3 the edges connecting, respectively, the points $P_1 - P_2, P_2 - P_3, P_3 - P_1$. An n–degree Bernstein–Bézier polynomial has the form (see e.g. [2], [26])

$$b(x, y; n) = b(u, v, w; n) := \sum_{i+j+k=n} \frac{n!}{i!j!k!} l_{i,j,k} u^i v^j w^k, \qquad (13)$$

where, setting $P = (x,y) \in T$, $u = u(x,y)$, $v = v(x,y)$, $w = w(x,y)$, are the barycentric coordinates defined by $P = uP_1 + vP_2 + wP_3$; $u + v + w = 1$, and $l_{i,j,k}$, $i + j + k = n$, are the *Bernstein ordinates* of $b(\cdot, \cdot, \cdot; n)$. If we take $x = x(u,v,w)$, $y = y(u,v,w)$ and connect in $\mathbb{R}^3$ the points (called *control points*)

$$L_{i,j,k} := \left(x\left(\frac{i}{n}, \frac{j}{n}, \frac{k}{n}\right), y\left(\frac{i}{n}, \frac{j}{n}, \frac{k}{n}\right), l_{i,j,k} \right), \quad i, j, k \geq 0, \quad i+j+k = n,$$

$$(14)$$

with triangular linear patches, we get the *control net*, $L = L(u,v,w;n)$. It is well known that a polynomial of the form (13) has the same values and gradients of the function $l(u,v,w;n)$ at the vertices.

Let R be a point internal to T, typically the barycenter (for proving the results below, some technical restrictions on the position of R have been required in [18]; these are here omitted to not add a formalism useless for the comprehension). We recall that the so called *Clough–Tocher cubic macro-element* ([2], [6], [65]) is obtained dividing $T := P_1 P_2 P_3$, called in this context *macro–triangle*, in three *mini–triangles*

$$T^{(1)} := P_1 P_2 R \ , \ T^{(2)} := P_2 P_3 R \ , \ T^{(3)} := P_3 P_1 R \ ,$$

and then in taking three polynomials (13) of degree 3 for each mini triangle. In order to get the internal C^1 continuity many of the control points must lye on the same plane (see fig. 3), and it is immediate to check that there are still 12 free control points to determine. The typical choice is to use function values, gradients and normal derivatives, as explained in fig. 4; the number of parameters is then reduced to 9 assuming, for the value of the normal derivative in the middle of any edge, the average of the (known) normal derivatives at the vertices.

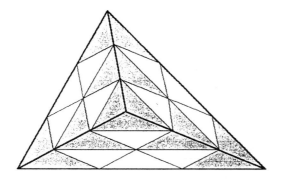

Figure 3. The Clough-Tocher macro-element. The shaded triangles correspond to planar regions.

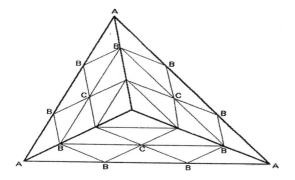

Figure 4. The determination of the Clough-Tocher control points.
A: from data values; B: from gradients; C: from normal derivatives.

The new macro–element introduced in [18], and hereafter called Basic Macro–Element (BME), consists in using the Clough–Tocher split and then in taking a polynomial of degree n for any of the three mini-triangles. We reduce the number of free parameters imposing that some of the control points are on the same plane, and others (along the edges) on the same bilinear function, as explained in figs. 5.a and 5.b. We see that, as n increases, the central plane tends to invade the macro–triangle, and, therefore, the corresponding BME tends to the plane interpolating the three data points $f(P_1), f(P_2), f(P_3)$. In other words the degree plays again the role of a tension parameter, exactly as in the previous schemes.

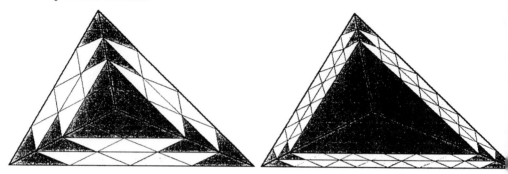

Figure 5.a The Basic Macro Net for $n = 4$.   **Figure 5 b** The Basic Macro Net for $n = 8$.

It is simple to see that the BME does coincide with the classical cubic Clough–Tocher element in the case $n = 3$ and has, for $n_i > 3$, 15 free parameters, obtainable from function's values, gradients and normal derivatives as shown in fig. 6. The number of free control points can again be reduced to 9, assuming, as in the cubic case, that the normal derivative of the control net along the edge e_r is a linear function of the normal derivatives at P_r and P_{r+1}. We refer to [18] or [19] for the detailed formulation.

Let us denote with $\mathcal{B} = \mathcal{B}(x, y; n) = \mathcal{B}(u(x,y), v(x,y), w(x,y); n)$ our BME of degree n; in [18] the following results are proven.

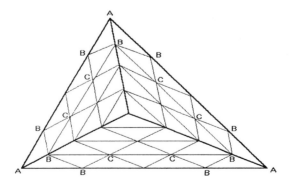

Figure 6. The determination of the Basic Macro Net's control points.
A: from data values; B: from gradients; C: from normal derivatives.

Theorem 4. Let $\pi = \pi(x,y)$ the plane interpolating the data (P_r, f_r), $r = 1, 2, 3$. Then

$$\lim_{n\to\infty} \|\mathcal{B}(\cdot,\cdot;n) - \pi\|_{T,\infty} = \lim_{n\to\infty} \|\mathcal{L}(\cdot,\cdot;n) - \pi\|_{T,\infty} = 0 \ .$$

Definition. The data are INC (DEC) along the edge e_r if

$$f_r < f_{r+1}, \ \langle \nabla f_r, e_r \rangle > 0, \ \langle \nabla f_{r+1}, e_r \rangle > 0,$$

$$(f_r > f_{r+1}, \ \langle \nabla f_r, e_r \rangle < 0, \ \langle \nabla f_{r+1}, e_r \rangle < 0) \ .$$

Definition. The data are CVX (CNC) along the edge e_r if

$$\langle \nabla f_r, e_r \rangle < (f_{r+1} - f_r) < \langle \nabla f_{r+1}, e_r \rangle,$$

$$(\langle \nabla f_r, e_r \rangle > (f_{r+1} - f_r) > \langle \nabla f_{r+1}, e_r \rangle) \ .$$

Theorem 5. Let the data be increasing (decreasing) and/or convex (concave) along the edge e_r. There exists a threshold degree, $\tilde{n}^{(r)}$, such that for any $n \geq \tilde{n}^{(r)}$ the interpolating piecewise Bézier polynomial $\mathcal{B}(\cdot,\cdot;n)$ is increasing (decreasing) and/or convex (concave) in the sub–triangle $T^{(r)}$ w.r. to a direction parallel to the edge e_r.

For a better comprehension of some forthcoming ideas, we emphasize that in theorem 5 the three mini–triangles are considered separately, and the given asymptotic results (similar to those already discussed) are also stated separately for the three mini–nets.

Theorem 6. Let $f \in C^2(T)$. Then for any n,

$$\|f(\cdot,\cdot) - \mathcal{B}(\cdot,\cdot;n)\|_{T,\infty} = O(h^2) \ , \quad h := \max_{r=1,2,3} \|P_{r+1} - P_r\| \ .$$

Let now come back to the set of scattered, triangulated data, of (12). If we take two adjacent triangles T_α, T_β of the given triangulation with vertices $P_{\alpha_1} P_{\alpha_2} P_{\alpha_3}$ and $P_{\beta_1} P_{\beta_2} P_{\beta_3}$ sharing the common edge $e_{\alpha,\beta} := P_{\alpha_1} P_{\alpha_2} = P_{\beta_1} P_{\beta_2}$, the corresponding two BME $\mathcal{B}_\alpha(\cdot,\cdot;n)$ and $\mathcal{B}_\beta(\cdot,\cdot;n)$ will produce a C^1 function across $e_{\alpha,\beta}$ because they share the same interpolatory conditions and *the same degree*. On the other hand, theorem 5 provides us with shape-preserving, automatic criteria for computing admissible degrees, and we can sketch the following algorithm.

Algorithm 3.

1. For $\nu = 1, \ldots, N_p$

    1.1. Check the triangle T_ν and compute the three threshold degrees $\tilde{n}_\nu^{(r)}$; $r = 1, 2, 3$ according to Theorem 5.

2. Set $n := \max\{\tilde{n}_\nu^{(r)} \ ; \ \nu = 1, \ldots, T_p \ ; \ r = 1, 2, 3\}$.

The algorithm above is very simple, but suffers of the serious drawback of the global choice of the tension parameter. It is in fact clear that a large value of n, even if suitable for some "sharp" subset of data, does force the other triangles to accept useless strong tension factors, and therefore, the resulting surface could, in some cases, not be visually pleasing. In addition we have the obvious disadvantage that all the data have to be processed for a single evaluation of the interpolating function.

In [19] a better, local scheme is presented. The first problem we must solve if we want that large $n_n^{(r)}$ don't spread on the whole domain D, is to find a trick permitting to smoothly connect BMEs of different degree. This trick is the construction of a *Modified Macro Element* (MME) which is rigorously explained in [19] with a heavy formalism, but which relies upon an idea not difficult to understand.

The starting points are: (a) the degree elevation process – any p-degree net of the form (14) can be transformed into a q-degree net, $q > p$, such that the coresponding p and q polynomials of the form (13) are identical, using the *degree elevation algorithm* (see, e.g. [2] or [26]), and (b) the cross boundary normal derivatives along an edge e_r do depend only on the two first rows of control points. A trivial consequence of (b) is the following. If we have two Bèzier polynomials, defined over two adjacent triangles sharing a common edge e_r, which are C^1 across e_r, we can arbitrarily modify all the control points which are not on the first two lines parallel to e_r without destroying the C^1 continuity across the edge.

For solving the problem of a local construction of the degrees, let us recall the two adjacent triangles T_α, T_β, let's define R_α, R_β their internal points determining the Clough–Tocher split and let $T_\alpha^{(\beta)}$ ($T_\beta^{(\alpha)}$) be the mini-triangle of the split determined by $P_{\alpha_1} P_{\alpha_2} R_\alpha$ ($P_{\beta_1} P_{\beta_2} R_\beta$) that is the sub-triangle of T_α (T_β) adjacent to T_β (T_α). (To help the comprehension of the notation, we advise that subscripts refer to triangles and superscripts to mini-triangles). Suppose that we have decided to use a degree n_α for T_α and n_β for T_β, with, say, $n_\alpha < n_\beta$. If we take the corresponding Clough–Tocher like BME of degrees n_α and n_β, the composite surface will not be C^1, because the two BME share the same interpolatory conditions, but not the same degree. On the other hand, if we had taken n_α even in T_β, we would have obviously obtained the desired C^1 continuity. The trick consists in accepting a compromise (which will be shown to have no negative consequence in tensioned interpolation) between n_α and n_β. We compute the *Basic Macro–Nets* (BMNs) of degree n_α both in T_α and in T_β, say $\text{BMN}_\alpha(n_\alpha)$ and $\text{BMN}_\beta(n_\alpha)$, and then we form, in T_β, a new macro–net, say $\text{MN}_\beta(n_\alpha \to n_\beta)$, elevating the degree of $\text{BMN}_\beta(n_\alpha)$ from n_α to n_β. Note that $\text{BMN}_\alpha(n_\alpha)$ and $\text{MN}_\beta(n_\alpha \to n_\beta)$ form a C^1 surface across $e_{\alpha,\beta}$. We now construct in T_β a new BMN of degree n_β, $\text{BMN}_\beta(n_\beta)$, and note that we have at this point two n_β-degree objects in T_β: $\text{BMN}_\beta(n_\beta)$, which is good for the tension but not for the continuity, and $\text{MN}_\beta(n_\alpha \to n_\beta)$ which is good for the continuity but not for the tension. We finally construct our solution, the *Modified Macro–Net* in T_β, MMN_β taking the control points of $\text{BMN}_\beta(n_\beta)$ but substituting those on the first two rows parallel to $e_{\alpha,\beta}$ with the analogous control points of $\text{MN}_\beta(n_\alpha \to n_\beta)$. The tension is maintained in the most part of T_β and the smoothness is preserved across $e_{\alpha,\beta}$.

The scheme involving the construction of the MMN and the corresponding MME described above permits us to have different degrees in each triangle – in practice the first two rows of the BMNs act as shock absorbers between different normal derivatives. We are now going to introduce a local automatic algorithm for computing degrees which satisfy shape–preserving criteria; the main idea (see the remark after theorem 5) is that we associate a good degree to any edge of a triangle, and then we take the maximum (to be precise: a number greater or equal to the maximum, as we will see immediately) as the good degree for the triangle itself. Before giving the details we need for the MME a theorem analogous to theorem (5) for the BME.

Theorem 7. *Let S be any non-empty subset of $\{1,2,3\}$. Suppose the data $f_r, \nabla f_r, r = 1,2,3$, are increasing and/or convex along the edge $P_s P_{s+1}$ of the triangle T, $s \in S$. Moreover let $\tilde{n}^{(1)}, \tilde{n}^{(2)}, \tilde{n}^{(3)}$, be such that, if $s \in S$, then $\tilde{n}^{(s)}$ are the threshold degrees of theorem 5 for which the BMN of degree $\tilde{n}^{(s)}$ is increasing and/or convex in $T^{(s)}$ along parallels to $P_s P_{s+1}$. Then there exists $n \geq \max_{s \in S}\{\tilde{n}^{(s)}\}$ such that the MMN and the MME are increasing and/or convex in $T^{(s)}$ along parallels to $P_s P_{s+1}$, $s \in S$.*

Given a point $Q = (x,y) \in T_k \subset D$ the following algorithm computes the degree of the corresponding MME.

Algorithm 4.
1. Find the triangle T_k such that $Q \in T_k$, and denote with T_α, T_β, T_γ the neighboring triangles.
2. For $s = \alpha, \beta, \gamma$
    2.1. Set $n_k^{(s)} = 3$.
    2.2. If the data are monotone and/or convex along the edge $e_{k,s}$ between T_k and T_s then
        2.2.1. Compute the threshold degrees according to theorem 5 $\tilde{m}_k^{(s)}$, $\tilde{m}_s^{(k)}$ such that $\mathrm{BMN}_k(m_k^{(s)})$ and $\mathrm{BMN}_s(m_s^{(k)})$ are shape–preserving.
        2.2.2. Set $n_k^{(s)} = \max(\tilde{m}_k^{(s)}, \tilde{m}_s^{(k)})$.
3. Set $n_k = \max(n_k^{(\alpha)}, n_k^{(\beta)}, n_k^{(\gamma)})$.
4. Increase n_k as necessary until the MMN applied to all the mini-triangles $T_k^{(\alpha)}, T_k^{(\beta)}, T_k^{(\gamma)}$, is shape preserving (see theorem 7).

We note that the application of the above algorithm to further triangles does not effect the computation in T_k because the "boundary degrees" $n_k^{(\alpha)}, n_k^{(\beta)}, n_k^{(\gamma)}$, and the internal one given in step 4 are uniquely defined and cannot be modified.

We close this section observing that algorithms 3 and 4 require knowing the values of the gradients at the interpolation points, but, in practice, this information is often unavailable. It is clear that our resulting surface, as well as any other two dimensional shape preserving interpolant of gridded data, will depend heavily on the method we have adopted to recover the gradients from the data points. The problem of a good choice of the partial derivatives, as far as we know, has been investigated in the scattered data setting only in [36], where alternatives to the classical weighted least squares approximation of the neighboring data slopes have been proposed. Other criteria, possibly based upon shape–preserving properties, would be very desirable, and researches in this direction are under current work.

Similarly, algorithms 3 and 4 require the data have been organized in a triangulation, and it is widely known how significantly the triangulation method affects the shape of the interpolant. The most famous is the so called Delaunay scheme, ([45]), but new algorithms, based on data–dependent strategies have been recently developed ([24], [50], [54], [61]). Obviously, we would like to have a data dependent triangulation especially tailored for our macro–elements and for the goals we want to achieve; in this connection, we believe the method proposed in [25] can produce nice results when our functions are used for shape–preserving interpolation.

§ 5. Variable degree polynomial splines in curve design

From the methods surveyed in the previous sections, and from the "C^2 approach" quoted in the introduction, we can infer that variable degree polynomial splines constitute a powerful tool for functional constrained interpolation, which is an important argument of CAGD and one of the goals of the *FAIR-SHAPE* project. Since they have also had direct applications in the construction of shape–preserving curves (see P.D. Kaklis's paper on this volume and the references therein), we are driven to consider them as a possible "building box" in CAGD applications. The aim of this section is to present some new simple results on direct applications in free–form curve (and surface) design which enforce the importance of variable degree polynomial splines.

We start rewriting polynomials of the form (1) in a Bèzier like form. Let $u \in [0,1]$, let $\mathbf{b}_r = [x_r y_r z_r]^T$; $r = 0, 1, 2, 3$ points of $\mathbb{R}^3$ and let us define $\mathbf{b}(u) = [x(u)\ y(u)\ z(u)]^T$ as

$$\mathbf{b}(u) := \sum_{r=0}^{3} \mathbf{b}_r B_r(u; n) , \qquad (15)$$

where B_r are our four "Bernstein polynomials" defined by (1.a) with the underlying linear function $l = l(u; n)$ connecting in $[0, 1]$ respectively the four sequences of points: $\{(0,1),(1/n,0),((n-1)/n,0),(1,0))\}$; $\{(0,0),(1/n,1),((n-1)/n,0),(1,0))\}$; $\{(0,0),(1/n,0),((n-1)/n,1),(1,0))\}$; $\{(0,0),(1/n,0),((n-1)/n,0),(1,1))\}$. It is worthwhile to remark that we are dealing with an n-degree curve, which has $n + 1$ control points; however, from the second to the second last they lye on the same straight line and it isn't important to write them down explicitly.

It is immediate to see that the basis functions B_r have the same properties of cubic Bernstein polynomials, and that curves of the form (15) have the same properties of cubic Bèzier curves, except for subdivision – an n-degree curve can be splitted in two curves of the form (15) only for $n = 3$. We can use the control points to roughly describe the curve in the usual way (remember that the computational cost is independent of n) having in addition the possibility of "playing" with the tension parameter n. Because of the component wise tension properties described in section 2 and of the convergence of the net to the curve as the degree tends to infinity, we have that the curve (15) tends to the control points $\mathbf{b}_1$ and $\mathbf{b}_2$, as shown in figs. 7 and 8.

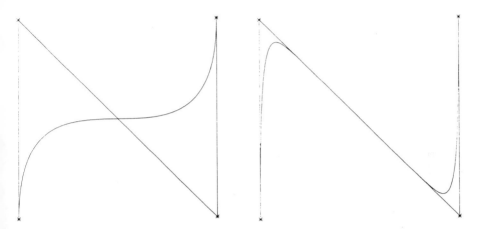

Figure 7. The Bèzier polynomial for $n = 3$.

Figure 8. The Bèzier polynomial for $n = 30$.

Although the above result is useful for curve design (we have a strong control on the shape of the curve without additional costs and changes in the main structure of design tools – the space of n-degree curves (15) is "the same" of cubics' one), for real CAGD applications we need something similar to B-spline curves.

¿From the theoretical point of view, we may ground our developments on the results of [44], because the space of polynomials of the form (1) fits in the general class described there. Let $\{t_i;\ i = -3, \ldots, N+3\}$ be a suitable sequence of knots with $h_i := t_{i+1} - t_i$, let $\{n_i;\ i = -3, \ldots, N+2\}$ be a suitable sequence of degrees and let us define the following space of *generalized splines*

$$GS := \{s \in C^2[t_0, t_N] \text{ s.t. } s|_{[t_i, t_{i+1}]} \in span\{B_r,\ r = 0, 1, 2, 3\}\} \quad (16)$$

where we have assumed for the Bernstein basis functions in (16) $n = n_i$ and $u = (t_{i+1} - t_i)/h_i$.

In [44] is given a constructive proof for the existence of a generalized B-spline basis, that is of a set of functions $N_i = N_i(t) = N_i(t; n_{i-2}, n_{i-1}, n_i, n_{i+1})$ which belong to GS, are positive, have the compact support $[t_{i-2}, t_{i+2}]$, form a partition of unity and have all the standard properties of B-splines. From this result we may describe and construct B-spline like curves using the B-spline control polygon; however we would have a more clear geometric interpretation and a more direct control on the curve if we were able to reproduce the direct computation of cubic Bèzier control points from the B-spline control points. We remind (see, e.g. [26]) that, for the cubic case, the Bèzier control points are taken on the B-spline polygonal legs with distances which are proportional to the knots' spacing; we should in the present case expect proportionality factors in which also the degree is taken into account.

Let $\{\mathbf{d}_i;\ i = -1, 0, \ldots, N + 1\}$ the sequence of (generalized) B-spline control points. We will denote with $\mathbf{b}_{3i}, \mathbf{b}_{3i+1}, \mathbf{b}_{3i+2}, \mathbf{b}_{3i+3}$, the four significant Bèzier control points for the i-th n_i-degree polynomial piece of the spline. We recall that the first and last control points $\mathbf{d}_{-1}, \mathbf{d}_0, \mathbf{d}_N, \mathbf{d}_{N+1}$ are in general treated separately to permit a stronger control on the shape at end interval, and we omit the corresponding (straightforward) details for the sake of brevity. From algebraic considerations we have the following result.

Theorem 8. *Let the knots t_i and the B-spline control points $\mathbf{d}_i$ be given. For $i = 1, \ldots, N - 1$ let the Bèzier control points $\mathbf{b}_{3i}, \mathbf{b}_{3i+1}, \mathbf{b}_{3i+2}, \mathbf{b}_{3i+3}$, be defined by the formulas below:*

$$\mathbf{b}_{3i+1} = \frac{(\tau_i + \sigma_i)\mathbf{d}_i + \rho_i \mathbf{d}_{i+1}}{\rho_i + \tau_i + \sigma_i} \ ;\ \mathbf{b}_{3i+1} = \frac{\sigma_i \mathbf{d}_i + (\rho_i + \tau_i)\mathbf{d}_{i+1}}{\rho_i + \tau_i + \sigma_i},$$

where

$$\rho_i := \left(\left(\frac{h_{i-1}}{n_{i-1}} + \frac{h_i}{n_i}\right)\frac{h_{i-1}}{n_{i-1} - 1}\right) \Big/ \left(\frac{h_{i-1}}{n_{i-1} - 1} + \frac{h_i}{n_i - 1}\right),$$

$$\sigma_i := \left(\left(\frac{h_i}{n_i} + \frac{h_{i+1}}{n_{i+1}}\right)\frac{h_{i+1}}{n_{i+1} - 1}\right) \Big/ \left(\frac{h_i}{n_i - 1} + \frac{h_{i+1}}{n_{i+1} - 1}\right),$$

$$\tau_i := (n_i - 2)h_i/n_i\ ;$$

$$\mathbf{b}_{3i} = \left(\frac{h_i}{n_i}\mathbf{b}_{3i-1} + \frac{h_{i-1}}{n_{i-1}}\mathbf{b}_{3i+1}\right) \Big/ \left(\frac{h_{i-1}}{n_{i-1}} + \frac{h_i}{n_i}\right),$$

$$\mathbf{b}_{3i+3} = \left(\frac{h_{i+1}}{n_{i+1}}\mathbf{b}_{3(i+1)-1} + \frac{h_i}{n_i}\mathbf{b}_{3(i+1)+1}\right) \Big/ \left(\frac{h_i}{n_i} + \frac{h_{i+1}}{n_{i+1}}\right).$$

Then the piecewise polynomial defined for any $i = 1, \ldots, N - 1$ by (15) with $n = n_i$ and $u = (t_{i+1} - t_i)/h_i$ belongs to the space (16).

We can repeat the same comments on the Bèzier form; we have an extension of cubic B-spline curves with the same structure and the same cost, with the possibility of a strong control on the curves' shape. The effect of enlarging the degrees is clearly shown in figs. 9, 10 and 11. We simply put in evidence that the possibility of stretching the curve between several B-spline control points and of pushing it towards one of them provides a shape-control tool more powerful than the NURBS' one; it may be compared with the effect of non-uniform β parameters of β-splines [1], which are, however, only G^2 continuous and by far more complicate in their mathematical structure.

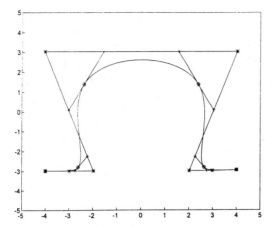

Figure 9. A cubic B-spline curve.

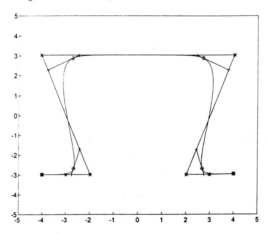

Figure 10. One degree is increased to 15.

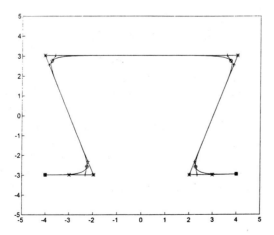

Figure 11. Three degrees are increased to 15.

We conclude this section observing that tensor–product and rational extension are straightforward. For the tensor–product case we have a regular grid of knots in the parameter space $\{(u_i, v_j)\}$ and the sequences of knots $\{n_i\}$, $\{m_j\}$. We can construct, for example applying for u and v the formulas of theorem 8, a tensor–product, variable degree B-spline surface, given, for any sub rectangle $[u_i, u_{i+1}] \times [v_j, v_{j+1}]$ by a polynomial of the form

$$\mathbf{b}_{i,j}(u,v) := \sum_{r,s=0}^{3} \mathbf{b}_{r,s} B_r(u; n_i) B_s(v; m_j) \ .$$

The effect of increasing the degrees is to stretch the central part of the surface; we have however the same problems discussed in section 3 for the functional case: the same tension factor is propagated along all the patches of the corresponding strips.

Rational curves (and surfaces) can be generated by functions of the form

$$R_i(t) := \frac{w_i N_i(t)}{\sum_{r=i-2}^{i+1} w_r N_r(t)}$$

or directly applying the formulas of theorem 8 in the rational case; we haven't the space to deep into details, but we point out that the combination of degrees and weights is an extremely powerful tool for shape–control.

§6. Conclusions and remarks

The collection of results presented in the above sections, in addition to the "global C^2 interpolation" ones, are an experimental proof of the importance that variable degree polynomial splines could have in CAGD environments. We again put in evidence that the success in the several applications relies in the simple structure of the net, which permits us to work with a "cubic like" polynomial space, looking at constraints and extensions in an intuitive, geometric manner.

We conclude the paper anticipating that new developments seem possible and are under current study of some researchers. The new results could include theoretical extensions to larger spline spaces and new points of view, like in the blossoming approach. Another simple extension, in which we have *two* degrees (related to shape at the left and right part of the interval) for any polynomial piece, is possible and seems to have very promising practical applications in CAGD.

References

1. Barsky, B.: *Computer graphics and geometric modeling using Beta–splines.* Springer Verlag, 1988.
2. Böhm, W., G. Farin, and J. Kahman: *A survey of curves and surfaces methods in CAGD.* Computer Aided Geom. Design , **1** (1984), 1–60.
3. de Boor, C.: *A practical guide to splines.* Springer, Berlin, 1978.
4. Clements J.C.: *Convexity preserving piecewise rational cubic interpolation.* SIAM J. Numer. Anal., **27** (1990), 1016–1023.
5. Cline, A.K.: *Scalar and planar valued curve fitting using splines under tension.* Commun. A.C.M., bf 17 (1974), 218–223.
6. Clough, R.W. and J.L. Tocher: *Finite element stiffness matrices for analysis of plates in bending.* in *Proc. Conf. on Matrix Methods in Structural Mechanics*, Wright–Patterson A.F.B., Ohio 1965.
7. Costantini, P.: *On monotone and convex spline interpolation.* Math. Comp., **46** (1986), 203–214.
8. Costantini, P.: *Co–monotone interpolating splines of arbitrary degree. A local approach.* SIAM J. Sci. Statist. Comput., **8** (1987), 1026–1034.
9. Costantini, P.: *An algorithm for computing shape–preserving interpolating splines of arbitrary degree.* J. Comp. Appl. Math., **22** (1988), 89–136.
10. Costantini, P.: *SPBSI1: a code for computing shape–preserving bivariate spline interpolation.* Universitá di Siena, Report no. 216, 1989.
11. Costantini, P.: *On some recent methods for shape–preserving bivariate interpolation.* In *Multivariate Interpolation and Approximation*, W. Haussmann and K. Jetter (eds.) International Series in Numerical Mathematics, Vol. 94, Birkhäuser Verlag, Basel 59–68, 1990.
12. Costantini, P.: *A general method for constrained curves with boundary conditions.* In *Multivariate Approximation: From CAGD to Wavelets*, K. Jetter and F.I. Utreras (eds.), World Scientific Publishing Co., Inc., Singapore, 1993.
13. Costantini, P.: *Boundary valued shape–preserving interpolating splines.* Preprint, 1996, submitted to ACM Trans. Math. Software .
14. Costantini, P. and F. Fontanella: *Shape–preserving bivariate interpolation.* SIAM J. Num. Anal., **27** (1990), 488–506.
15. Costantini, P. and C. Manni: *A local scheme for bivariate co–monotone interpolation.* Computer Aided Geometric Design, **8** (1991), 371–391.
16. Costantini, P. and C. Manni: *A bicubic shape-preserving blending scheme.* Computer Aided Geom. Design , **13** (1996), 307–331.
17. Costantini, P. and C. Manni: *Monotonicity–preserving interpolation of non–gridded data.* Preprint, 1995, to appear in Computer Aided Geometric Design.
18. Costantini, P. and C. Manni: *On a class of polynomial triangular macro–elements.* Preprint, 1995, to appear in proceedings of the *International Conference on Scattered Data Fitting*, J. Comp. Appl. Math..
19. Costantini, P. and C. Manni: *A local tension scheme for scattered data interpolation.* Preprint, 1996, submitted to SIAM J. Numer. Anal. .

20. Delbourgo, R.: *Accurate C^1 rational interpolants in tension*. SIAM J. Numer. Anal. , **30** (1993), 595–607.
21. Delbourgo R. and J.A. Gregory:*Piecewise rational quadratic interpolation to monotonic data*. IMA J. Numer. Anal., **2** (1982), 123–130.
22. Delbourgo R. and J.A. Gregory: *C^2 rational quadratic spline interpolation to monotonic data*. IMA J. Numer. Anal., **3** (1983), 141–152.
23. Delbourgo R. and J.A. Gregory: *The determination of derivative parameters for a monotonic rational quadratic interpolant*. IMA J. Numer. Anal., **5** (1985), 397–406.
24. Delbourgo R. and J.A. Gregory: *Shape preserving piecewise rational interpolation*. SIAM J. Sci. Stat. Comput., **6** (1985), 967–976.
25. Dyn, N., D. Levin and S. Rippa: *Data dependent triangulations for scattered data interpolation*. Appl. Numer. Math, **12** (1993), 89–105.
26. Farin, G.: *Curves and Surfaces for Computer Aided Geometric Design*, Academic Press, London, 1988.
27. Foley T.A.: *Interpolation with interval and point tension control using cubic weighted ν-splines*. ACM Trans. on Math. Softw., **13** (1987), 68–96.
28. Foley T.A.: *Local control of interval tension using weighted splines*. Computer Aided Geom. Design **13** (1986), 281–294.
29. Foley T.A.: *A shape preserving interpolant with tension controls*. Computer Aided Geom. Design **5** (1988), 105–118.
30. Fritsch, R.E. and R.E. Carlson: *Monotone piecewise cubic interpolation*. SIAM J. Numer. Anal. , **17** (1980), 238–246.
31. Goodman T.N.T.: *Shape preserving interpolation by parametric rational cubic splines*. In *International Series in Numerical Mathematics*, vol 86, Birkhäuser Verlag, Basel, 1988, 149–158.
32. Goodman T.N.T.: *Inflection of curves in two and three dimensions*. Computer Aided Geom. Design **8** (1991), 37–50.

33. Goodman T.N.T. and B.H. Ong: *Shape–preserving interpolation by space curves*. University of Dundee, Report AA/963, 1996.
34. Goodman T.N.T. and K. Unsworth: *Shape–preserving interpolation by parametrically defined curves*. SIAM J. Numer. Anal. **25** (1988), 1–13.
35. Goodman T.N.T. and K. Unsworth: *Shape–preserving interpolation by curvature continuous parametric curves*. Computer Aided Geom. Design **5** (1988), 323–340.
36. Goodman, T.N.T., H.B. Said and L.H.T. Chang: *Local derivatives estimation for scattered data interpolation*, Appl. Math. and Comp. **68** (1995), 41–50.
37. Gregory J. A.: *Shape preserving rational spline interpolation*. In *Rational approximation and interpolation*, Graves-Morris, Sarfaz and Varga (eds.), Springer, 1984, 431–441.
38. Gregory, J.A.: *Shape–preserving spline interpolation*. Comput. Aided Design , **18** (1986), 53–57.
39. Gregory J.A. and M. Sarfraz: *A rational cubic spline with tension*. Computer Aided Geom. Design **7** (1990), 1–13.
40. Heß ,W. and J.W. Schmidt: *Convexity preserving interpolation with exponential splines*. Computing **36** (1986), 335–342.

41. Kaklis, P.D. and D.G. Pandelis: *Convexity preserving polynomial splines of non uniform degree.* IMA J. Numer. Anal. **10** (1990), 223–234.
42. Kaklis, P.D. and N.S. Sapidis: *Convexity–preserving interpolatory parametric splines of non–uniform polynomial degree.* Computer Aided Geometric Design **12** (1995), 1–26.
43. Kaufman, E.H. and G.D. Taylor: *Approximation and interpolation by convexity–preserving rational splines.* Constr. Approx., **10** (1994), 275–283.
44. Kvasov, B.I.: *GB–splines and algorithms of shape–preserving approximation.* Preprint, 1994, to appear on Int. J. on CAGD.
45. Lawson, C.L.: *Software for C^1 surface interpolation.* In J.R. Rice (ed.) *Mathematical Software III*, Academic Press, New York, 161–194, 1977.
46. Lorentz, G.G.: *Bernstein Polynomials.* University of Toronto Press, Toronto, 1953.
47. Nielson G.M.: *Some piecewise polynomial alternatives to splines under tension.* In *Computer Aided Geomatric Design*, R.E. Barnhill and R.F. Riesenfeld (eds.), Academic Press, 1974, 209–235.
48. Pruess, S.: *Properties of splines in tension.* J. Approx. Theory, **17** (1976), 86–96.
49. Pruess, S.: *Alternatives to exponential splines in tension.* Math. Comp., **33** (1979), 1273–1281.
50. Quak, E. and L.L. Schumaker: *Cubic spline fitting using data dependent triangulations.* Computer Aided Geometric Design **7** (1990), 293–301.
51. Renka, R.J.: *Algorithm 716. TSPACK: Tension spline curve fitting package.* ACM Trans. Math. Software **19** (1993), 81–94.
52. Rentrop, P.: *An algorithm for the computation of the exponential spline.* Numer. Math., **35** (1980), 81–93.

53. Rentrop P. and U. Wever: *Computational strategies for the tension parameters of the exponential spline.* In: *Lecture Notes in Control and Information Sciences*, Vol. 95 , 1987, 122–134.
54. Rippa, S.: *Long thin triangles can be good for linear interpolation.* SIAM J. Numer. Anal. **20** (1992), 257–270.
55. Sapidis N.S. and P.D. Kaklis: *An algorithm for constructing convexity and monotonicity–preserving splines in tension.* Computer Aided Geom. Design , **5** (1988), 127–137.
56. Sapidis N.S. and P.D. Kaklis: *A method for computing the tension parameters in convexity–preserving spline in tension interpolation.* Numer. Math., **54** (1988), 179–192.
57. Schmidt, J.W.: *On shape–preserving spline interpolation: existence theorems and determination of optimal splines.* Approximation and Function Spaces, Banach Center Publications, Volume 22, PWN-Polish Scientific Publishers, Warsaw, 1989.
58. Schmidt, J. W.: *Rational biquadratic C^1 splines in S–convex interpolation.* Computing **47** (1991), 87–96.
59. Schmidt J. W. and W. Heß: *Convexity preserving interpolation with exponential splines.* Computing, **36** (1986), 335–342.

60. Schmidt, J.W. and W. Heß: *Quadratic and related exponential splines in shape-preserving interpolation.* J. Comp. Appl. Math. **18** (1987), 321–329.
61. Schumaker, L.L.: *Computing optimal triangulations in polygonal domains.* Computer Aided Geometric Design **10** (1993), 329–345.
62. Schweikert, D.G.: *Interpolatory tension splines with automatic selection of tension factors.* J. Math. Phis. **45** (1966), 312–317.
63. Späth, H.: *Exponential spline interpolation.* Computing, **4** (1969), 225–233.
64. Späth, H.: *Spline algorithms for curves and surfaces.* Utilitas Mathematica Publishing Incorporated, Winnipeg, 1974.
65. Strang, G. and G. Fix: *An Analysis of the Finite Element Method.* Prentice-Hall, Englewood Cliffs, NJ, 1973.
66. Wever, U.: *Non-negative exponential splines.* CAD, **20** (1988), 11–16.

Author

Paolo Costantini
Dipatimento di Matematica,
Università di Siena
via del Capitano 15, 53100 Siena
Italy
E–mail: costantini@unisi.it

Fairing of Surfaces

Functional Aspects of Fairness

Malcolm I.G. Bloor, Michael J. Wilson

The University of Leeds

Abstract: This paper will describe work in which the physical properties of objects defined in terms of PDE surfaces have been calculated and optimized against their functional requirements, subject to pre-specified constraints. The surface properties of such objects are also analysed in terms of some conventional measures of fairness based upon surface curvature, and the results presented. A limited discussion of application areas where fairnesss, or indeed lack of it, may be desirable is also given.

1 Introduction

This paper discusses the influence of surface fairness on the design of shape for function, with particular reference to the use of the PDE (partial differential equation) Method. The PDE method defines the shape of an object by specifying what might be called characterlines on its surface which act as boundaries between the individual patches which comprise the whole surface. This approach reduces the constructional data inherent in the conventional patch based methods which currently dominate surface representation, and is thus a basis for the functional design of shape, since it reduces the number of parameters (both shape and functional) that describe the physical properties of an object—an essential feature for any shape optimization.

The method was developed as a means of generating blend surfaces, and this it achieves by viewing blending essentially as a boundary-value problem ([1]). At the heart of the technique is a mapping from a two dimensional region of (u, v) parameter space to a surface in real space, obtained from the solution of an elliptic partial differential equation (PDE). The mapping satisfies boundary conditions derived from the continuity conditions at the edge of the blend surface. Being obtained from the solutions of partial differential equations the surfaces are intrinsically 'fair'.

The technique has been extended to cover free-form surface design where the aim has been to generate functionally useful surfaces (e.g. propellers, ship-hulls) from patches of PDE surface ([2]). An important feature of the method is that complex free-form surfaces can be specified by a relatively small set of shape parameters, which arises by virtue of its boundary-value approach. The method forms a basis for functional design, since with it the numerical optimization of complex curved surfaces is feasible ([4, 5, 3]).

2 The PDE Method

In the PDE approach we view u and v as coordinates of a point in a region Ω of a 2D parameter space, and the function $\underline{X}(u,v)$ as a mapping from that point in Ω to a point in 3-space. To satisfy these requirements we regard $\underline{X}$ as the solution of a partial differential equation

$$D_{uv}^m(\underline{X}) = \underline{F}(u,v) \tag{1}$$

where $D_{uv}^m()$ is a partial differential operator of order m in the independent variables u and v, while $\underline{F}$ is a vector function of u and v. Since we are considering boundary-value problems, it is natural to consider the class of elliptic partial differential equations.

In much of the surface generation work, the equations that have been considered are based upon the following fourth-order elliptic PDE:

$$\left(\frac{\partial^2}{\partial u^2} + a^2 \frac{\partial^2}{\partial v^2}\right)^2 \underline{X} = 0 \tag{2}$$

To find a solution to this equation we can specify boundary conditions along $\partial\Omega$ on the function $\underline{X}$ and on its normal derivative $\frac{\partial \underline{X}}{\partial \mathbf{n}}$; this provides sufficient control to achieve a tangent plane continuous blend. Note that $\frac{\partial \underline{X}}{\partial \mathbf{n}} = \underline{X}_u$ on a region of $\partial\Omega$ parallel to the v axis, and $\frac{\partial \underline{X}}{\partial \mathbf{n}} = \underline{X}_v$ on a region of $\partial\Omega$ parallel to the u axis. The three components of the function $\underline{X}$ are the Euclidean coordinate functions of points on the surface given parametrically in terms of the two surface parameters u and v which define a coordinate system on the surface; note that in the simplest case the equation is solved independently for the x,y and z coordinates.

The partial differential operator in equation (5) represents a smoothing process in which the value of the function at any point on the surface is, in a certain sense, an average of the surrounding values. In this way a surface is obtained that is a smooth transition between the boundary conditions on both the function and its normal derivative. We can see intuitively that this process achieves the basic requirement of surface fairness, namely that short length scales cannot be introduced inadvertantly, in other words length scales are determined by the boundary conditions and, where used, by a non-zero right hand side of the governing PDE. A fuller discussion of these principles is given elsewhere. The parameter a controls the relative rates of smoothing between the u and v parameter directions, and for this reason has been referred to as the smoothing parameter ([6]).

3 Blending

The problem of blend generation in computer-aided design is essentially that of being able to generate a smooth surface to act as a bridging transition between neighbouring 'primary' surfaces ([7]). This may be necessary in order to satisfy aesthetic requirements, or for more functional reasons such as the need to relieve stress concentrations or for manufactuing considerations. The first requirement is clearly a case where the smoothness of the surface and the freedom from blemishes is the primary consideration, and such is the case for part of a car body. Here we see the essential feature of fairness, that is a lack of surface variations on length scales inappropriate to the overall length scale of the surface.

Functional Aspects of Fairness

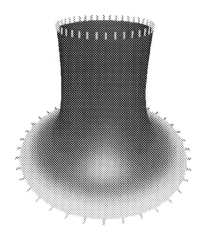

Fig. 1 A simple blend between a circular cylinder and a flat plane

Mathematically the calculation of a blending surface may be formulated as a classic boundary-value problem, in that what we require a function $\underline{X}$ defined over a domain Ω, that satisfies specified boundary data around the edge the region $\partial\Omega$. In the case of blending, this boundary data will typically be in the form of $\underline{X}$ and a number of its derivatives specified on $\partial\Omega$. The number of derivatives specified will depend on the required degree of continuity between the blend and the surfaces to which it joins.

A simple example of a blend is given in Fig. (1), which shows a blend between a circular cylinder and a flat plane at right angles to each other. It is important to understand the nature of the control of the surface afforded by the boundary conditions and the parameters in the problem. The PDE has been solved over the rectangular region $[0, 1] \times [0, 1]$ in the (u, v) parameter plane. The boundary conditions on $\underline{X}$ give the shape of the curves bounding the surface in E^3, and the parameterization of these curves in terms of u and v is something that must be chosen. In this example the surface is taken to be periodic in the v-direction, hence the 'function' boundary conditions amount to specifying $\underline{X}$ along the iso-lines $u = 0$ and $u = 1$, i.e. $\underline{X}(0, v)$ and $\underline{X}(1, v)$. The derivative boundary conditions specify the variation of the coordinate vector $\underline{X}_u$ along the iso-lines $u = 0$ and $u = 1$. The geometric interpretation of the derivative boundary condition is straightforward, for the direction of this vector determines the direction of approach of the surface to the trimline, while its magnitude determines the 'speed' with which the surface moves away from the trimline. These effects are illustrated in Fig. (2) where the effect of changing the derivative conditions is shown.

The surface shape, and indeed its fairness, can also be influenced by the choice of the parameter a. As mentioned above this can be interpreted as a smoothing parameter in the sense that it controls the relative smoothing of the dependent variables between the u and v directions. This point is illustrated in Fig. (3) where is shown a blend between a cylinder of 'fluted' cross-section and a plane perpendicular to the generators. By adjusting the value of

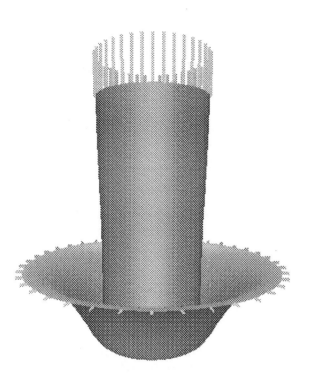
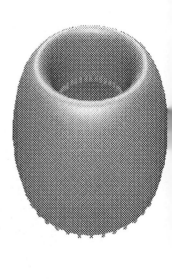

Fig. 2 The effect of derivative boundary conditions

Functional Aspects of Fairness

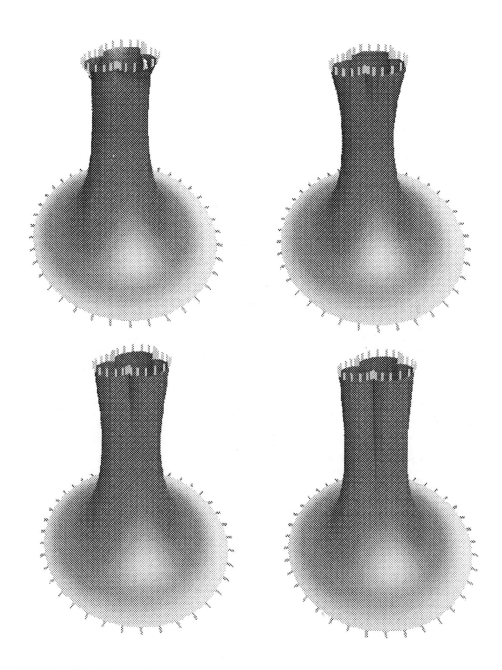

Fig. 3 The effect of the smoothing parameter a

a one can change the properties of the blending surface, either propagating the effects of the boundary curves over substantial parts of the blend, or confining their effects to regions near the edges of the blend. For large a, changes in the u direction occur over a relatively short length scale, namely $1/a$ times the length scale in the v direction over which similar changes take place. Note that it is possible to associate different values of a with features on the surface of different length scales. In fact, this is what is happening in Fig. (3), where the fluting effect is produced in the boundary conditions by the addition of a higher frequency mode to the basic Fourier mode describing a cylinder of circular cross-section. The decay of these modes as one moves away from the boundary is controlled by the value of a, which is different for the two modes. Thus we see a mechanism by which the short length scales introduced by the boundary conditions can be almost eliminated from a large part of the surface, their effect being confined to 'boundary layers'.

4 Freeform Design

The defining characteristic of a free-form surface is that its shape can be influenced by a designer. Such surfaces occur in computer-aided geometric design in a variety of contexts ranging from the representation of existing objects through to shape design. It is conventional to represent free-form curves or surfaces in terms of simple polynomial functions, of a single parameter in the case of curves or two parameters in the case of surfaces. The type of polynomial functions used gives the various classic forms of curve and surface representations ([8]).

In contrast to polynomial surfaces, PDE surfaces, being generated as solutions to boundary-value problems, are manipulated by changing the boundary conditions imposed at the edges of the patch. Here, we will indicate how the PDE method can be used to generate free-form surfaces. The surfaces generated by this method are expressed parametrically, often in terms of transcendental functions of the surface parameters rather than simple polynomial expressions, and as we shall see again that the resulting surfaces are extremely smooth.

When generating a free-form surface one has considerably more freedom with which to choose the boundary conditions in order to achieve a desired shape. The generating equation, being an elliptic PDE, has a solution that is sensitive to the choice of boundary conditions, and these can be used by the designer as a tool for surface manipulation. The boundary conditions on the function $\underline{X}$ are obviously chosen so that the curves forming the edges of the surface patch have the desired shape. To get a feel for the way in which the boundary values of $\underline{X}_u$ and $\underline{X}_v$ affect the shape, one should remember that the directions of these vectors are tangential to the isoparametric u and isoparametric v lines. Therefore, by altering the values specified for $\underline{X}_u$ and $\underline{X}_v$ along the boundaries one can affect the direction in which the surface moves away from the edges of the patch and also how how far it will move in space before it begins to respond to the boundary conditions on other parts of the patch.

Fig. (4) shows the hull of a yacht that has been generated using the PDE method. The design has been carried out as a boundary value problem with boundary conditions specified along curves in the (u,v) plane which correspond to closed curves in E^3. The $u = 0$ boundary has been taken to be the plan outline of the hull at the deck level, and this curve is given parametrically in terms of v. The $u = 1$ boundary is mapped to a curve lying at the bottom of the yacht's keel, which is again given in terms of v. Derivative conditions at the base of

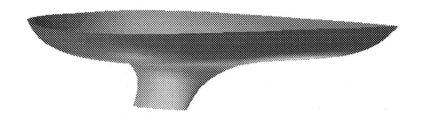

Fig. 4 Yacht hull generated from single boundary-value problem

Fig. 5 Ship hull generated from single boundary-value problem

the boat can be used to ensure that we get a fin-like shape for the keel providing that the vertical derivative is sufficiently large and positive. This basic geometry will be used later in a functional optimisation example.

Fig. (5) shows another example of a ship hull where each half of the hull has been constructed from a non-periodic solution. The boundary conditions on $\underline{X}(u,v)$ have been chosen so as to give suitable outlines for the deck, keel, bow and stern profiles for the vessel, which were mapped to the iso-lines $u = 1$, $u = 0$, $v = 0$ and $v = 1$ respectively. The mapping between points along the boundary of the (u,v) parameter domain and points in E^3 has been chosen so that the boundary curves of the hull have arc-length parameterization. The derivative boundary conditions were chosen so that the interior surface of the ship's hull had a realistic shape.

Fig. (6) shows what can be achieved using multiple PDE patches joined together to form a complex surface: it is the swirl-inlet port to a diesel engine. A further example of multiple-patch geometry is given in Fig. (7) which shows a 4-valve cylinder head for a 4-stroke petrol engine. Note that in both cases, the boundary-value nature of the surface generation process means that adjacent surfaces meet exactly, since they share a common set of function boundary-conditions.

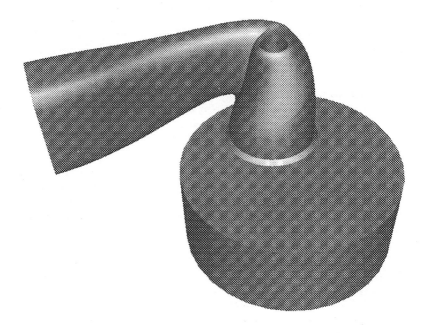

Fig. 6 swirl-inlet port for diesel engine

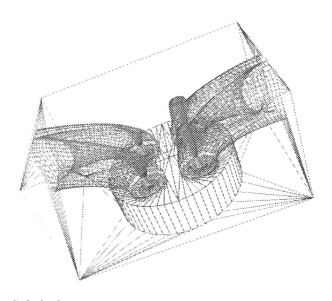

Fig. 7 4-stroke cylinder head

5 Variational Approaches to Surface Fairness

A similar approach has been used by Celniker and Gossard ([9]) who, by a physical analogy, use, in their 'energy-based' method, the equations describing the time-dependent deformation of an 'elastic' surface under the action of an externally applied load as a means of generating and designing free-form surfaces. This process involves minimizing an energy functional, and tends to produce surfaces which, in some ways, resemble those generated by the present approach. Unlike the PDE method, however, Celniker and Gossard achieve their shape design by varying their loading function rather than by varying the boundary conditions imposed on the surface patch. There is also a link with the work of Williams who considers a structural analogy ([10]).

Also related are the variational methods for surface design that have been adopted by a number of authors ([11, 12, 13, 14, 15, 16, 17, 18, 19, 20]) In such techniques, the basic idea is to find a fair surface that is an approximation to a given set of data points. This involves minimizing a functional which contains basically two contributions, one of which determines how closely the surface that is produced approaches the data points, while the other controls the fairness properties of the surface. The resulting surface is viewed as a compromise between these two competing requirements. There is a certain analogy to the PDE approach in that the Euler-Lagrange equation associated with the integral above is a PDE of the type which we have been considering, except it has a forcing term on the right-hand side which is a combination of generalized functions ([21, 22]).

However, unlike the PDE method, variational methods specify the mapping between parameter and physical space in the sense that discrete points in parameter space are assigned to specified points in physical space; and the result of the variational process is an approximating surface. Another point of difference is that the PDE approach designs a surface using data distributed around the edges of the surface patch, whereas these variational techniques use data distributed across the surface.

6 Physical Aspects of Surface Fairness

It is well known that surface fairness based upon, or related to, surface curvature, is a desirable feature for surfaces whose primary function is to appeal to the eye. A great deal of effort goes into designing surfaces for car bodies in order that their appearance under various lighting systems is pleasing ([23, 24]). This leads to reflection line analysis which, by the very nature of the exercise, must be subjectively judged. Very often, however, we are concerned with more easily quantifiable measures of performance relating to some measure of functional efficiency. First of all, let us consider some physical applications of design and analysis where the role of fairness can be assessed qualitatively without reference to detailed analyses.

If we consider the area of heat transfer, it is clear from everyday experience that surface fairness in the design of heat exchangers is not a particularly desirable feature. Witness the 'dimpled' surface of some household radiators, or fin-like structures on heat sinks attached to electronic devices. In both cases, the surface has associated with it a length-scale of a different order of magnitude from the length scale of the object overall. Simply designing a 'fair' surface to contain hot material within it would not lead to functional efficiency. It is clear that a large surface area is required, as would be the case in any process governed by diffusion from the surface. Furthermore, as soon as convection begins to play a role the situation is a good deal more complicated, as fluid mechanics begins to have a significant effect.

In the case of designing objects to have an appropriate strength for the function they must perform, yet minimising the amount of material used in manufacture, we have already indicated that blends might be used to remove excessive stress concentrations at 'junctions' between different patches of surface of the object. Thus fair surfaces are used to link other primary surfaces to achieve this end. On the other hand, structures may frequently have reinforcing structures such as flanges, which are used for strengthening whilst keeping the weight down. In these circumstances, short length-scales (associated with the flanges) are introduced into the design to satisfy functional design requirements.

In neither of the last two areas discussed is the requirement for fairness at all straightforward, and any criteria based on purely geometric reasoning would not represent the whole story.

In certain areas of fluid mechanics, the position is somewhat clearer, although even here a great deal of care must be exericed. In the design of ship hulls using spline-based representations, it is often advantageous to carry out an optimisation exercise in order to maximise some geometric measure of surface fairness whilst meeting the original design constraints as closely as possible. This approach is of value because it avoids expensive fluid flow calculations at an early stage and yet experience has shown it to be of value ([26]). The cautionary note that needs to be sounded is that certain effects in fluid mechanics are counter-intuitive. It is well known that the dimples on a golf ball reduce its resistance for its particular flight regime. In other words, in this area of fluid mechanics a very geometrically unfair surface is desirable for functional reasons.

To understand the situation more fully, it is instructive to discuss at a very basic level some fluid mechanics. Let us consider the flow past a bluff body with fluid properties and speed such that the flow is dominated by inertia effects, in other words the Reynolds' number, which measures the relative importance of inertia to viscous effects overall, is high. Under these circumstances, because the fluid is certainly viscous and must therefore satisfy a no-slip condition at the surface of the body, the viscous effects are confined to a boundary layer adjacent to the body surface. In this region the kinetic energy of the fluid is sapped, with the result that the flow separates from the body, and this failure of the pressure to recover at the rear of the body gives rise to form-drag. In addition, the friction of the fluid flowing past the body causes, naturally enough, skin friction. These two contributions to the drag experienced by the body need to be understood. A further complication is the onset of turbulence which causes momentum transfer between adjacent layers of fluid on a macroscopic scale rather than molecular, which is the case for viscosity.

Separation of the boundary layer is caused by flow into a region of increasing pressure and this adverse pressure gradient is determined by the flow outside the boundary layer. An increase in pressure in this inviscid flow region is associated with a decrease in fluid speed, which in turn may result from curvature in the surface of the body causing the streamlines to diverge. Thus local curvature in the surface has a strong influence on separation and thus on drag. Here surface fairness is a critical factor and in most cases, fair surfaces are desirable in designs where fluid resistance needs to be kept to a minimum. Turbulence is also affected by surface fairness, and the increased length scale for momentum interchange associated with

Functional Aspects of Fairness

this phenomenon causes increased skin friction. The reduction in drag of a golf ball is caused by a secondary effect of the turbulence in that particular flow regime when the form drag is reduced.

7 Functional Design

As mentioned above, the PDE method is not primarily a method for surface representation, but for surface generation. It is envisaged that objects whose surfaces are generated by the method are designed to serve some function, and hence it is crucial that the design of an object's geometry is integrated with functional considerations. What makes the PDE approach especially suitable for this is that it can parameterize complicated free-form surfaces in terms of a relatively small number of parameters. This is possible because of its boundary-value approach: surfaces are specified in terms of data distributed around curves, rather than across the surface itself. In practice, this means that the number of shape parameters specifying an object's surface is often small enough for the task of optimizing its shape to be computationally feasible; yet at the same time the method is sufficiently flexible for a wide range of shapes to be accessible to it.

Work has been carried out on the integration of an object's geometric design with its functionality in a number of applications, ranging from heat transfer ([25]), stress minimization ([4, 5]), and the design of a surface for its hydrodynamic properties ([3]). In each case the approach has been to take an object whose surface is described as a solution to a PDE, and then to calculate (usually numerically) the physical properties of the object on the basis of a mathematical model. Thus, the object's shape is described by a PDE, and the object's physical properties are described by a (different) PDE. How well the object performs its desired function is quantified by some merit function: for instance, in the case of a stress problem this may be the maximum shear stress that occurs in a blend; or in a hydrodynamic problem this may be the wave drag experienced by a yacht hull-form. The merit function is then regarded as a function of those shape parameters describing the surface that are allowed to vary, as well as the various physical parameters in the problem. Then, starting from an initial design some method for optimization (such as Fletcher-Powell minimization) can be used to modify the values of the shape parameters so that the performance of the design is improved. This involves moving towards some (local) minimum in the parameter space defined by the parameters in the problem, and, even if a minimum is not reached, this procedure usually results in a new design with better properties than the original. Furthermore, this design process is fully automatic once the initial design and the design variables have been specified. Note that very often the minimization is performed subject to certain constraints which result from requirements that the eventual design must satisfy. For instance, where the wave drag on a hull is minimized it is sometimes a good idea to fix the vessel's displacement if the final design is to have a certain volume of cargo space.

Fig. (8) shows an example where a PDE yacht hull has been designed so as to minimize the wave drag it experiences: Fig. (8a) shows the initial design, while Fig. (8b) shows the final design. The shape parameters that have been allowed to vary are essentially a number of the Fourier coefficients in the representation of the derivative boundary conditions. The minimization of the wave drag has been carried out subject to the constraints of fixed lateral

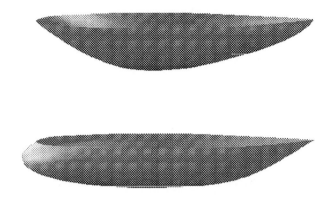

Fig. 8 Yacht hull optimised to reduce wave drag: (a) Starting Design; (b) Final Design

stability, draught and displacement. An interesting point to note that in terms of the one of the conventional measures of fairness:

$$\int\int \left(\kappa_1^2 + \kappa_2^2\right) |\underline{X}_u \times \underline{X}_v| \, dudv, \tag{3}$$

the optimised surface ('fairer' from the point of view of wave-drag) is actually worse than the intial design.

8 Conclusions

In many physical applications the influence of surface fairness is extremely important. We have seen that in the area of fluid flow a fair surface is generally speaking a desirable feature. The usual fairness measures based on curvature can be a useful tool in developing an initial design. How this compares with a functionally optimal design remains to be seen, but in applications such as drag reduction, the computationally economic approach based on geometry is worth further investigation.

References

[1] Bloor, M.I.G. & Wilson, M.J., "Generating Blend Surfaces using Partial Differential Equations", *CAD*, Vol. 21, No 3, pp. 165-171, 1989.

[2] Bloor, M.I.G. & Wilson, M.J., "Using Partial Differential Equations to generate Free-form Surfaces", *CAD*, Vol. 22, pp. 202-212. 1990.

[3] Lowe, T. W., Bloor, M. I. G. and Wilson, M. J., "The Automatic Functional Design of Hull Surface Geometry", Journal of Ship Research, Vol 38, No.4, pp. 319-328, 1994.

[4] Doan, N., Bloor, M. I. G. and Wilson, M. J., "A Strategy for the Automated Design of Mechanical Parts", eds Rossignac, J., Turner, J. and Allen, G., A., *Second Symposium of Solid Modelling and Applications*, ACM Press, New York, pp.15-21, 1993.

[5] Wilson, D. R., Bloor, M. I. G. and Wilson, M. J., "An Automated Method for the Incorporation of Functionality in the Geometric Design of a Shell", eds Rossignac, J., Turner, J. and Allen, G., A., *Second Symposium of Solid Modelling and Applications*, ACM Press, New York, pp.253-259, 1993.

[6] Bloor, M.I.G. & Wilson, M.J., 'Blend Design as a Boundary-Value Problem', eds. Straber,W. & Seidel H.-P., in *Theory and Practise of Geometric Modelling*, Springer-Verlag, Berlin, pp.221-234, 1989.

[7] Woodwark, J.R., "Blends in Geometric Modelling", in The Mathematics of Surfaces II, ed. Martin,R R , Oxford University Press, Oxford, UK, pp. 255-297, 1987.

[8] Mortenson, M.E., "Geometric Modeling", Wiley-Interscience, New York, 1985.

[9] Celniker, G. and Gossard, D. "Energy-based models for free-form surface shape design", *ASME Design Automation Conference*, 107-112. 1988.

[10] Williams, C.J.K, "Use of structural analogy in generation of smooth surfaces for engineering purposes", *CAD*, Vol. 19, pp. 310-322, 1987.

[11] Nowacki, H. and Reese, D., "Design and Fairing of Ship Surfaces", in: Barnhill, R.E. and Boehm, W., eds., *Surfaces in CAGD*, North-Holland, pp. 121-134, 1983.

[12] Hagen, H., and G. Schulze, "Automatic Smoothing with geometric surface patches", Computer Aided Geometric Design 4, pp. 231-235, 1987. 1987.

[13] Terzopoulos, D. and Witkin, A., "Physically Based Models with Rigid and Deformable Components", *IEEE Computer Graphics & Applications*, pp. 41-51, 1988.

[14] Platt, J.C. and Barr, A.H., "Constraint Methods for Flexible Models", *Computer Graphics*, Vol. 22, No. 4, pp. 279-288, 1988.

[15] Nowacki, H., Dingyuan, L., and Xinmin, L., "Mesh Fairing GC Surface Generation Method", in : Straber, W. and H.-P. Seidel, H.-P., eds., *Theory and Practise of Geometric Modeling*, Springer-Verlag, Berlin, pp. 93-103, 1989.

[16] Hagen, H. and Schulze, G., "Variational Principles in Curve and Surface Design", in: Hagen, H. and Roller, D., eds., *Geometric Modelling*, Springer, pp. 161-184, 1990.

[17] Hagen, H. and Santarelli, P., "Variational Design of Smooth B-Spline Surfaces", in: Hagen, H., ed., *Topics in Surface Modelling, SIAM*, pp. 85-94, 1992.

[18] Moreton, H. and Sequin, C., "Functional Optimization for Fair Surface Design", ACM Computer Graphics 26, pp. 167-176, 1992.

[19] Hagen, H. and Bonneau G.-P. (1993), "Variational Design of Smooth Rational Bezier Surfaces", Computing Suppl. 8, pp. 133-138, 1993.

[20] Welch, W. and Witkin, A., "Variational Surface Modeling", *Computer Graphics*, Vol. 26, 157-166.

[21] Zipps, J.M., "Untersuchung der analyytischen Beziehungen zwischen Variationsformuliezungen und elliptischen Differential gleichungen zur Generierung von Freiformflachen", Diplomarbeit un Fache Schiffstheorie, T.U.Berlin, 1991.

[22] Bloor, M.I.G., Wilson, M.J. and Hagen, H., "The Smoothing Properties of Variational Schemes for Surface Design", *Computer Aided Geometric Design*, Vol 12, No. 4, pp. 381-394, 1995.

[23] Klass, R., Correction of local surface irregularities using reflection lines, *Computer Aided Design*, Vol. 12, pp. 73-77, 1980.

[24] Kaufmann, E., Klass, R., "Smoothing surfaces using reflection lines for families of spline", *Computer Aided Design*, Vol. 20, pp. 312-316, 1988.

[25] Lowe, T., Bloor, M.I.G. & Wilson, M.J., "Functionality in Blend Design", *CAD*, Vol. 22, pp. 655-665, 1990.

[26] Nowacki, H., in *Computational Geometry for Ships*, eds. Nowacki, H., Bloor, M.I.G. and Oleksiewicz, B., World Scientific, Singapore, 1995.

Authors

Prof. Dr. M. Bloor
Dept. of Applied Mathematical Studies
The University of Leeds
Woodhouse Lane
LEEDS (UK)

Prof. Dr. Mike Wilson
Dept. of Applied Mathematical Studies
The University of Leeds
Woodhouse Lane
LEEDS (UK)
E–mail: mike@amsta.leeds.ac.uk

Surface design based on brightness intensity or isophotes–theory and practice

Roger Andersson

Mid Sweden University

Abstract: In this paper we will consider surface design through interactive improvements of image intensities of a current surface. Systems supporting this kind of design are immensely efficient tools, enabling operators quickly achieve surface shapes that may be unattainable with other means.

The problem that must be solved in this process is that of finding a surface from its image intensity and some boundary conditions. It is a non-linear boundary value problem that, unfortunately, may very well lack a smooth solution. The main issue of the paper is to find conditions that can be used in practice, ensuring useful solutions

1 Introduction

In the process of styling objects, visual properties play an important role. This holds in particular in the automotive industry, where a designer may spend days at the computer to produce surfaces, reflecting light as desired. Families of reflection lines as a design tool were first introduced and analysed by Klass in [4] and practical means for surface modifications based on them were presented in [3]. We refer to [2] for an overview of methods for fairing of surfaces.

Shaded pictures have a long tradition in art and styling as a medium to communicate shape. Some sources trace it back at least as long as to the work of Leonard da Vinci in the fourteenth century. As a result, stylists have often a very clear idea on a desirable variation of the brightness over the surfaces. Despite these good prerequisites, the efforts to produce surfaces with a desired appearance, aided by current styling systems, are enormously time-consuming. It is also often frustrating and many times have to stop before one has reached the goal.

To improve this situation, we created a prototype system a few years ago that, in principle, allows a stylist to prescribe a desired image intensity of a current surface, leaving the task to supply a surface having this image intensity to the computer. Such systems, ran on powerful computers, have an enormous impact on the process, releasing the stylists the very hard work of producing such a surface by the current trial-and-error process.

This process has an immense power when successful, being capable to produce the surface looked for in a few seconds. It is not free from problems, however, being also capable of producing self-intersecting, creased surfaces, like the one shown in Fig. 1 and Fig. 2. In the present paper, we will discuss some of the problems inherent with any system of this kind.

Since some of the key problems are common to all property-driven design, we will start in this more general setting.

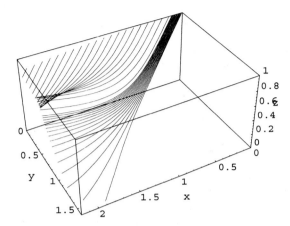

Fig. 1 Curves in a surface with modified image intensity.

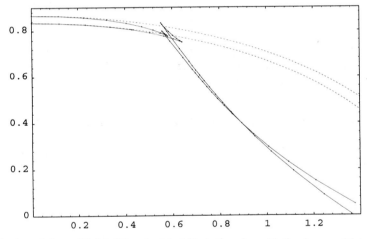

Fig. 2 Sections of the modified (solid) and original (dashed) surface with the planes x=2.1 and x=2.2.

2 Design based on properties.

In general terms, the design goes on as follows: the operator prescribes a property he wants fulfilled on the surface and the system returns a surface having the property. This is a new procedure, with a vast potential. Its interactive form is just going to be practical due to the affordable big increase in computer power now emerging.

Most shape properties are expressed through partial derivatives. To find a surface having a specified shape thus calls for the solution of a partial differential equation. Like equations arising in other areas of the design process, as for instance in mechanics of materials, they are non-linear. For their numerical solution one should expect a similar amount of work as for the latter equations.

Having much in common with other computational problems in the construction process, there are at least two very significant differences. The first concerns smoothness. In shape problems, the solution to the equation is the desired surface and needs to be smooth. Required smoothness is at least G^1, in several cases G^2 and for some reasons even higher.

To the best of our knowledge, no currently available FEM software gives that smooth solutions. Still the piecewise linear element seems to be the bread-and-butter method, giving only C^0 solutions. The p-method, developed during the 80's, rely on higher degree piecewise polynomials, but the pieces are not connected in a manner giving more than C^0 solutions. For the numerical solution with FEM, one has thus to implement smooth basis functions oneself.

The other, even more important difference, relate to existence of a solution. In most other computational problems, this is rarely a concern, since one may often infer existence of a useful solution for physical or other reasons.

In shape problems, on the other hand, this is probably the most important question. Here, we cannot rely on physical evidence or similar arguments to ensure a useful solution. After all, why should there be any physically realisable surface having some property prescribed by the user? Sometimes, this will not be the case.

The main issue of the present paper is to elucidate this lack of solutions concerning image intensities or isophotes. We will enlighten the reasons behind, suggest how to detect the problems early in the computations and, most important, will indicate guidelines to set up the problem in ways that will admit smooth solutions.

3 Surface design based on image intensities or isophotes.

Here we want to highlight the main steps to interactively improve image intensities or isophotes of surfaces and briefly discuss the steps and their requirements.

PROCEDURE

1. Form a shading or a set of isophotes of the current surface.
2. Modify the image or the family of isophotes until good enough.
3. Order the system to modify the current surface into one having the image or isophotes just formed.

The first step is very well known since years, easy to implement and available in several commercial systems. For big objects, like a hood or even more pronounced a uni-side, the comparatively small shaded image on the screen is insufficient to properly reveal the intensity variation of the full-size surface.

Here and in other aspects of the problem as well, it is important to realise that image intensity is basically a function defined on the surface. Thus we are free to illustrate this function in the way we find most suitably for the problem at hand. For instance, we may multiply it with a suitable, possibly nonlinear, scaling function and present the outcome as an enhanced shading of the surface. For other aspects, it may be better to present the image intensity to the user just as a graph of function.

The second step still calls for many improvements. Convenient means for image editing, editing of graphs of functions or for families of isophotes all belong to the new generation of user interfaces that have to be developed to efficiently support property driven design.

In the third step, the input is the modified image intensity function. The way to create it has no consequence here. This fact is significant for surface design based on isophotes. Since isophotes are just level curves of the image intensity function, families of them may be considered just as a device to estimate or modify the image intensity function.

We may benefit from this remark both for the analysis and the synthesis. First, it may often be more easy and natural to modify families of isophotes through modification of the image intensity function than directly. Recall it is not a single or just a few curves to improve. Rather it is the interplay between the curves in the family that counts. Experts on surface design utilise simplified variants of this method to resolve particularly difficult cases today.

Second, it enables a valuable shortcut in designing with isophotes, in alleviating the computation intense step in several design cycles. Since the changes in position of the surface caused in changes of its isophotes is small, one may set up an inner design loop for the improvement through isophotes. After this inner loop has led to a distribution of isophotes good enough, one continues to recalculate the surface.

Having identified the role of isophotes as a means to set up a desired image intensity function, they will have no further impact on the problem at hand. We will not consider them anymore in this paper, with the concluding remark that the changes imposed on the image intensity function through them is usually rather small. Here, small does not mean unimportant!

4 Image irradiance equation.

To gain understanding on the problem to find a surface with an image intensity prescribed by the user, we need to formulate the question precisely and identify what kind of problem it is. Let us start with the easy problem to find the image intensity of a smooth existing surface, illuminated by a single distant light source.

Avoiding to hide the essens of the problem behind long and complicated formulas, we assume the current surface may be considered to be a function surface S_0. Thus $S_0 = \{(x, y, u^{(0)}(x,y)) \,|\, (x,y) \in \Omega\}$, where $u^{(0)}$ is a smooth function defined over the domain Ω in the plane. The upward unit normal to the surface in the point $P = (x, y, u^{(0)}(x,y))$ of S_0 is given by

$$N_0(x,y) = \frac{(-u_x^{(0)}(x,y), -u_y^{(0)}(x,y), 1)}{\sqrt{1 + (u_x^{(0)}(x,y))^2 + (u_y^{(0)}(x,y))^2}}.$$

Since we will not consider light sources located in the plane, the unit vector $\mathbf{l}$ pointing toward the light source may conveniently be given as $\mathbf{l} = \frac{(a,b,1)}{\sqrt{1+a^2+b^2}}$. The unit vector $\mathbf{l}$ is determined by the user, but the system may assist in obtaining a good choice.

The image intensity $I^{(0)}(x,y)$ in the point P is given by $I^{(0)} = \cos \alpha$, where α is the angle between l and the unit normal $N_0(x,y)$. Since $\cos \alpha$ equals the inner product between l and $N_0(x,y)$, we find that the image intensity function $I^{(0)}$ is given by

$$I^{(0)} = \frac{-a\,u_x^{(0)} - b\,u_y^{(0)} + 1}{\sqrt{1+a^2+b^2}\sqrt{1+(u_x^{(0)})^2+(u_y^{(0)})^2}}.$$

This is the expression one have to evaluate in several points $(x,y) \in \Omega$ to form e g a Gourand- or Phong shaded picture of the surface S_0. In the problem we are considering, one turns the things around. Here, one modifies the image intensity function $I^{(0)}$ of the present surface S_0 into a new function I. The user supplies the function I as indicated in the preceding section. We now look for a smooth function surface S having this image intensity I. Usually, one also wants to keep as much as possible of the boundary. In practice, this implies that one does not change the intensity on or in the immediate vicinity of the part of the boundary one wants to fix.

So we are looking for a smooth function u such that

$$\frac{1 - a\,u_x(x,y) - b\,u_y(x,y)}{\sqrt{1+a^2+b^2}\sqrt{1+(u_x(x,y))^2+(u_y(x,y))^2}} = I(x,y) \qquad \text{in } \Omega \qquad (1)$$

$$u(x,y) = u_0(x,y) \qquad \text{on } \Sigma,$$

where Σ the part of the boundary one wants to fix. The user must indicate this part, but as will turns out later, he needs aid from the system to make this choice properly.

This is a boundary value problem for a first order non-linear partial differential equation, known as the image irradiance equation. As a result, our surface modification problem has a solution if and only if the boundary value problem has. Unfortunately, as has been well known in the theory of partial differential equations in many years, it is not generally the case!

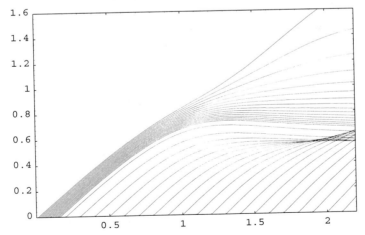

Fig. 3 Projections of the curves in Fig. 1 into the plane.

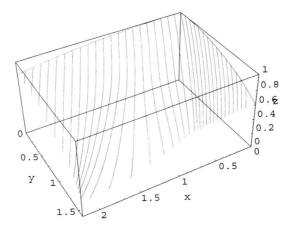

Fig. 4 Curves in a surface with modified image intensity.

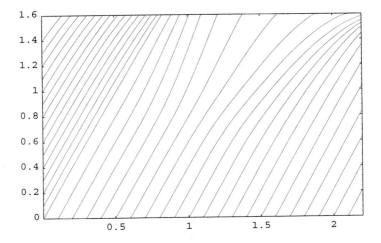

Fig. 5 Projections into the plane of the curves in Fig. 4.

To cite Lions et al, [5] smooth solutions ... are in fact "isolated accidents" for the general equation of this kind. Indeed, Fig. 1 and Fig. 3 illustrate a situation in which our problem lacks a smooth solution. Fig. 4 and Fig. 5 on the other hand indicate one of several cases in which the problem does admit a smooth solution.

What is the mechanisms governing when there is a smooth solution and where there is not? This is one of the key questions one is facing if one tries to create reasonably reliable software for this kind of surface design. Furthermore, insight in this question is also needed in assisting a user in setting up problems admitting a smooth solution.

In the rest of the paper we will consider some aspects of this question. Our main tool in the analysis will be the classical method of characteristics. In the next section, we will briefly outline how the method works, first for the simple first order linear case. We will then indicate the modifications needed to adapt it for the image irradiance equations.

5 Smooth solvability

5.1 Method of characteristics for linear equations.

Our purpose is not to give an introduction to the method but just to present a few facts that are indispensable for the understanding of the solvability conditions of the image irradiance equation. Our approach will be pragmatic. We will use figures that display its use on our problem for the illustration. This is achieved by numerical solutions of the systems of ordinary differential equation arising in our problem. For the numerical solution, we use the function **NDSolve** in Mathematica.

We first consider the partial differential equation

$$A(x,y)u_x + B(x,y)u_y = D(x,y)$$

in a domain in the plane, where A and B are C^1 - functions on a simply connected domain $\tilde{\Omega}$ that contains Ω and its boundary $\partial\Omega$. In Fig. 6, Ω is formed by all points (x,y) in the plane with $0 < x < 2.2$ and $0 < y < 1.6$.

This figure also shows a field of unit vectors over the rectangle. It is the vector field $\zeta = (A, B)$, normalised to have length 1. As turns out of the figure, no one of the vectors along the boundary is parallel to the boundary. Let ν be the outwards directed normal to the boundary, that exists except in the four corners.

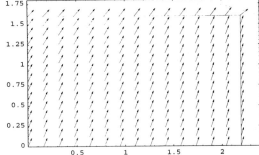

Fig. 6 The vector field $\zeta = (A, B)$ over $\overline{\Omega} = [0, 2.2] \times [0, 1.6]$.

By inspection of the figure, we find that the angle between ν and the vector field $\zeta = (A, B)$ lies between $\frac{\pi}{2}$ and π along the lower and left sides and between 0 and $\frac{\pi}{2}$ along the upper and right sides. This may conveniently be expressed by the sign of the inner product $<\nu, \zeta> = |\nu|\,|\zeta|\cos\beta$, where β is the angle between ν and ζ and $|\;|$ denotes length of a vector. We denote by Σ_- and Σ_+ the parts of the boundary where ν exists and $<\nu, \zeta>$ is negative and positive respectively.

For the further discussion, it is convenient to describe the closure $\overline{\Sigma}_-$ of Σ_- as a parametric curve. Thus we will represent $\overline{\Sigma}_-$ as the curve $\sigma(s)$, for $s_1 \leq s \leq s_2$. In the figure $\overline{\Sigma}_-$ is the lower and left side of the rectangle and we may e g chose σ as follows

$$\sigma(s) = \begin{cases} (2.2 - s, 0) & for\ 0 \leq s \leq 2.2 \\ (0, s - 2.2) & for\ 2.2 \leq s \leq 3.8 \end{cases}.$$

Now let s be any point in $[s_1, s_2]$ and (x_s, y_s) be the corresponding point $\sigma(s)$ in $\overline{\Sigma}_-$ and consider the initial value problem

$$\begin{aligned} x(t) &= A(x(t), y(t)),\ x(0) = x_s \\ y(t) &= B(x(t), y(t)),\ y(0) = y_s. \end{aligned} \qquad (2)$$

Its solution is called a characteristic base curve for the problem.

In the example we are considering, the functions A and B are obtained through linearization of the image irradiance equation around the current surface for a light source direction corresponding to $a = 0.64, b = 1.8$. The initial value problem above was solved numerically for several different s-values and the corresponding solution curves are displayed in Fig. 7.

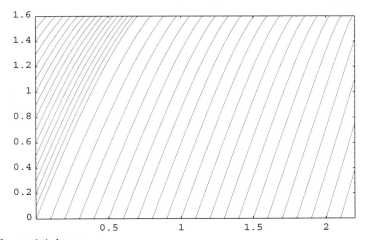

Fig. 7 Characteristic base curves.

As turns out of the figure, the characteristic base curves are smooth, start on Σ_- and leave the domain on Σ_+ after some finite time. The figure also suggests that the curves changes in a smooth way with s. It seems also likely from the figure that given any point (x_0, y_0) in the interior of the rectangle, we can find a unique $s = s_0$ such that the corresponding characteristic base curve passes through the point (x_0, y_0). Moreover, the figure suggests that different characteristic base curves never cross.

All these claims are true in general, provided that there is a simply connected neighbourhood of $\overline{\Omega}$ on which both components A and B of the vector field never vanish simultaneously. It may be proved by the existence- and uniqueness theory for systems of ordinary differential equations in the plane, together with the Poincare-Bendixon theory, see e g [1].

Assuming this, it is not hard to solve the boundary value problem

$$A(x,y)u_x + B(x,y)u_y = D(x,y) \text{ in } \Omega, \, u(\sigma(s)) = \varphi(s) \text{ for } s_1 \leq s \leq s_2. \quad (3)$$

for a smooth function $\varphi : [s_1, s_2] \to \mathbf{R}$ and conclude that the solution is smooth. It depends on a simple observation, which has been utilised within computational fluid dynamics since long ago: along a characteristic base curve, the solution of the boundary value problem is obtained by a simple integration!

If u is a C^1-function and $t \mapsto (x(t), y(t))$ is a characteristic base curve, then by the chain rule

$$\frac{d}{dx}[u(x(t), y(t))] = u_x(x(t), y(t)) \cdot x'(t) + u_y(x(t), y(t)) \cdot y'(t)$$

Since $t \mapsto (x(t), y(t))$ satisfies (2), the right hand side may be written as $A(x(t), y(t)) \cdot u_x(x(t), y(t)) + B(x(t), y(t)) \cdot u_y(x(t), y(t))$, and if u is a solution to (3), this equals $D(x(t), y(t))$. Thus by integration we get

$$u(x(t), y(t)) = u(x(0), y(0)) + \int_0^t D(x(\tau), y(\tau)) \, d\tau.$$

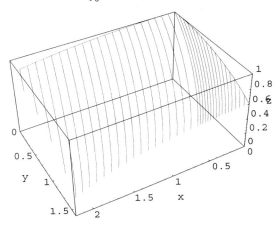

Fig. 8 Characteristics for a linear equation.

The space curves in Fig. 8 are obtained in this way. They are called characteristics for the partial differential equation. The boundary value function φ over $\overline{\Sigma}_-$ is indicated by the curve pieces in the xz- and yz-planes above the lover and left sides of the rectangle. Thus we get the solution u to (3) as the union of the characteristics through the boundary value curve.

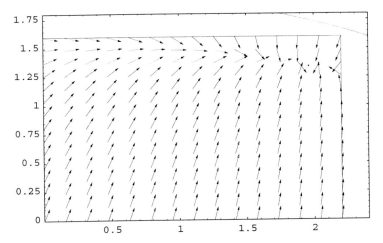

Fig. 9 The vector field $\zeta = (A, B)$ when both components vanish at a point (marked with the small dot).

5.2 First set of consequences.

We see that the starting point of a characteristic uniquely determines its further course. As a consequence, we cannot prescribe boundary values of any more part of the boundary $\partial\Omega$. Of course, we could have given the boundary values over $\overline{\Sigma}_+$ instead, but in this case, we cannot prescribe any boundary values outside $\overline{\Sigma}_+$.

This phenomenon carries over to the non-linear problem as well, which implies that we must be careful when fixing parts of the boundary for the image irradiance equation not to ruin smooth solvability by accident. Where the sets Σ_- and Σ_+ are located depend on an interplay between the outward normal to the boundary and the position of the light source.

To make this choice correctly, a user needs support from the system, making the necessary analysis for him. The same applies to fulfil the related condition that the vector field $\zeta = (A, B)$ not be a tangent to the boundary.

Recall our assumption that both the components of the vector field may not vanish. In Fig. 9 we display the same unit vector field as in Fig. 6, but with the light source direction changed, by making the change $a = 0.22, b = 0.6$. This causes both components of the vector field to vanish in a point $(x, y) \approx (1.96333, 1.33863)$, indicated by a small dot in the figure.

The effects on the characteristic base curves are severe. We will not display the curves obtained by numerical solution because **NDSolve** cannot produce a correct solution in this case. The indicated point is what is know as a strictly stable point for the system, see e g [6], pp 121 - 129. The qualitative behaviour of the solution of the system close to such points is well known, as also indicated in e g [6].

We cannot expect a smooth solution of (3) in this case and have to exclude it. In our image intensity problem such exceptional points are those in which the upward surface normal is parallel to the light source direction.

5.3 Method of characteristics for the image irradiance equation.

We will now look at how to enhance the above method into one for the problem (1). To keep the formalism simple, it is appropriate to introduce the function

$$g(p,q) = \frac{1 - ap - bq}{\sqrt{1 + a^2 + b^2}\sqrt{1 + p^2 + q^2}}.$$

In this notation, setting $p = u_x$ and $q = u_y$ the equation in (1) becomes $g(u_x, u_y) - I = 0$. Furhermore, the coefficients A and B in the linearized equation considered above are then given by $A = g_p(u_x, u_y)$ and $B = g_q(u_x, u_y)$, where g_p and g_q denotes the $p-$ and $q-$ partial respectively.

In this terminology the equations for the characteristic base curve considered above are

$$x'(t) = g_p(u_x(x(t), y(t)), u_y(x(t), y(t)))$$
$$y'(t) = g_q(u_x(x(t), y(t)), u_y(x(t), y(t))).$$

In writing $p(t) = u_x(x(t), y(t))$ and $q(t) = u_y(x(t), y(t))$, it takes the simply form

$$x'(t) = g_p(p(t), q(t)) \tag{4}$$
$$y'(t) = g_q(p(t), q(t)).$$

Moreover, by the chain rule

$$p'(t) = u_{xx}(x(t), y(t))\, x'(t) + u_{xy}(x(t), y(t))\, y'(t) \tag{5}$$
$$q'(t) = u_{xy}(x(t), y(t))\, x'(t) + u_{yy}(x(t), y(t))\, y'(t),$$

provided $u \in C^2$.

Similarly, writing $u(t) = u(x(t), y(t))$, we find that $u'(t) = u_x(x(t), y(t))\, x'(t) + u_y(x(t), y(t))\, y'(t)$ which, recalling the definitions of $p(t)$ and $q(t)$ above, simplifies to

$$u'(t) = p(t)\, x'(t) + q(t)\, y'(t) = p(t)\, g_p(p(t), q(t)) + q(t)\, g_q(p(t), q(t)). \tag{6}$$

By taking $x-$ and $y-$ partials of the equation $g(u_x, u_y) - I = 0$ we get also

$$g_p(u_x, u_y)\, u_{xx} + g_q(u_x, u_y)\, u_{xy} - I_x = 0 \tag{7}$$
$$g_p(u_x, u_y)\, u_{xy} - g_q(u_x, u_y)\, u_{yy} - I_y = 0.$$

Along a characateristic base curve for the linearized equation (4) holds. Substituting the expressions for $x'(t)$ and $y'(t)$ into (5) and using (7), it follows that (5) takes the form

$$p'(t) = I_x(x(t), y(t)) \tag{8}$$
$$q'(t) = I_y(x(t), t(t)).$$

Together, equations (4) and (8) form a system of four ordinary differential equations with no reference to the function u that was used in deriving it. Forgetting how $p(t)$ and $q(t)$ were defined, we have thus reached a system that, supplemented with suitable initial conditions, determines four functions x, y, p, q uniquely. Having determined the functions x, y, p, q of t from this initial value problem, by prescribing an initial value $u(0)$ for the function u of t, we then immediately get $u(t)$ by a straightforward integration as we did for the linear equation.

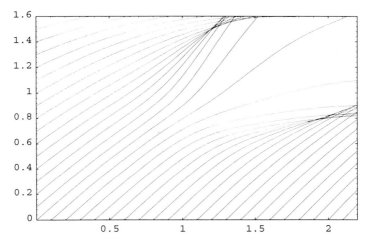

Fig. 10 Projections of characteristics into the plane for a moderately sized but steep perturbation.

This is the method of characteristic for (1)! In practice we have $I = I^{(0)}$ in a vicinty of $\overline{\Sigma}_-$ – we do not expect the surface remains the same in a region where we change its image intensity. Then the natural inital values for the functions x, y, p, q, u of t are $x(0) = x_s$, $y(0) = y_s$ where $\sigma(s) = (x_s, y_s)$ used earlier, together with $u(0) = u^{(0)}(\sigma(s))$ and $p(0) = u_x^{(0)}(\sigma(s))$, $q(0) = u_y^{(0)}(\sigma(s))$.

Each solution, one for each s, is usually called a characteristic strip. The name deduces from the fact that while the characteristics $t \mapsto (x(t), y(t), z(t))$ for different $s \in [s_1, s_2]$ are curves in the solution surface, the functions $t \mapsto (p(t), q(t))$ determine the tangent planes of the solution surface along the curves. It is a valuable property for practical numerical solution, enabling Hermite- rather than Lagrange interpolation to get a surface definition between the curves.

5.4 Second set of consequences.

With all this behind us, we may now look more closely into when our equation has a smooth solution and when it fails. From the above discussion one may expect that in favorable cases, the solution to the boundary value problem

$$\frac{1 - a\, u_x(x,y) - b\, u_y(x,y)}{\sqrt{1 + a^2 + b^2}\sqrt{1 + (u_x(x,y))^2 + (u_y(x,y))^2}} = I(x,y) \qquad \text{in } \Omega \qquad (9)$$

$$u(x,y) = u_0(x,y) \qquad \text{on } \overline{\Sigma}_-$$

may be obtain as a union of characteristic, similar to the linear case.

There is one very significant difference, causing lack of smooth solutions of nonlinear problems in many cases. The characteristic base curves in the linear problem are each a solution of an initial value problem and as such cannot cross. The corresponding curves $t \mapsto (x(t), y(t))$ in the nonlinear problem are only projections into the plane of solutions of initial value problems and may very well cross. Fig. 3 and 10 are both examples of such crossings, that demolish smooth solvability. In Fig. 5 on the contrary, no crossing does occur and, as Fig. 4 suggests, the solution surface is smooth.

There is one practically very important general condition ensuring that the curves $t \mapsto (x(t), y(t))$ obtained for different s-values do not cross and that they cover all of Ω. The condition is that the difference between the intensity I_0 of the current surface and the desired intensity I, *and their partials of the first two orders*, is small enough. When it holds, as shown in [1], one may deduce a sufficiently regular behaviour of the curves $t \mapsto (x(t), y(t))$ from that of the characteristic base curves of the linearised problem.

It is significant to notice that the difference of the derivatives must be taken into account. The depence is illustrated in Fig. 5, in which the C^2-norm of the perturbation producing the crossing is about 5.39 and with C^0-norm 0.14. A similar perturbation of the same intensity function I_0 and same C^0-norm, but with a C^2-norm of about 2.87 does not produce any crossing.

References

[1] Andersson, R.K.E.: Smooth solvability and solutions of the image irradiance equation. Report 1996:1, Preprint Mid Sweden University.

[2] Hoschek, J., Lasser, D.: Grundlagen der geometrischen Datenverarbeitung. 2 Auflage Teubner 1992.

[3] Kaufmann, E., Klass, R. Smoothing surfaces using reflection lines for families of splines. Computer-Aided Design 20 (1988) 312-316.

[4] Klass, R.: Correction of local surface irregularities using reflection lines. Computer-Aided Design 12 (1980) 73-77.

[5] Lions, P.L., Rouy, E. and Tourin, A.: Shape-from Shading, viscosity solutions and edges. Numer. Math. 64 (1993), 323-353.

[6] Birkhoff, G., Rota, G-C.: Ordinary differential equations, 3 ed, Wiley 1978.

Author

Prof. Dr. R. Andersson
Dept. of Technique and Natural Science
Midsweden University
OESTERSUND
Schweden
E-mail: mg17813@gaia.swipnet.se

Fair surface blending, an overview of industrial problems

Alain Massabo

MATRA DATAVISION

Abstract: Aesthetics, the commercial impact of which is continuously increasing, is not a synonym for freedom as the "free form" term leads us to suppose. The blending of surfaces whether complex or ... less complex is an important mode of generating CAD/CAM geometric entities not only because it involves several domains of mathematical expertise but also because it often conditions the quality (or at least the aesthetic quality) of the resulting product.
The presentation, based on a survey published in the May '94 of CAD edition, brings an industrial point of view to the theme, illustrated by some examples and results.
Key Words: *surface blending, filleting, contour filling, thin plate.*

1. Definitions

The geometrical description of an industrial object is rarely done in a single piece. Often the current computer aided design process defines main parts, **connected** or **not**, satisfying functional and/or aesthetic criteria, that should be linked by transition areas, *the blendings*. Their generation needs a **reference** geometry, and because they contribute to the object description, they need to be **accepted without restriction** in all downstream operations. This leads the CAD/CAM system developers to prefer an explicit parametric representation, hence *approximations,* the accuracy stability of which should be carefully controlled in downstream treatments.
Blendings are very important in domains where the quality transition is the criterion to be targeted. For instance, they influence aerodynamic properties or mechanical characteristics and, in Style design, they constitute smart *highlight* or *isophote* transitions between areas, hence participating in the global aesthetic quality of the object.
While the ease of control and the predictability of the blendings are important [1] for the methods of implementation, industrial reality needs to satisfy the criteria which have led the user to request a blending. Therefore, a small number of parameters is not sufficient if the requested blending quality cannot be reached. The exact nature of the blending must be respected when it is a design constraint. This sometimes leads to over-constrained problems.
In general, blendings are G^k-continuous (k=0,1 rarely 2, and even more rarely or never 3 or more) with the blended geometry along *contact curves* (also called *trimlines*). These curves may be either input constraints or results of the blending. When they are input constraints and make a closed contour the blending is *a filling*[1]. When they make two branches it is *a blend*. If the blending is generated by a motion envelope of *profile* curves or surfaces (constant or variable) then the contact curves are a result and blending is *a fillet*. The motions are defined by *spine* or *guide* curves (*trajectories*).
The contact order (*external continuity*) and the blending and reference geometry *internal continuities* are determinant components of the final quality. They are coupled and may even be *incompatible.* Their "uncoupling" leads to singularities (Gregory's patch for instance) which have to be accepted in downstream algorithms. Likewise *physically invalid* blendings (self-intersection for instance) have to be accepted and treated or avoided (i.e. forbidden).

---

[1] A filling may also occur when there are several closed contours "inside" a given closed contour.

As mentioned in [1], the above definitions do not constitute either a partition of construction cases and problems or their exhaustive list. However they fit rather well with generation methods and problems encountered in industry. It is why they have been used as guidelines for the presentation hereafter summarized.

2. Methods

2.1. Blends
In general they are based on Hermite's interpolations of weighted boundary osculating constraints.

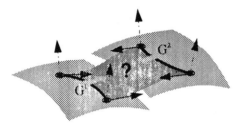

In our system, the weights may be variable along the contact curves and the directions of tangent transversal derivative are computed as an interpolation of extremity values projected in 3D on the tangent osculating surface.
Assignment (cf. 3.1) is supposed to be performed before the blend is requested.

2.2. Fillings
Few systems are able to fill a "more-than-4-sided" hole with a continuity higher than G^1.

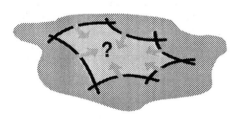

When a Rational mathematical representation is used:
- for n-sided G^0 fillings and n=<4 algebraic solutions may be exact
- G^k fillings are often approximated due to osculating constraint approximations and the degree "explosion".

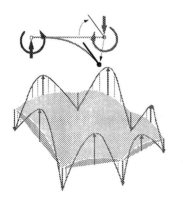

The "thin plate" solution [2] is (mainly) used to fill a "more-than-4-sided" contour. It is based on the simulation of a thin plate behavior under a curvilinear effort distribution (force, moment for G^1 continuity, etc.) applied on the boundary so that the contact constraints are satisfied. The behavior is given by the minimization of the $(k+2)^{th}$ order energy when a G^k contact is requested.
The solution is always approximated if the target mathematical representation is a Rational one (the theoretical solution includes Logarithms).
One problem is the choice of the 2D contour corresponding to the 3D boundary.

It can be noticed that Differential Partial Equation solutions are similar to the thin plate method.

2.3. Fillets
These are generated by the motions of curves (profiles) or envelopes of surfaces (mainly spheres).

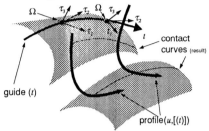

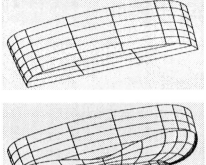

Profiles are not necessary planar. Contact curves are often results, when this is not the case the profile may vary to satisfy the contact order.

If there is a profile deformation

$\text{Profile} = \{\Pi^i(u[,t]) | i = 1, \cdots 3\}$

Laws ∩ (guide[s] ∪ support geometry)

$\text{Motion}(\tau) = \{\vec{\Omega}(\tau), \vec{\tau}_i(\tau) | i = 1, \cdots 3\}$

$\vec{S}(u,t) = \vec{\Omega}(t) + \Pi^i(u[,t])\vec{\tau}_i(t), i = 1, \cdots 3$

The motion can be defined by several curves (guides) that may be contact curves
In the figure on the left, Einstein's convention is used and "[ ]" means optional
For sphere envelopes, the locus of their center is the Voronoï surface of the geometry to blend. Sphere motions may also be used to initialize contact curves and osculating constraints especially when profile(s) is(are) planar, then the fillet can be generated either as a circle motion or as a blending with adequate weight law (using Rationals). This generation constitutes a better numerical solution.

The envelopes are only G^1 with the geometry to blend but provide a better freedom versus the parameterization.
Cyclides do not provide enough controllability (freedom) for free form blending nor do the truncation of Fourier's expansion methods

3. Problems

3.1. Assignment
The problem is to set a correspondence between the two sides of the blend. It has numerous solutions, this is one of the reasons of the great number of blending methods.

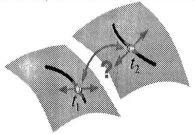

Assignment involves point to point and transversal derivatives correspondence.
In general it leads to re-parameterization of contact curves, hence approximation and/or degree increase.
A way to make the assignment is to use plane motions (cf. 2.3.).
The solution to this problem may generate invalid shapes and therefore should be treated carefully.

3.2. Compatibility

Internal continuities of blendings and reference geometry and their external continuities are interdependent.

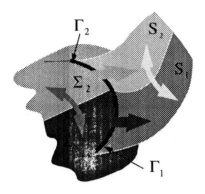
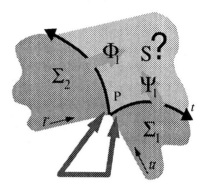

The compatibility conditions are **sufficient conditions**. It means that there are necessary conditions that are less constraining but that limit the controllability (degree of freedom) of the blending shape.

Roughly speaking, it is sufficient that the internal continuity order of the geometry to be blended is the double of the requested external one to get a blend internal continuity equal to the external one.

Internal vs external continuities:
if

$Continuity(\vec{\Sigma}_i, \vec{S}_i) = G^k$

for

$Continuity(\vec{S}_1, \vec{S}_2) = G^k$

excepted specific cases, it is required that

$Continuity(\vec{\Sigma}_1, \vec{\Sigma}_2) >= G^{2k}$

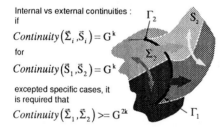

The condition described on the left occurs in Hermite's blend generations as well as in curve motion fillets.

When possible, necessary conditions, such as the use of the common tangent plane directions as the directions of transversal derivatives, solve the problem.

Schwartz's compatibility:
example with a $G^1 \otimes G^1$ filling at the corner P

to get a S solution, excepted specific cases, it is required that

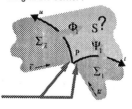

The condition described on the left, is found in filling.

It exists in addition to the one due to that of the requested order of contact.

For example in case of a "$G^1 \otimes G^1$" filling is requested, the two tangent planes at the corner must be identical **and** the reference geometry has to be G^2 continuous.

When possible and/or allowed, a forced compatibility around a vertex solves the problem.

In both of the above described conditions, the addition of singularities such as Gregory's patches, bi or triangular patches, helps to set a solution by "uncoupling" the continuities. However, the singularities are encountered afterwards in downstream treatments.

At each blending operation the intrinsic continuity decreases. Approximation of osculating constraints must be done carefully (i.e. with the right order at the extremities of what is approximated).

Compatibility "justifies" high degrees of Rational mathematical representation in the sense that they delay the problems because they allow high order continuity.

Fair Surface Blending

3.3. Boundary transitions
This problem is encountered at G^k discontinuities (G^1 examples: edges, vertices) when filleting. Intersecting geometry is not required.

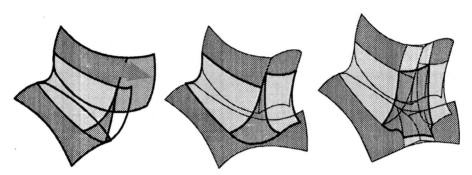

The G^k extension (for an external G^k continuity => G^{2k} extension) *or even C^∞ extension* of the reference geometry is one of the solution.
The best extension method is the C^∞ one, but it is not always possible because of self-intersection (due to high curvatures or cusp spine).
Any curve lying on the blending geometry, and passing through the contact points (before the extension), suffices. Intersection with the surrounding geometry is often used to trim the extension.
Contact switch is an other solution. It consists of managing the kind of contact between the blending element and the reference geometry:

surface (face)			the blending element			surface (face)
curve (edge)	X	i.e. (profile, sphere)	X	curve (edge)
point (vertex)			point (vertex)

The result is rather predictable and often leads to a loss of one order of continuity.
Filling is often (but not necessarily) used at vertices where several blends and/or fillets meet. The area where the blends or fillets are replaced by the filling has to be defined.
Boundary curves (hence contact curves) may belong to the geometry to be blended. This happens when the contact curves have no intersection.
For G^1 blending it should be expected (except in particular cases) not more than G^0 between the blending and the filling (cf. previous compatibility problems in 3.2.).

3.4. Validity
This consists of non aesthetic and/or self-intersecting shapes (see the following example with a rolling sphere).

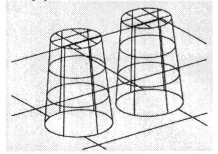 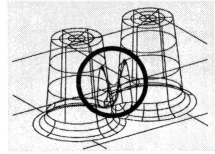

Self-intersection does not only occur in sphere motions, but also in backward curves motions, blending with bad weights or wrong assignment and filling with a wrong parameter distribution.

- validity

e.g.: "variable" sphere

$|\dot{r}| \leq \|\dot{\vec{\Gamma}}\|$

$f(r,\dot{r},\ddot{r},\dot{\vec{\Gamma}},\ddot{\vec{\Gamma}}) \leq 0$

e.g.: $r\kappa_\Gamma - 1 \leq 0$

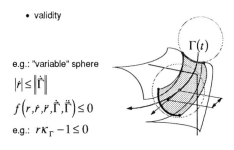

The two columns on the left present sufficient conditions to avoid a self-intersecting fillet generated from a variable sphere motion envelope.

The radius law and sphere center curve (it is supposed that its first derivative does not vanish) are not independent. The sphere center curve must belong to the Voronoï surface of the geometry to be blended.

The problem is a little bit more complex if the contour center curve is closed.

For the variable radius sphere fillet (see on the left), the sufficient condition is obtained by stating that two neighboring contact planes do not intersect "closer than the radius distance".

$$r^2 - \left(\frac{r\dot{r}}{V_\Gamma}\right)^2 - \left(\frac{V_\Gamma^3 - V_\Gamma r\ddot{r} - V_\Gamma \dot{r}^2 + \dot{V}_\Gamma r\dot{r}}{\kappa_\Gamma V_\Gamma^3}\right)^2 \leq 0$$

$V_\Gamma = \text{"speed" of } \vec{\Gamma} \Leftrightarrow V_\Gamma = \|\dot{\vec{\Gamma}}\|, \dot{V}_\Gamma = \frac{\dot{\vec{\Gamma}} \bullet \ddot{\vec{\Gamma}}}{\|\dot{\vec{\Gamma}}\|}$

$\kappa_\Gamma = \text{curvature of } \vec{\Gamma} \Leftrightarrow \kappa_\Gamma = \frac{\|\dot{\vec{\Gamma}} \times \ddot{\vec{\Gamma}}\|}{\|\dot{\vec{\Gamma}}\|^3}$

The self-intersection must be detected and treated. When self-intersection occurs outside the domain of interest it is not a problem. When self-intersection belongs to the domain and if not rejected then NMT[2] acceptance (for internal edges) is a better solution than the splitting of the resulting trimmed blend. However, singularity acceptance (due to some singular points in the self-intersecting curves) is also requested in downstream algorithms.

3.5. Singularities

Degenerated contacts (see figure on the left) appear when the whole profile is the contact solution or when the contact edge between the surface in motion (e.g. the sphere) and its envelope is also contact solution between the surface and the geometry to be blended. *Singular points* are due (among other reasons) to compatibility solutions presented previously (cf. 3.2.). They are encountered in downstream treatments such as blending of blending or offsetting or NC machining etc.

4. Conclusion

There is a great variety of blending methods. Nevertheless in industrial applications few are fully automatic and when they are, either their domains of use and/or their user's satisfaction are limited. This leaves the door open for further research. Good luck to whoever is willing to accept the challenge.

5. Bibliography

Reference [1] proposes a rather exhaustive list of previous works, it covers the theme more broadly than this presentation does.

[1] J. Vida, R.Martin, T. Varady. "A survey of blending methods that use parametric surfaces"
Computer-Aided Design Volume 26 Number 5 May 1994.

[2] G. Durand, A. Lieutier, A. Massabo. "Accuracy incompatible modelers: are complex data exchange possible ?", 28th ISATA (18th -22nd , September 1995) 95ME063.

Author

Dr. Alain C. Massabo
MATRA DATAVISION GmbH
53, avenue de l'Europe
13082 Aix–en–Provence CEDEX 2
Tel.: ++33 42 52 20 00 Fax.: ++33 42 52 21 99

Multivariate Splines with Convex B-Patch Control Nets are Convex

Hartmut Prautzsch

Fakultät für Informatik, Universität Karlsruhe

Abstract: In this paper results from a forthcoming paper are presented concerning the convexity of multivariate spline functions built from B-patches. Conditions are given under which it is possible to define a control net for such spline functions. The control net is understood as a piecewise linear function. If it is convex, then so is the underlying spline.

Keywords: multivariate splines, B-patches, convexity, control nets, Greville-abscissae.

1 Introduction

For the Bézier representation of a bivariate polynomial over some triangle $\triangle$ it is well-known that the convexity of the Bézier net implies the convexity of the polynomial over the triangle $\triangle$. This fact was first proved by Chang and Davis [1984] and later generalized to multivariate polynomials and their Bézier representations over a simplex [DM88, Bes89, Pra95].

Here it is shown that this property is, more generally, even shared by multivariate polynomials and their B-patch representations. Moreover it is also possible to extend the proof to multivariate spline functions and their B-patch control nets.

2 Multivariate B-splines

This paper is based on the B-splines constructed by Dahmen, Micchelli and Seidel [1992] from B-patches. To begin with let us recall the relevant properties and thereby introduce the notation used in this paper:

For any set of **knots** $\mathbf{u}_0, \ldots, \mathbf{u}_k \in \mathbf{R}^s$ the **simplex spline** $M(\mathbf{x}|\mathbf{u}_0 \ldots \mathbf{u}_k)$ is defined as the solution of the functional equation

$$\int_{\mathbf{R}^s} f(\mathbf{x}) M(\mathbf{x}|\mathbf{u}_0 \ldots \mathbf{u}_k) d\mathbf{x} = k! \int_{\sigma} f([\mathbf{u}_0 \ldots \mathbf{u}_k]\mathbf{t}) dt$$

for all continuous functions $f(\mathbf{x})$ where

$$\sigma = \{\mathbf{t} \in \mathbf{R}^{k+1} | \mathbf{o} \leq \mathbf{t}, |\mathbf{t}| = 1\} \ , \quad |\mathbf{t}| = \text{sum of all coordinates of } \mathbf{t}$$

denotes the **standard k-simplex**.

Thus the above normalization implies that

$$\int_{\mathbf{R}^s} M(\mathbf{x}|\mathbf{u}_0 \ldots \mathbf{u}_k)d\mathbf{x} = 1 \ .$$

Now for any $s+1$ knot clusters $\mathbf{u}_\beta^\alpha$, $\alpha = 0, \ldots, s$, $\beta = 0, \ldots, n$, consider the simplices $\sigma_{\mathbf{i}}$ with vertices $\mathbf{u}_{i_\alpha}^0, \ldots, \mathbf{u}_{i_s}^s$ where $\mathbf{i} = (i_0, \ldots, i_s) \in \mathbf{N}_0^{s+1}$ and $|\mathbf{i}| = n$. Then the corresponding splines

$$B_{\mathbf{i}}(\mathbf{x}) = \frac{\mathrm{vol}_s \sigma_{\mathbf{i}}}{\binom{n+s}{s}} M(\mathbf{x}|\mathbf{u}_0^0 \ldots \mathbf{u}_{i_\alpha}^0 \ldots \mathbf{u}_0^s \ldots \mathbf{u}_{i_s}^s)$$

are the **multivariate B-splines** which were introduced in [DMS92] with the name B-weights. Throughout the paper we will assume that bold indices $\mathbf{i}, \mathbf{j}, \ldots$ are in $\mathbf{N}_0^{s+1}$ and that the intersection Ω of all simplices $\sigma_{\mathbf{i}}$, $|\mathbf{i}| \leq n$, is non-empty. Then one has the following crucial property:

Theorem 2.1 *Let $p(\mathbf{x})$ be any s-variate polynomial of total degree n and let $p[\mathbf{x}_1 \ldots \mathbf{x}_n]$ be the unique symmetric multiaffine polynomial with the diagonal property $p[\mathbf{x} \ldots \mathbf{x}] = p(\mathbf{x})$. Then for all $\mathbf{x} \in \Omega$ one has*

$$p(\mathbf{x}) = \sum_{|\mathbf{i}|=n} p[\mathbf{u}_0^0 \ldots \mathbf{u}_{i_\alpha-1}^0 \ldots \mathbf{u}_0^s \ldots \mathbf{u}_{i_s-1}^s] B_{\mathbf{i}}(\mathbf{x}) \ .$$

For the proof one can use the properties of the so-called polar form $p[\mathbf{x}_1 \ldots \mathbf{x}_n]$ and the recurrence relation of simplex splines to evaluate the left and respectively the right hand side of the equation recursively. A comparison then reveals the identity above.

A dimension count futher shows that the $\binom{n+s}{s}$ B-splines $B_{\mathbf{i}}$ are linearly independent (over Ω).

Remark 2.2 *Theorem 2.1 also shows that for $s = 1$ the $B_i(x)$ are the common univariate B-splines. Further if $\mathbf{u}_0^\alpha = \cdots = \mathbf{u}_n^\alpha$ for all α, then the $B_{\mathbf{i}}(\mathbf{x})$ are the truncated Bernstein polynomials over Ω*

3 Control nets

In order to describe the control net of a polynomial

$$p(\mathbf{x}) = \sum c_{\mathbf{i}} B_{\mathbf{i}}(\mathbf{x}) \ , \quad \mathbf{x} \in \Omega \ ,$$

we need the B-spline representation of $\mathbf{x}$. From Theorem 2.1 we obtain

$$\mathbf{x} = \sum \mathbf{x}_{\mathbf{i}} B_{\mathbf{i}}(\mathbf{x}) \ , \quad \text{where } \mathbf{x}_{\mathbf{i}} = \frac{1}{n} \sum_{\alpha=0}^{s} \sum_{\beta=0}^{i_\alpha-1} \mathbf{u}_\beta^\alpha \ .$$

In particular, if $s = 1$, then the $\mathbf{x}_{\mathbf{i}}$ are the so-called **Greville abscissa** and if

$$\mathbf{u}_\beta^\alpha = \mathbf{u}^\alpha \quad \text{for all } \alpha \text{ and } \beta$$

then the $\mathbf{x_i}$ lie on a regular grid, i.e.

$$\mathbf{x_i} = (i_0\mathbf{u}^0 + \cdots + i_s\mathbf{u}^s)/n \ .$$

Next we will triangulate the abscissae $\mathbf{x_i}$ and define the control net of p as the piecewise linear function $c(\mathbf{x})$ which is linear over each simplex of this triangulation and which interpolates the $c_\mathbf{i}$ at the $\mathbf{x_i}$.

If the $\mathbf{x_i}$ are not too far away from the vertices of a regular grid, then we can obtain a triangulation from a triangulation of the regular grid. Therefore we will first describe a triangulation for the case $\mathbf{u}_\beta^\alpha = \mathbf{u}^\alpha$. Then we change the triangulation by moving the $\mathbf{u}_\beta^\alpha$ independently from each into general positions and present conditions under which the triangulation remains a triangulation with disjoint simplices.

For the construction of a Bézier net Dahmen and Micchelli [1988] used a triangulation due to Allgower and Georg:

Let π be the simplex $\mathbf{u}^0 \ldots \mathbf{u}^s$ and ρ the subsimplex whose vertices have the barycentric coordinates

$$\frac{1}{n}\begin{bmatrix} n \\ 0 \\ 0 \\ \vdots \\ 0 \end{bmatrix}, \frac{1}{n}\begin{bmatrix} n-1 \\ 1 \\ 0 \\ \vdots \\ 0 \end{bmatrix}, \ldots, \frac{1}{n}\begin{bmatrix} n-1 \\ 0 \\ 0 \\ \vdots \\ 1 \end{bmatrix}$$

with respect to π. Let $\mathbf{a}_0, \ldots, \mathbf{a}_s$ be these vertices in any arbitrarily fixed order. Counting indices modulo $s+1$ the vertex $\mathbf{a}_0$ and the ordered sequence of vectors

$$\mathbf{v}_0 = \mathbf{a}_1 - \mathbf{a}_0, \ldots, \mathbf{v}_s = \mathbf{a}_{s+1} - \mathbf{a}_s$$

describe a simple closed path through all vertices of ρ. Note that $\mathbf{a}_i$ and $\mathbf{v}_i, \ldots, \mathbf{v}_{i+s}$ describe the same path. Now if any two successive vectors, say $\mathbf{v}_i$ and $\mathbf{v}_{i+1}$ are interchanged, then $\mathbf{a}_i; \mathbf{v}_{i+1}, \ldots, \mathbf{v}_{i+s+1}$ describes a path around a simplex ρ' which shares an $s-1$-dimensional face with ρ. By further transpositions of successive vectors one gets paths around successively adjacent simplices. All the simplices obtained in this way form a triangulation of the entire space $\mathbf{R}^s$. This triangulation is also formed by all hyperplanes spanned by the knot $\mathbf{u}_0$ and any $s-1$ vectors out of $\{\mathbf{v}_0, \ldots, \mathbf{v}_s\}$ and translates of these hyperplanes by integer multiples of the $\mathbf{v}_i$. Thus this triangulation respects the simplex σ and can be restricted to σ.

Remark 3.1 *If the $\mathbf{a}_i$ denote the vertices of σ in a different order, then the construction above results in a different triangulation.*

4 Conditions on the knot clusters

Assume that all knots in every cluster coincide, i.e. $\mathbf{u}_\beta^\alpha = \mathbf{u}^\alpha$ for all α and β. Then the above triangulation has the following property:

Lemma 4.1 *The union of all simplices with vertex $\mathbf{a}_0$ forms the set of all points*

$$\mathbf{x} = \mathbf{a}_0 + \mu_0 \mathbf{v}_0 + \cdots + \mu_s \mathbf{v}_s \ , \quad \text{where } \mu_i \in [0,1] \ .$$

Proof

Let $\mu_0 \geq \cdots \geq \mu_s$. Then since $\mathbf{v}_0 + \cdots + \mathbf{v}_s = \mathbf{o}$, we can write $\mathbf{x} = \mathbf{a}_0 + \mu_0 \mathbf{v}_0 + \cdots + \mu_s \mathbf{v}_s$ as

$$\mathbf{x} = (1 - \mu_0 + \mu_s)(\mathbf{a}_0 + \mathbf{v}_0 + \cdots + \mathbf{v}_{s-1} + \mathbf{v}_s)$$
$$+ (\mu_{s-1} - \mu_s)(\mathbf{a}_0 + \mathbf{v}_0 + \cdots + \mathbf{v}_{s-1})$$
$$\vdots$$
$$+ (\mu_0 - \mu_1)(\mathbf{a}_0 + \mathbf{v}_0)$$

which is a convex combination of the vertices of the simplex given by the loop $\mathbf{a}_0, \mathbf{v}_0 \ldots \mathbf{v}_s$. Similarly any ordering of the μ_i corresponds to a loop $\mathbf{a}_0, \mathbf{w}_0 \ldots \mathbf{w}_s$ where $(\mathbf{w}_0, \ldots, \mathbf{w}_s)$ is a permutation of $(\mathbf{v}_0, \ldots, \mathbf{v}_s)$ and vice versa. This completes the proof since all these loops describe all the simplices with vertex $\mathbf{a}_0$. ∎

Now we move the $\mathbf{u}_\beta^\alpha$ independently from each other into general positions. This will also change the positions of the $\mathbf{x_i}$ and the shape and positions of the simplices of the triangulation given in Section 3. The new triangulation is still feasible under the following mild restrictions on the knot positions:

Theorem 4.2 *If for all $\alpha = 0, \ldots, s$ and $\beta = 0, \ldots, n$*

$$\mathbf{u}_\beta^\alpha \in \mathbf{u}^\alpha + [\mathbf{v}_0 \ldots \mathbf{v}_s][0, 1/2)^s \ ,$$

then any two simplices of the new triangulation have disjoint interiors.

We omit the full proof here and derive only the crucial property on which the proof is based:

$$\mathbf{x_i} = \frac{1}{n} \sum_{\alpha=0}^{s} \sum_{\beta=0}^{i_\alpha - 1} \mathbf{u}_\beta^\alpha$$

$$\in \frac{1}{n} \sum_{\alpha=0}^{s} \sum_{\beta=0}^{i_\alpha - 1} \mathbf{u}^\alpha + [\mathbf{v}_0 \ldots \mathbf{v}_s][0, 1/2)^s$$

$$= \frac{1}{n} \sum_{\alpha=0}^{s} i_\alpha \mathbf{u}^\alpha + [\mathbf{v}_0 \ldots \mathbf{v}_s][0, 1/2)^s \ .$$

Thus different $\mathbf{x_i}$ lie in disjoint convex regions.

5 B-patches with convex control nets

Consider the control net of a single B-patch. It is a piecewise linear function defined over some triangulation of the abscissae $\mathbf{x_i}$. In general, this triangulation does not form a convex domain for the control net. Therefore we need to explain what is meant by a convex net:
First let $\mathbf{q}(\mathbf{x}) = [\mathbf{x} \ q(\mathbf{x})]$ be the graph of a quadratic polynomial and let $\mathbf{c_i} \in \mathbf{R}^{s+1}$, $|\mathbf{i}| = 2$, be its B-spline control points with respect to the knots $\mathbf{u}_\beta^\alpha$, $\alpha = 0, \ldots, s; \beta = 0, 1, 2$, and further let $\mathbf{b_i}$ be the Bézier points of $\mathbf{q}(\mathbf{x})$ over the simplex $\mathbf{u}_0^0 \ldots \mathbf{u}_0^s$. Then it follows from Theorem 2.1 that

$$\mathbf{c_i} = \mathbf{b_i} \quad \text{for all } \mathbf{i} \leq (1, \ldots, 1)$$

and furthermore that the points $\mathbf{b_i}$ and the points $\mathbf{c_i}$, for $\mathbf{i} = \mathbf{e}_i + \mathbf{e}_j$, i fixed, $j = 0, \ldots, s$, span the same plane. Thus we have the following property:

Lemma 5.1 *The Bézier and the B-spline control nets of the quadratic polynomial $q(\mathbf{x})$ above are identical over the intersection of their domains.*

Hence we say that the B-spline control net of the quadratic polynomial $p(\mathbf{x})$ is convex if the associated Bézier net of $p(\mathbf{x})$ is convex.

Next consider again a polynomial of degree n

$$p(\mathbf{x}) = \sum_{|\mathbf{i}|=n} c_{\mathbf{i}} B_{\mathbf{i}}$$

given by its B-spline representation over the knot clusters $\mathbf{u}_\beta^\alpha$, $\alpha = 0, \ldots, s; \beta = 0, \ldots, n$. Let $p[\mathbf{x}_1 \ldots \mathbf{x}_n]$ be the polar form of $p(\mathbf{x})$. Then the quadratic polynomials

$$p_{\mathbf{i}}(\mathbf{x}) = p[\mathbf{x} \ \mathbf{x} \ \mathbf{u}_0^0 \ldots \mathbf{u}_{i_0-1}^0 \ldots \mathbf{u}_0^s \ldots \mathbf{u}_{i_s-1}^s] \ , \quad |\mathbf{i}| = n-2 \ ,$$

have the B-spline representations

$$p_{\mathbf{i}}(\mathbf{x}) = \sum_{|\mathbf{j}|=2} c_{\mathbf{i}+\mathbf{j}} B_{\mathbf{j}}(\mathbf{x})$$

over the knots $\mathbf{u}_{i_\alpha+\beta}^\alpha$, $\alpha = 0, \ldots, s; \beta = 0, 1, 2$. Now we can state the main result of this section.

Theorem 5.2 *If the control nets of all quadratic polynomials $p_{\mathbf{i}}(\mathbf{x})$, $|\mathbf{i}| = n-2$, are convex, then $p(\mathbf{x})$ is convex over the intersection Ω of all simplices $\mathbf{u}_{i_\alpha}^0 \ldots \mathbf{u}_{i_s}^s$, $|\mathbf{i}| \leq n$.*

Let us sketch the proof: Let $D_\mathbf{v}^2 f(\mathbf{x})$ be the second derivative of the function f with respect to the direction $\mathbf{v}$. Then one can use, e.g., the multidimensional analog of Proposition 8.2 in [Ram87] to derive

$$D_\mathbf{v}^2 p(\mathbf{x}) = \frac{n(n-1)}{2} \sum_{|\mathbf{i}|=n-2} (D_\mathbf{v}^2 p_{\mathbf{i}}) B_{\mathbf{i}}(\mathbf{x}) \ .$$

Since the $p_{\mathbf{i}}$ have a convex Bézier net, they are convex functions, see e.g. [DM88]. Hence the second directional derivatives $D_\mathbf{v}^2 p_{\mathbf{i}}$ are non-negative which implies that $D_\mathbf{v}^2 p(\mathbf{x})$ is non-negative and thus the convexity of $p(\mathbf{x})$ over Ω.

6 Splines with convex control nets

The results above for a single B-patch can be extended to splines:
Let $\mathbf{u}^\alpha$, $\alpha \in \mathbf{Z}$, be the vertices of some triangulation $\mathcal{T}$ covering the entire space $\mathbf{R}^s$. Here we think of $\mathcal{T}$ as a subset of $\mathbf{Z}^{s+1}$ such that the simplices $\mathbf{u}^{a_0} \ldots \mathbf{u}^{a_s}$, $\mathbf{a} = (a_0, \ldots, a_s) \in \mathcal{T}$ form the triangulation. In the following we will always assume that $\mathcal{T}$ contains each simplex only once, i.e. for any $\mathbf{a} \in \mathcal{T}$ there is no other permutation of $\mathbf{a}$ in $\mathcal{T}$. Further let $\mathbf{u}_\beta^\alpha$, $\beta = 0, \ldots, n$

be associated knot clusters and assume that the intersections $\Omega_{\mathbf{a}}$ of all simplices $\mathbf{u}_{i_0}^{a_0} \ldots \mathbf{u}_{i_s}^{a_s}$, $|\mathbf{i}| \leq n$, are non-empty for all $\mathbf{a} \in \mathcal{T}$. Then consider the spline

$$s(\mathbf{x}) = \sum_{\mathbf{a} \in \mathcal{T}} \sum_{|\mathbf{i}|=n} c_{\mathbf{i}}^{\mathbf{a}} B_{\mathbf{i}}^{\mathbf{a}}(\mathbf{x})$$

where $B_{\mathbf{i}}^{\mathbf{a}}$ is the B-spline over the knots $\mathbf{u}_\beta^\alpha$, $\alpha = a_0, \ldots, a_s; \beta = 0, \ldots, i_\alpha$. In order to define the control net of $s(\mathbf{x})$ as a piecewise linear function we need the abscissae

$$\mathbf{x}_{\mathbf{i}}^{\mathbf{a}} = \frac{1}{n} \sum_{\alpha=a_0, \ldots, a_s} \sum_{\beta=0}^{i_\alpha} \mathbf{u}_\beta^\alpha .$$

Then for each $\mathbf{a} \in \mathcal{T}$ we triangulate the abscissae $\mathbf{x}_{\mathbf{i}}^{\mathbf{a}}$ as described in Section 4 using the loops

$$\mathbf{v}_0^{\mathbf{a}} = \mathbf{u}^{a_1} - \mathbf{u}^{a_0}, \quad \ldots \quad , \quad \mathbf{v}_s^{\mathbf{a}} = \mathbf{u}^{a_0} - \mathbf{u}^{a_s} .$$

In order to obtain a correct triangulation of all $\mathbf{x}_{\mathbf{i}}^{\mathbf{a}}$, $\mathbf{a} \in \mathcal{T}$, we need to restrict the positions of the knots. Such a condition is given by the following extension of Theorem 4.2:

Theorem 6.1 *Let Ω_α be the intersections*

$$\Omega_\alpha = \cap \{[\mathbf{v}_0^{\mathbf{a}} \ldots \mathbf{v}_s^{\mathbf{a}}][0, 1/2)^s \mid \mathbf{a} \in \mathcal{T}, \ \alpha \text{ is a coordinate of } \mathbf{a}\}$$

and for all $\alpha \in \mathbb{Z}$ and $\beta = 0, \ldots, n$ let $\mathbf{u}_\beta^\alpha \in \mathbf{u}^\alpha + \Omega_\alpha$. Then any two simplices of the triangulation of the $\mathbf{x}_{\mathbf{i}}^{\mathbf{a}}$, $\mathbf{a} \in \mathcal{T}$, $|\mathbf{i}| = n$, have disjoint interiors.

Theorem 6.1 enables us to define the **control net** of $s(\mathbf{x})$ as the piecewise linear function which is composed of the control nets of the patches

$$s_{\mathbf{a}}(\mathbf{x}) = \sum_{|\mathbf{i}|=n} c_{\mathbf{i}}^{\mathbf{a}} B_{\mathbf{i}}^{\mathbf{a}}(\mathbf{x}) , \quad \mathbf{x} \in \Omega_{\mathbf{a}} .$$

Note that the control nets of the patches over the sets $\Omega_{\mathbf{a}}$ are always continuous, but the entire control net of $s(\mathbf{x})$ is continuous only if $c_{\mathbf{i}}^{\mathbf{a}} = c_{\mathbf{j}}^{\mathbf{b}}$ whenever $\mathbf{x}_{\mathbf{i}}^{\mathbf{a}} = \mathbf{x}_{\mathbf{j}}^{\mathbf{b}}$. Now, for this control net of $s(\mathbf{x})$ we can state the main result presented in this paper:

Theorem 6.2 *Let the control net of $s(\mathbf{x})$ be continuous and such that the subnets for all patches $s_{\mathbf{a}}(\mathbf{x})$, $\mathbf{x} \in \Omega_{\mathbf{a}}$, satisfy the conditions of Theorem 5.2. Then the spline function $s(\mathbf{x})$ is convex for all $\mathbf{x} \in \mathbf{R}$.*

References

[Bes89] M. Beska. Convexity and variation diminishing property of multidimensional Bernstein polynomials. *Approximation Theory and its Applications*, 5:59–78, 1989.

[CD84] G. Chang and P.J. Davis. The convexity of Bernstein polynomials over triangles. *Journal of Approximation Theory*, 40:11–28, 1984.

[DM88] W. Dahmen and C.A. Micchelli. Convexity of multivariate Bernstein polynomials and box spline surfaces. *Studia Sci. Math., Hungary*, 23:265–287, 1988.

[DMS92] W. Dahmen, C.A. Micchelli, and H.-P. Seidel. Blossoming begets B-splines built better by B-patches. *Mathematics of computation*, 59(199):97–115, 1992.

[Pra92] H. Prautzsch. On convex Bézier triangles. *Mathematical Modelling and Numerical Analysis*, 26(1):23–36, 1992.

[Pra95] H. Prautzsch. On convex Bézier simplices. In *Notas de Matematicas, revista de Departamento de Matematicas*, pages 1–12. Universidad de los Andes, 1995.

[Ram87] L. Ramshaw. Blossoming: a connect-the-dots approach to splines. Technical report, Digital Systems Research Center, Palo Alto, Ca, June 21 1987.

Author

Prof. Dr. H. Prautzsch
Universität Karlsruhe
Fakultät für Informatik
D–76128 Karlsruhe
E–mail: prau@ira.uka.de

Shape Preserving of Surfaces

Parametrizing Wing Surfaces using Partial Differential Equations

Malcolm I.G. Bloor, Michael J. Wilson

The University of Leeds

Abstract: A method is presented for generating three-dimensional surface data given two-dimensional section data. The application on which this paper concentrates is that of producing wing surfaces through a set of airfoil sections. It is an extension of a new method for the the efficient parametrization of complex three-dimensional shapes, called the PDE Method. The method views surface generation as a boundary-value problem, and produces surfaces as the solutions to elliptic partial differential equations.

1 Introduction

Aircraft design is, traditionally, divided into three phases: (a) conceptual design, (b) preliminary design, and (c) detail design [1, 2]. Raymer [2] characterizes conceptual design as being the process whereby basic questions concerning the aircraft's configuration, size and weight, and performance are answered. During preliminary design, design and analysis of such aspects as structures, landing gear, and control systems takes place; the precise mathematical description of the outside skin of the aircraft also takes place at this point; testing is begun in areas such as aerodynamics, propulsion, structures, stability and control. Finally, detail design involves the design of the actual parts out of which the aircraft will be built and the planning of the production process itself.

The increasing power of computer-aided technologies has meant that the boundaries between these three phases have become somewhat blurred. For instance, computational fluid dynamics (CFD) and computer hardware have both advanced to the point where not only is it feasible to carry out an analysis of the aerodynamic properties of a particular design [3, 4] but also, to some extent, carry out numerical optimization to improve the design [5, 6, 7, 8, 9, 10, 11, 12]. Thus it is becoming increasingly possible for operations that traditionally formed part of preliminary design to be carried out during concept design.

In earlier work it was shown how generic aircraft geometry could be produced from the solutions to elliptic partial differential equations using a technique known as the PDE Method ([13],[14],[15]). In particular it was shown how the method was capable of producing many of the large-scale features associated with the double-delta shape characteristic of a high-speed civil transport (HSCT)[13]. The method was able to parametrize this geometry using only a small number of design variables, and by altering these variables it was possible to rapidly produce a wide variety of radically different alternative designs.

However, in certain circumstances, it is desirable to represent certain standard geometric features that are, traditionally, used by designers. For instance, conceptual design often starts with a basic planform and airfoils from standard families distributed at certain span-stations along a wing [1, 2]. The choice of this starting design will be based upon the prior experience and expertize of the designer, coupled with the mission priorities for the aircraft. The precise geometry of the airfoils and the plan-form may need to be optimized to suit the particular aircraft [7, 9, 10, 11]. However, in many situations it is unlikely that they will alter drastically in character during the design process, since the broad configurations required for different mission-profiles are by now well established.

The object of this paper is to show how the PDE Method can accurately represent certain geometric features that are typically used by designers when specifying an initial design. In this way we aim to demonstrate that a designer using the PDE Method can, in principle, start the process of conceptual design with the same type of geometry specification as in the past. The particular feature this paper will concentrate upon are the wings since these are, in many ways, the key-features of an aircraft, although the technique described here is applicable to other parts of the aircraft.

2 Outline of the PDE Method

When designing using the PDE Method, a complex surface, such as that of an aircraft, is broken up into constituent patches which meet at space-curves ('characterlines') which form the boundaries between adjacent patches. In this respect it is similar to conventional Boundary-Representation (BRep) CAD systems [16] which define an object by a description of its bounding surface. However, unlike conventional systems, the method is based upon a view of surface generation as a boundary-value problem in which each surface patch is defined by data defined along the characterlines which form the patch boundaries. Adjacent surface patches share common boundary conditions and thus continue to meet exactly throughout any changes to the overall geometry that may occur during the design process. This is in constrast to conventional BRep systems which typically use surface patches generated from polynomial spline functions, and tend to require 'trimming' at the boundaries between adjacent patches and the further 'stitching' together for them to meet continuously. Also, far fewer surface patches are usually required using the PDE method than a spline-based approach.

Past work has concentrated upon solutions to the following equation,

$$\left(\frac{\partial^2}{\partial u^2} + a^2 \frac{\partial^2}{\partial v^2}\right)^2 \underline{X} = 0 \tag{1}$$

This equation is solved over some finite region Ω of the (u, v) parameter plane, subject to boundary conditions on the solution which usually specify how $\underline{X}$ and its normal derivative $\frac{\partial \underline{X}}{\partial \mathbf{n}}$ vary along $\partial\Omega$. The three components of the function $\underline{X}$ $((x(u,v), y(u,v), z(u,v))$ are the Euclidean coordinate functions of points on the surface, given parametrically in terms of the two parameters u and v which define a coordinate system on the surface. Note that in the simplest cases (1) is solved independently for the x,y and z coordinates.

The boundary conditions on $\underline{X}$, which we shall refer to as function boundary conditions, determine the shape of the curves bounding the surface patch in physical space, or, more specifically, their parametrization in terms of u and v. The boundary conditions on $\frac{\partial X}{\partial \mathbf{n}}$, which we shall refer to as derivative boundary conditions, basically determine the direction in physical space in which the surface moves away from a boundary and how 'fast' is does so.

The problem addressed in this paper is similar in intention to previous work on describing the blades of marine propellers ([17]), in that we are seeking a smooth surface that interpolates (or at least closely approximates) specified airfoil cross-sections. In the propeller work the specified cross-sections where approximated by an appropriate choice of boundary-conditions, an approach which was not straightforward to automate. In this paper a somewhat different method is used, in that the PDE surface that interpolates specified airfoil cross-sections will be obtained by solving a high-order partial differential equations.

3 Interpolating Wing Sections

Consider the problem of generating a smooth wing-surface that passes through 2N two-dimensional wing sections which are specified. Suppose for the moment that each wing section $\underline{W}^i(u)$ is given parametrically as a vector-valued function of a periodic coordinate v that runs around the wing,

$$\underline{W}^i(v) = (W_x^i(v), W_y^i(v), W_z^i) \tag{2}$$

where the sections $i = 1, 2N$ correspond to the ends of the wing, and W_z^i is the specified span-station of the ith section. In what follows below we will assume that the x coordinate is approximately aligned with the wing chord and the y coordinate is aligned with the wing thickness.

In previous work [13] the order of the partial differential equation was kept as low as possible consistent with the requirements of surface control and continuity at patch boundaries. Here we consider solutions of the equation

$$\left(\frac{\partial^2}{\partial u^2} + a^2 \frac{\partial^2}{\partial v^2} \right)^{(N+1)} \underline{X} = 0 \tag{3}$$

Now consider the solution $\underline{X}(u,v)$ of equation (3) over the (u,v) region Ω: $[0,1] \times [0, 2\pi]$, subject to periodic boundary conditions in the v direction. The $u = 0$ and $u = 1$ isoparametric lines correspond to the wing-sections $\underline{W}^1(v)$ and $\underline{W}^{2N}(v)$ respectively, which form the boundary curves for the surface patch.

If we assume that the boundary conditions for equation (3) take the form

$$\underline{X}(0,v) = \underline{W}^1(v), \tag{4}$$
$$\underline{X}(1,v) = \underline{W}^{2N}(v),$$
$$\underline{X}_u(0,v) = \underline{s}^0(v), \tag{5}$$
$$\underline{X}_u(1,v) = \underline{s}^1(v),$$

where $\underline{W}^1(v)$ and $\underline{W}^{2N}(v)$ are given by an expression of the form (2), and the derivative functions $\underline{s}^0(v)$ and $\underline{s}^1(v)$ are specified, then, by using the method of separation of variables [18], the solution to equation (3) may be written in closed-form thus

$$\underline{X}(u,v) = \underline{A}_0(u) + \sum_{n=1}^{M} \{\underline{A}_n(u) \cos(nv) + \underline{B}_n(u) \sin(nv)\} \tag{6}$$

where, depending on the boundary conditions, M may be infinite. The 'coefficient' functions $\underline{A}_n(u)$ and $\underline{B}_n(u)$ are of the form

$$\underline{A}_0(u) = \underline{a}_{00} + \underline{a}_{01}u + \underline{a}_{02}u^2 + \ldots\ldots + \underline{a}_{0(2N+1)}u^{2N+1} \tag{7}$$

$$\underline{A}_n(u) = \underline{a}_{n(2N+2)}u^{2N+1}e^{anu} + \underline{a}_{n(2N+1)}u^{2N+1}e^{-anu} + \underline{a}_{n(2N)}u^{2N}e^{anu}$$
$$+\underline{a}_{n(2N-1)}u^{2N}e^{-anu} + \ldots\ldots + \underline{a}_{n4}ue^{anu} + \underline{a}_{n3}ue^{-anu}$$
$$+\underline{a}_{n2}e^{anu} + \underline{a}_{n1}e^{-anu} \tag{8}$$

$$\underline{B}_n(u) = \underline{b}_{n(2N+2)}u^{2N+1}e^{anu} + \underline{b}_{n(2N+1)}u^{2N+1}e^{-anu} + \underline{b}_{n(2N)}u^{2N}e^{anu}$$
$$+\underline{b}_{n(2N-1)}u^{2N}e^{-anu} + \ldots\ldots + \underline{b}_{n4}ue^{anu} + \underline{b}_{n3}ue^{-anu}$$
$$+\underline{b}_{n2}e^{anu} + \underline{b}_{n1}e^{-anu} \tag{9}$$

and $\underline{a}_{n(2N+2)}, \underline{b}_{n(2N+2)}$, etc. are constant vectors.

We then make the assumption that the wing-sections and the derivative boundary conditions can be written thus,

$$\underline{W}^i(v) = \underline{C}_0^i + \sum_{n=1}^{M} \left(\underline{C}_n^i \cos(nv) + \underline{S}_n^i \sin(nv)\right) + \underline{R}_w^i(v) \qquad i = 1, \ldots, 2N \tag{10}$$

$$\underline{s}^m(v) = \underline{c}_0^m + \sum_{n=1}^{M} \left(\underline{c}_n^m \cos(nv) + \underline{s}_n^m \sin(nv)\right) + \underline{R}_s^m(v) \qquad m = 0, 1 \tag{11}$$

i.e. as the sum of a finite Fourier Series to M terms, plus 'remainder' functions $\underline{R}_w^i(v)$ or $\underline{R}_s^m(v)$.

Then the approach is to choose a value for M in equations (10) and (11) and approximate the wing-sections and boundary conditions by a finite Fourier Series, i.e. ignoring the remainder functions, using a wing-surface $\tilde{\underline{X}}(u,v)$ of the form (6) that interpolates these approximate sections (and approximately satisfies the boundary conditions). Note that this function can be found from the values of the Fourier coefficients of the boundary conditions and wing-sections. Then define the following $(2N+2)$ 'difference' functions

$$\underline{dW}^i(v) = \underline{W}^i(v) - \tilde{\underline{X}}(u_i, v) \quad \text{for i = 1,...,2N} \tag{12}$$

$$\underline{ds}^m(v) = \underline{s}^m(v) - \tilde{\underline{X}}_u(m, v) \quad \text{for m = 0,1.} \tag{13}$$

To obtain a surface $\underline{X}(u,v)$ that interpolates the actual wing-sections, the following function is defined

$$\underline{R}(u,v) = \underline{r}_{(2N+2)}(v)u^{2N+1}\exp(\omega u) + \underline{r}_{(2N+1)}(v)u^{2N+1}\exp(-\omega u) + \underline{r}_{(2N)}(v)u^{2N}\exp(\omega u)$$
$$+\underline{r}_{(2N-1)}(v)u^{2N}\exp(-\omega u) + \ldots\ldots + \underline{r}_4(v)u\exp(\omega u) + \underline{r}_3(v)u\exp(-\omega u)$$
$$+\underline{r}_2(v)\exp(\omega u) + \underline{r}_1(v)\exp(-\omega u) \tag{14}$$

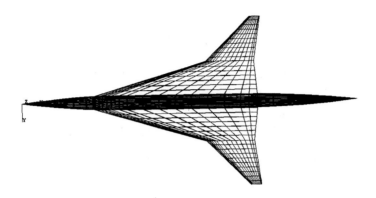

Fig. 1 Wire-frame model of initial wing geometry

which is required to satisfy the conditions

$$\underline{dW}^i(v) = \underline{R}(u_i, v) \quad \text{for i = 1,...,2N} \tag{15}$$

$$\underline{ds}^m(v) = \underline{R}_u(m, v) \quad \text{for m = 0,1} \tag{16}$$

of which there are sufficient to determine the $(2N + 2)$ functions $\underline{r}_1(v), \ldots, \underline{r}_{(2N+2)}(v)$ in equation (14). Finally $\underline{X}(u,v)$ is given by

$$\underline{X}(u,v) = \underline{\tilde{X}}(u,v) + \underline{R}(u,v) \tag{17}$$

This approach can be viewed as a means of generating an approximate solution (17) to equation (3) that exactly satisfies the boundary conditions and interpolates the specified wing-sections (to within machine accuracy).

In the results below it will be shown that the choice of a can have an important influence on the smoothness of the interpolating surface.

In some circumstances it is more appropriate to specify airfoils as data points distributed around the wing sections, rather than as closed-form expressions. These cases can be treated using essentially the method above, either by first interpolating (or approximating) each wing-section using some sort of spline function to obtain a closed-form expression, or by using discrete fourier transforms and a discrete version of equation (14), which is the procedure used in this paper.

4 Results

We will demonstrate some features of the method using data from a double-delta wing from a model HSCT supplied by R.E. Smith of NASA's Langley Research Center. The data consists

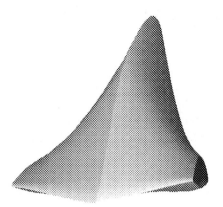

Fig. 2 PDE representation of wing sections 1 → 8 (inner wing). $a = 0.5$

of 18 planar wing-sections spaced at equal span intervals from the wing-root to the wing-tip. Each wing-section is specified discretely in terms of 41 ordered points, distributed around the wing surface from the leading edge along the upper-surface to the trailing edge and then back along the lower-surface to the leading edge. The distribution of wing-sections along the wing's span and data points around each wing-section is shown in Fig. (1): each line around the wing lies along one of the specified wing-sections, while each line down the wing passes through equivalent data-points (same value of the parameter v) on each wing-section.

The results shown below are in the form of shaded surfaces produced on a Silicon Graphics workstation. A representation for the PDE surfaces has been obtained in the form of equation (17). To visualize a surface, it is broken down into a collection of four-sided facets which are then used to produce the shaded image using standard computer graphics techniques. To aid comparison with the original wing-data, each PDE wing representation is shown alongside a shaded surface constructed from the original wing-sections, where the surface facets for the 'original' wing are those with edges as shown in Fig. (1). The original data has been put next to the corresponding PDE wing by 'reflecting' the data about a vertical plane passing through the lowest numbered wing-section. Note that although we have a paramaterization of the entire surface of a PDE wing, in the case of the original wing we only have the wing surface defined at discrete points. Note also that enough surface facets are used to render the PDE wing in order to show up any variations in the PDE surface that are not present in the original data.

In what follows, to make the data easier to visualize, the wing thickness has been uniformly increased by a factor of 10.

A value of 5 for M has been used in all the results presented below.

Fig. (2) shows a PDE surface that has been made to pass through sections 1 → 8. Section

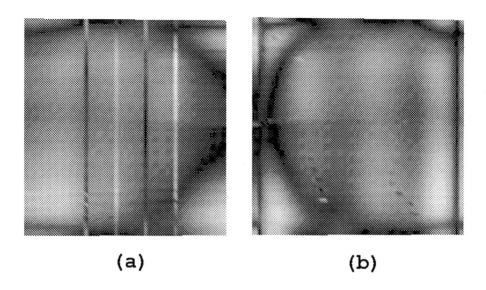

(a)   (b)

Fig. 3 Curvature variation over (u,v) plane: (a) original wing-data; (b) PDE wing surface

1 lies nearest to the fuselage, while Section 8 lies at the start of the wing-crank. Thus we are considering the inner wing of the double-delta. A value of 0.5 has been used for a and a value of 5 has been used for ω. The wing has been parametrized by assuming that each wing-section lies at a constant value of u and that there is an equal increment in u between successive wing-sections. The surfaces have been parametrized in v by assuming that the point at the leading edge of each wing-section corresponds to $v = 0$, 2π and that successive data points along a section are equally spaced in v.

One can see that the PDE representation (right-hand surface) of the wing-data is acceptably smooth, showing no variations in shape that are not present in the original data (left-hand surface).

An alternative method of comparing the smoothness of the PDE wing surface with that of the original data is by considering the variation of curvatures over the surface (see Fig. (3)). Since the original wing-data is given discretely, the surface curvature has been calculated numerically on a fine grid of points distributed evenly over the wing-surface.

However, since the original data is given on a grid of points that is relatively coarse in the u-direction, the original section data have been interpolated onto 64 new wing sections uniformly spaced along the wing, i.e. uniformly spaced in u; 64 ($= 2^6$) sections were chosen for reasons that will become apparent. Furthermore, each new section has been reparametrized in the v direction so that it is given in terms of 64 points, uniformly spaced in arclength. Note that to obtain the new sections cubic interpolation has been used in the u direction; linear interpolation to reparametrize in v.

In this way a wing given discretely on a mesh of points, 64 × 64 in the (u,v) plane, has been created from the original data. Note that because each wing-section is parametrized in arc-length, and successive wing-sections are uniformly spaced along the wing-span, the mesh is approximately uniform in physical space as well as in (u,v) space. At each point of this mesh the surface curvature has been calculated using finite-differences. A similar procedure was adopted for the PDE wing, which was initially calculated on a grid of 50 points in the u direction (41 in v), and interpolated onto a 64 × 64 grid as described above.

Although not ideal, the above procedure is forced upon us by the fact that the original data is given relatively sparsely in the u direction, and also by the need to obtain a convenient grid for the wavelet analysis that follows.

When it comes to choosing a measure for surface curvature there are plenty of alternative to choose from: Gaussian, Mean, Principal, to name but a few. In this example, a useful choice is the curvature of the isoparametric u lines. This is because a wing is a particularly awkward surface to assess, consisting as it does of relatively flat regions (the upper and lower surfaces) bounded by regions of very high curvature (the leading and trailing edges). Thus, unless we consider only sub-regions, 2D plots of the obvious curvature measures such as Gaussian or Mean over the whole surface are dominated by the values these quatities take in relatively small regions, making the variation in curvature very difficult to assess due to the limited dynamic range offered by colour-code plots. However, in this example, the curvature of the u isosparametric lines does not dramatically alter across the surface, and since the PDE surface generation method is basically interpolating in the u direction anyway, the variation in the u curvature is an appropriate measure with which to assess the smoothness of the surface produced.

Fig. (3) shows the colour-coded variation in u curvature across the 64 × 64 mesh in the (u,v) plane for both the original data and the PDE wing. The u direction is horizontal; the v is vertical; the origin is in the top left hand corner.

Note that since the mesh is approximately uniform in physical space, the rate of variation of curvature across the (u,v) mesh gives us a fairly good measure of the rate of variation of curvature in physical space. The important point to notice is that, although not identical, there is no 'structure' in the curvature variation across the PDE wing on shorter length-scales that any present in the original data. In fact, in the original data one can see very rapid curvature variations across several lines of constant v, where none is present in the PDE wing. These variations occur at the positions of the original wing-sections and are an artefact of the 'cheap-and-cheerful' interpolation procedure that has been used.

The curvature maps shown in Fig. (3) can be further analysed, as is shown in Fig. (4) which gives wavelet representations of the two (u,v) curvature maps. The wavelets representations have been generated by taking the 64 × 64 curvature 'images' of Fig (3) and generating multiresolution representations using, for convenience, the simplest of all 2D wavelets, tensor-product Haar wavelets. Since the original images contain $2^6 \times 2^6$ 'pixels', we use 7 (1 (coarse) $\rightarrow$ 7 (fine)) resolution levels. The 2D wavelet decompositions are arranged in the manner suggested by Mallat ([20]). The quantity plotted is the absolute values of the wavelet coeffcents.

The wavelet decomposition can be interpreted as a decomposition of the curvature map into a set of independent, spatially orientated, frequency channels. At each resolution level (J), the top right 'detail' image shows the sampled frequencies in the u-direction, the bottom left image shows the sampled frequencies in the v-direction, the bottom right image shows the

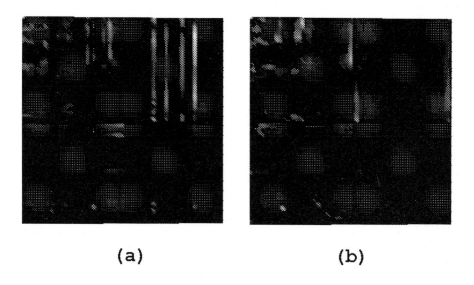

Fig. 4 Wavelet decomposition of curvature images of Fig. (4): (a) original wing data; (b) PDE wing surface

sampled frequencies in both u and v directions, while the image in the top left corner, the projection onto the next solution level $(J-1)$, is itself decomposed in a similar fashion.

Crudely speaking, at each resolution level the detail images show the spatial frequencies with wavelengths on the same scale as the mesh size at that resolution. Representing the curvature distribution in this fashion is a method of quantitatively assessing the length scales on which a surface is varying. Using Fig. (4) to compare the PDE wing surface with the surface generated from the original wing data, shows that there are no significant high frequency variations in the PDE wing not present in the original data. In fact, the highest frequency variations are present in the original wing data, at the locations of the given wing-sections, and are generated by the interpolation procedure.

Fig. (5) shows a PDE surface through the same wing sections, but now a value of 0.2 has been used for the smoothing parameter. Unlike the previous case, the PDE representation, although it interpolates the original wing sections, shows distinct 'ripples' or oscillations in the surface that are not present in the original data, and is clearly not an acceptable representation of a wing. The ripples are clearly visible in Fig. (6a) which shows the (u,v) curvature map for the PDE wing and also in Fig. (6b) which shows the wavelet decomposition of that map. The effect is closely akin to the well-known result that interpolating data points with high-order spline functions tends to give rise to unwanted oscillations in the interpolation function[19]. That we experience a similar phenomenon here is not surprising since as $a \to 0$, the PDE solution approaches a high-order polynomial in u. This tendency for a PDE surface to oscillate if required to pass through an large number of sections that show sufficient variation, can be reduced by increasing the value of the smoothing parameter a.

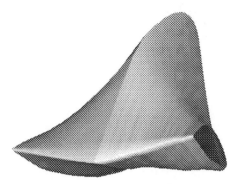

Fig. 5 PDE representation of wing sections $1 \to 8$ (inner wing). $a = 0.2$

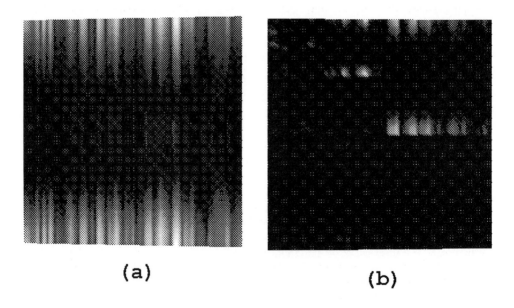

Fig. 6 (a)Curvature variation over (u,v) plane; (b) Wavelet decomposition

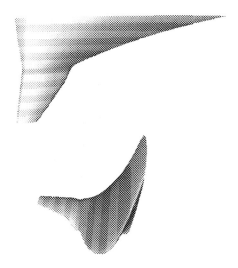

Fig. 7 PDE representation of entire wing with multiple surface patches

5 Discussion

We have seen how the method outlined above can produce PDE surface patches to represent various portions of the original wing. The important point is that this is a method for going from two-dimensional section data to a complete surface in three-dimensions. This is a crucial step, since a 3D geometric model is an essential prerequisite for mesh generation and CFD analysis, techniques which are becoming increasingly relied upon in modern aerodynamic design. Furthermore, unlike spline-based methods for surface generation, which typically use control points distributed over the entire surface between the wing-sections, the data set for a PDE wing representation is basically that used to represent the original wing-sections plus a few extra parameters, such as the smoothing parameter a, to control the behaviour of the PDE solution.

The PDE patches interpolate specified wing-sections and are free of unwanted oscillations, provided that the variation in the shape of the specified sections is not too large. These oscillations can be controlled by the choice of the smoothing parameter a, although in extreme cases this may not always be possible. However, this problem can be overcome by the simple expedient of representing the original data in terms of multiple surface patches. For example, Fig. (7) shows the double-delta wing represented in terms of multiple 4th-order PDE surface patches, each patch being generated between successive sections. Successive patches have been shaded in a different colour for the sake of clarity. The derivative boundary conditions imposed at the common boundaries between adjacent patches are such that the patches meet with tangent plane continuity, and, in this example, are calculated by finite-differencing the coordinates of corresponding points on the sections on either side of a particular patch boundary.

If higher-order continuity is required at the wing-sections, curvature continuity for instance, a higher-order PDE can be used to generated the individual patches with extra boundary conditions—on higher-order derivatives—being imposed at the wing-sections, e.g. boundary-conditions on second derivatives for curvature continuity. In terms of computational effort, there is little to choose between using multiple, low-order PDE patches or fewer, high-order PDE patches to generate a wing surface. With the data used in this paper, depending on the number of sections to be interpolated, both approaches took no more than a second or two of CPU time on a modestly powered workstation (a Silicon Graphics Indigo 2 with a MIPS R4400 processor). Thus, the number of surface patches used to represent a wing and the order of the generating PDEs, should be determined by the quality of the resulting surface. One could imagine a procedure, based upon some qualitative measure of surface 'fairness', which, given a set of wing-sections, automatically selects the appropriate combination of patches and their generating PDEs, to produce a composite surface that was free of oscillations, using the smallest number of patches.

6 Acknowledgments

This work was supported by NASA grant NAGW-3198. The authors would like to thank Dr R.E. Smith of the NASA Langley Research Centre for his support and encouragement of the work.

References

[1] Nicolai, L.M, 'Fundamentals of Aircraft Design', University of Dayton, 1975.

[2] Raymer, D. P., *Aircraft Design: A Conceptual Approach*, AIAA, Washington, 1989.

[3] van der Vooren, J. and van der Wees, A.J., 'Inviscid Drag Prediction for Transonic Transport Wings using a Full-Potential Method', *Journal of Aircraft*, Vol. 28, No. 12, 1991, pp. 869-875.

[4] Agrawal, S., Creasman, S.F. and Lowrie, R.B., 'Evaluation of Euler Solvers for Transonic Wing-Fuselage Geometries', *Journal of Aircraft*, Vol 28, No. 12, 1991, pp. 885-891.

[5] Hutchinson, M.G., Unger, E.R., Mason, W.H., Grossman, B. and Haftka, R.T., 'Variable-Complexity Aerodynamic Optimization of a High-Speed Civil Transport Wing', *J. Aircraft*, Vol 31., No. 1, 1994, pp. 110-116.

[6] Sloof, J. W., "Computational Methods for Subsonic and Transonic Aerodynamic Design", *AGARD*, Report 712, Paper 3, 1983.

[7] Lee, K.D. and Eyi, S., 'Aerodynamic Design via Optimisation', *Journal of Aircraft*, Vol 29, No. 6, 1992, pp. 1012-1019.

[8] Barthelemy, J.-F.M., Wrenn, G.A., Dovi, A.R., Coen, P.G. and Hall, L.E., 'Supersonic Transport Wing Minimum Weight Design Integrating Aerodynamics and Structures', *Journal of Aircraft*, Vol. 31, No. 2, 1994, pp. 330-338.

[9] Hager, J.O., Eyi, S. and Lee, K.D., 'Two-Point Transonic Airfoil Design using Optimization for Improved Off-Design Performance', *Journal of Aircraft*, Vol. 31, No. 5, 1994, pp. 1143-1147.

[10] Yiu, K.F.C, 'Computational Methods for Aerodynamic Shape Design', *Mathematical Computer Modelling*, Vol. 20, No. 12, 1994, pp. 3-29.

[11] Eyi, S., Hager, J.O. and Lee, K.D., 'Airfoil Design Optimization using the Navier-Stokes Equations', *J. Optimization Theory and Applications*, Vol 83, No. 3, 1994, pp. 447-461.

[12] El-banna, H.M. and Carlson, L.A., 'Aerodynamic Sensitivity Coefficients using the Three-Dimensional Full Potential Equation', *Journal of Aircraft*, Vol. 31, No. 5, 1994, pp. 1071-1077.

[13] Bloor, M. I. G. and Wilson, M. J., "The efficient parametrization of generic aircraft geometry", *Journal of Aircraft*, Vol. 32, No. 6, pp. 1269-1275, 1995.

[14] Smith, R.E., Bloor, M.I.G., Wilson, M.J. and Thomas, A.M., 'Rapid Airplane Parametric Input Design', *AIAA* Paper 95 1687, 1995.

[15] Sevant, N.M., Bloor, M.I.G., Lowe, T.W. and Wilson, M.J., "The Automatic Design of a Generic Wing/Body Fairing", eds. Thibault, P.A. and Bergeron, D.M., *CFD 95: Third Annual Conference of the CFD Society of Canada*, Banff, Canada, Vol. 1, pp. 163-170, 1995.

[16] Requicha, A., A., Voelcker, H.B., "Solid Modelling: Current Status and Research Directions", *IEEE Computer Graphics & Applications*, Vol. 3, No.7, 1983, pp. 25-37.

[17] Dekanski, C., Bloor, M.I.G., and Wilson, M.J., "The Representation of Marine Propeller Blades using the PDE method,', *Journal of Ship Research*, Vol 38, No.2, pp. 108-116, 1995.

[18] Zauderer, E, *Partial Differential Equations of Applied Mathematics*, Wiley-Interscience, New York, (1983).

[19] Gerald, C.F. and Wheatley, P.P., *Applied Numerical Analysis*, 3rd ed., Addison-Wesley, London, 1984.

[20] Mallat, S.G., "A Theory for Multiresolution Signal Decomposition: The wavelet representation", IEEE Trans. on Pattern Analysis and Machine Intelligence, Vol. 11, No. 7, pp. 674-693, 1989.

Authors

Prof. Dr. M. Bloor
Dept. of Applied Mathematical Studies
The University of Leeds
Woodhouse Lane
LEEDS (UK)

Prof. Dr. Mike Wilson
Dept. of Applied Mathematical Studies
The University of Leeds
Woodhouse Lane
LEEDS (UK)
E–mail: mike@amsta.leeds.ac.uk

Algorithms for convexity preserving interpolation of scattered data

J. M. Carnicer, Mikael S. Floater

Departamento de Matemática Aplicada,
Universidad de Zaragoza,
Spain

Abstract: All convex interpolants to convex bivariate Hermite scattered data are bounded above and below by two piecewise linear functions u and l respectively. This paper discusses numerical algorithms for constructing u and l and how, in certain cases, they form the basis for constructing a C^1 convex interpolant using Powell-Sabin elements.

Key words: Convexity, Shape preserving interpolation, Scattered data, Triangulation.

1 Introduction

This paper concerns the numerical computation of two piecewise linear functions which are fundamental to the problem of convexity preserving scattered data interpolation in two variables.

Suppose we are given a sequence $\mathbf{X} = (\mathbf{x}^1, \ldots \mathbf{x}^n)$ of distinct points in $\mathbb{R}^2$ and associated scalar values $\mathbf{f} = (f_1, \ldots, f_n)$. Then, the Lagrange data set $(\mathbf{X}, \mathbf{f})$ is called *convex* (resp., *strictly convex*) if there exists a convex (resp., strictly convex) function F satisfying

$$F(\mathbf{x}^i) = f_i, \quad \forall i \in \{1, \ldots, n\} \tag{1.1}$$

Given a finite sequence of distinct points $\mathbf{X} = (\mathbf{x}^1, \ldots, \mathbf{x}^n)$, $\mathbf{x}^i \in \mathbb{R}^s$, $i = 1, \ldots, n$, real values $\mathbf{f} = (f_1, \ldots, f_n)$ and vectors $\mathbf{G} = (\mathbf{g}^1, \ldots, \mathbf{g}^n)$, $\mathbf{g}^i \in \mathbb{R}^s$, $i = 1, \ldots, n$ we may pose a Hermite interpolation problem: find a continuous function F, which is differentiable in a neighbourhood of each $\mathbf{x}^i$ such that

$$F(\mathbf{x}^i) = f_i, \quad \nabla F(\mathbf{x}^i) = \mathbf{g}^i, \quad \forall i \in \{1, \ldots, n\}. \tag{1.2}$$

The Hermite data set $(\mathbf{X}, \mathbf{f}, \mathbf{G})$ is said to be *convex* (resp., *strictly convex*) if there exists a convex (resp., strictly convex) interpolant to the data.
In Proposition 4.2 of [3] it was shown that $(\mathbf{X}, \mathbf{f})$ are convex data if and only if

$$f_j \le \sum_{i \in I} \lambda_i f_i, \quad \text{for any } \lambda_i \ge 0 \, (i \in I) \text{ such that } \sum_{i \in I} \lambda_i = 1, \sum_{i \in I} \lambda_i \mathbf{x}^i = \mathbf{x}^j, \tag{1.3}$$

where I is any subset of $\{1, \ldots, n\}$ with exactly 3 elements.

When dealing with Hermite data we will avoid unnecessary complications by restricting ourselves to strictly convex data. In Corollary 3.5 of [1] it was shown that $(\mathbf{X}, \mathbf{f}, \mathbf{G})$ are strictly convex if and only if

$$\mathbf{g}^i \cdot (\mathbf{x}^j - \mathbf{x}^i) < f_j - f_i, \quad \forall i \neq j \in \{1, \ldots, n\}. \tag{1.4}$$

For given convex Lagrange data $(\mathbf{X}, \mathbf{f})$ we define the function

$$u(\mathbf{x}|\mathbf{X}, \mathbf{f}) := \min \left\{ \sum_{i=1}^{n} \lambda_i f_i \,\Big|\, \sum_{i=1}^{n} \lambda_i = 1, \lambda_i \geq 0, 1 \leq i \leq n, \sum_{i=1}^{n} \lambda_i \mathbf{x}^i = \mathbf{x} \right\}, \tag{1.5}$$

which we shall call the upper bound (or upper envelope) for all convex interpolants. In Lemma 4.1 and Proposition 4.1 of [3], it was shown that $(\mathbf{X}, \mathbf{f})$ is convex if and only if $u(\mathbf{x}|\mathbf{X}, \mathbf{f})$ is a convex interpolant to the data. Furthermore, if F is any convex interpolant to $(\mathbf{X}, \mathbf{f})$ then

$$F(\mathbf{x}) \leq u(\mathbf{x}|\mathbf{X}, \mathbf{f}), \quad \forall \mathbf{x} \in [\mathbf{X}],$$

where $[\mathbf{X}]$ denotes the convex hull of $\mathbf{X}$ (see [3], [1]). In [3] it is also shown that $u(\mathbf{x}|\mathbf{X}, \mathbf{f})$ is a piecewise linear function based on a triangulation. In section 2 we shall analyze the construction of the upper bound, by giving a construction of the corresponding triangulation.

Section 3 is devoted to the lower bound of all Hermite interpolants which was studied in [3], [1]. The lower bound is defined by

$$l(\mathbf{x}|\mathbf{X}, \mathbf{f}, \mathbf{G}) = \max\{f_i + \mathbf{g}^i \cdot (\mathbf{x} - \mathbf{x}^i) \,|\, i = 1, \ldots, n\},$$

and is linear on each tile of a tesselation of the plane. It is straightforward to see that if $(\mathbf{X}, \mathbf{f}, \mathbf{G})$ are strictly convex data, then $l(\mathbf{x}|\mathbf{X}, \mathbf{f}, \mathbf{G})$ is a convex interpolant to the given data. Moreover, for any other convex interpolant F,

$$l(\mathbf{x}|\mathbf{X}, \mathbf{f}, \mathbf{G}) \leq F(\mathbf{x}) \quad \forall \mathbf{x} \in \mathbb{R}^2.$$

We shall analyze some properties of the lower bound defined by strictly convex Hermite data, and construct the corresponding tesselation.

Since $l(\mathbf{x}|\mathbf{X}, \mathbf{f}, \mathbf{G})$ and $u(\mathbf{x}|\mathbf{X}, \mathbf{f})$ are lower and upper bounds of any convex interpolants, they provide constraints which must be taken into consideration when constructing some other interpolants. In fact, the construction of interpolants [2] based on Powell-Sabin elements [5] must take into consideration some properties of the structure of l and u.

We complete the paper with some numerical examples of l and u.

2 Construction of the upper bound

Definition 2.1. A triangulation $\mathcal{T}$ with set of vertices $\mathbf{X}$ is a set of triangles $\mathcal{T} = \{T_1, \ldots, T_N\}$ such that

(1) $|T_i \cap X| = 3$ and $T_i = [T_i \cap \mathbf{X}]$ for all $i \in \{1, \ldots, N\}$,
(2) $\cup_{i=1}^{N} T_i = [\mathbf{X}]$,
(3) $\text{Int}\, T_i \cap \text{Int}\, T_j = \emptyset$, for any $i \neq j$.

Given any data $(\mathbf{X}, \mathbf{f})$ (not necessarily convex) each triangulation $\mathcal{T}$ defines a unique continuous piecewise linear interpolant $F(\mathbf{x}|\mathbf{X}, \mathbf{f}; \mathcal{T})$ defined on $[\mathbf{X}]$ whose restriction to each triangle T_i coincides with the unique linear interpolant to $(\mathbf{x}^{\alpha(i)}, f_{\alpha(i)})$, $(\mathbf{x}^{\beta(i)}, f_{\beta(i)})$, $(\mathbf{x}^{\gamma(i)}, f_{\gamma(i)})$, where $\mathbf{x}^{\alpha(i)}, \mathbf{x}^{\beta(i)}, \mathbf{x}^{\gamma(i)}$ are the three vertices of T_i.

Let $\mathcal{T}$ be any triangulation of $\mathbf{X}$. A quadrilateral of $\mathcal{T}$ is the union of any two adjacent triangles, that is $T \cup S$, where $T, S \in \mathcal{T}$, $|T \cap S \cap \mathbf{X}| = 2$. The vertices of the quadrilateral are the four elements in $(T \cup S) \cap \mathbf{X}$. The sides of the triangles T, S which are not the common edge e are called the sides of the quadrilateral. The quadrilaterals of the triangulation can be classified into three types: a quadrilateral is *degenerate* if it contains any three collinear vertices, a quadrilateral is *convex* if it is nondegenerate and all interior angles are less than π, and a quadrilateral is *concave* if it is nondegenerate and contains an interior angle greater that π.

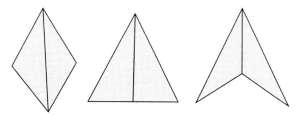

Fig. 1 Convex, degenerate and concave quadrilaterals

Let us describe now an elementary operation which allows us to define a new triangulation from a given one. Let $T \cup S$ be any convex quadrilateral. Any three vertices of the quadrilateral whose convex hull does not coincide with T or S determine precisely two adjacent triangles T', S' such that $T \cup S = T' \cup S'$. A *swap* is the operation performed on the triangulation $\mathcal{T}$ which produces the new triangulation $(\mathcal{T} \setminus \{T, S\}) \cup \{T', S'\}$. A swap cannot be applied to degenerate or concave quadrilaterals because, in these cases, the new set of triangles will not be a valid triangulation.

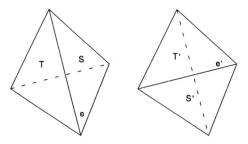

Fig. 2 A swap replaces T,S by T', S'. The edge e is replaced by e'

This observation suggests an algorithm for computing a triangulation $\mathcal{T}^*$, for convex data $(\mathbf{X}, \mathbf{f})$, such that the interpolant $F(\mathbf{x}|\mathbf{X}, \mathbf{f}; \mathcal{T}^*)$ is convex and coincides with the upper bound $u(\mathbf{x}|\mathbf{X}, \mathbf{f})$. This algorithm was first proposed by Scott [6] and has been further discussed by [4].

Algorithm 2.2. (Swapping algorithm)

1. **construct** a starting triangulation $\mathcal{T}_0$ and **let** $i := 0$.
2. **search** for a convex quadrilateral Q of $\mathcal{T}_i$ such that the restriction of $F(\mathbf{x}|\mathbf{X}, \mathbf{f}; \mathcal{T}_i)$ to Q is concave but nonlinear.
3. **if** such a quadrilateral Q exists **then** define $\mathcal{T}_{i+1}$ from $\mathcal{T}_i$ by swapping the diagonal edges of Q, **let** $i := i + 1$ and **go to** step 2.
4. **else** define $\mathcal{T}^* := \mathcal{T}_i$ and **output** $\mathcal{T}^*$.

end.

Let us remark that the algorithm has to stop in a finite number of steps. In fact, the sequence of functions obtained is decreasing:

$$F(\mathbf{x}|\mathbf{X}, \mathbf{f}; \mathcal{T}_0) \geq \cdots \geq F(\mathbf{x}|\mathbf{X}, \mathbf{f}; \mathcal{T}_{i-1}) \geq F(\mathbf{x}|\mathbf{X}, \mathbf{f}; \mathcal{T}_i) \geq \cdots$$

and since they take different values at at least one point, all the functions are distinct. Therefore all triangulations $\mathcal{T}_i$ produced by the algorithm must be different. Since there exists only a finite number of triangulations of $\mathbf{X}$, the algorithm stops.

3 An algorithm for the lower bound

Suppose now that we are given a strictly convex Hermite data set $(\mathbf{X}, \mathbf{f}, \mathbf{G})$. Define $h_i : \mathbb{R}^2 \to \mathbb{R}$ as

$$h_i(\mathbf{x}) = f_i + \mathbf{g}^i \cdot (\mathbf{x} - \mathbf{x}^i), \quad \mathbf{x} \in \mathbb{R}^2,$$

so that h_i is the linear function which coincides with the tangent plane defined by $\mathbf{x}^i$, f_i, and $\mathbf{g}^i$. Since $(\mathbf{X}, \mathbf{f}, \mathbf{G})$ are strictly convex data, (1.4) holds. Equivalently,

$$h_j(\mathbf{x}^i) < f_i, \quad \forall i \neq j, \tag{3.1}$$

that is to say, every tangent plane lies strictly below every data point other than the one it interpolates.

In this section we will discuss the lower bound l, an algorithm for computing it, and how a convex C^1 piecewise quadratic interpolant can, in certain cases, be constructed from it.

Let

$$\Omega_i = \{\mathbf{y} \in \mathbb{R}^2 : h_i(\mathbf{y}) \geq h_j(\mathbf{y}), \forall j = 1, \ldots, n\}.$$

It can easily be shown that each Ω_i is a convex polygon containing the point $\mathbf{x}^i$ and no other. Furthermore, since there are only a finite number of data points, there is in fact an open neighbourhood around each $\mathbf{x}^i$ which is contained in Ω_i. Moreover, the interiors of the Ω_i are pairwise disjoint and their union covers the whole of $\mathbb{R}^2$. Thus the set $\mathcal{L} = \{\Omega_i\}_1^n$ forms a *tiling* or *tesselation* of the plane.

Now we observe that l is piecewise linear, linear on each polygonal tile Ω_i, indeed

$$l(\mathbf{x}) = h_i(\mathbf{x}), \quad \mathbf{x} \in \Omega_i,$$

Algorithms for convexity preserving interpolation of scattered data

for all i. Furthermore, since there is an open neighbourhood around each $\mathbf{x}^i$ which is contained in Ω_i, l is a convex interpolant to $(\mathbf{X}, \mathbf{f}, \mathbf{G})$.

As usual with tilings, two points $\mathbf{x}^i$ and $\mathbf{x}^j$ are said to be *neighbours*, with respect to $\mathcal{L}$, if the boundaries of their polygons Ω_i and Ω_j intersect. In what follows, it will be important to distinguish between different kinds of neighbours.

Definition 3.1. *We will say that $\mathbf{x}^i$ and $\mathbf{x}^j$ are weak neighbours if the boundaries of their polygons Ω_i and Ω_j intersect at a single point. If their intersection contains a line segment, they will be called strong neighbours.*

The common part of the boundaries of the polygons of two neighbours $\mathbf{x}^i$ and $\mathbf{x}^j$ will be denoted by $e_{i,j}$. It can be either a point, a line segment, a semi-infinite line or, in extreme cases, an infinite line.

Note that if all neighbours are strong, then $\mathcal{L}$ forms a planar graph in which every node v is the intersection of three linear functions $h_i(v) = h_j(v) = h_k(v)$ and is incident on three edges $e_{i,j}$, $e_{j,k}$, and $e_{k,i}$, and the angle between any two of them is less than π.

The following concepts were partly discussed in [2].

Definition 3.2. *We will say that two strong neighbours $\mathbf{x}^i$ and $\mathbf{x}^j$ are direct neighbours if the line segment $[\mathbf{x}^i, \mathbf{x}^j]$ intersects the interior of $e_{i,j}$.*

Definition 3.3. *If all pairs of neighbours in $\mathcal{L}$ are direct neighbours we say that $\mathcal{L}$ is well-associated with $\mathbf{X}$.*

Although $\mathcal{L}$ is rarely well-associated with $\mathbf{X}$, there are always relatively many direct neighbours in $\mathcal{L}$. Every point $\mathbf{x}^i \in \mathbf{X}$ has at least one direct neighbour. In fact, if $j \in \{1, \ldots, n\}$ is such that $h_j(\mathbf{x}^i) = \max_{k \neq i} h_k(\mathbf{x}^i)$ then it can be shown that $\mathbf{x}^i$ and $\mathbf{x}^j$ are direct neighbours.

We now propose the following algorithm for finding the polygons of $\mathcal{L}$, assuming that all neighbours are strong (the algorithm can easily be extended to cover all strictly convex data sets). In the following we denote the unique point $\mathbf{x} \in \mathbb{R}^2$ for which $h_i(\mathbf{x}) = h_j(\mathbf{x}) = h_k(\mathbf{x})$, if it exists, as $\mathbf{p}_{i,j,k}$, for any distinct $i,j,k \in \{1, \ldots, n\}$. For any $i = 1, \ldots, n$, the algorithm constructs a list $\mathbf{P} = (\mathbf{p}_1, \ldots, \mathbf{p}_{|\mathbf{P}|})$ of the vertices of Ω_i in an anticlockwise direction. If Ω_i is closed, then $(\mathbf{p}_1, \ldots, \mathbf{p}_{|\mathbf{P}|})$ are its $|\mathbf{P}|$ vertices. If Ω_i is open, then $(\mathbf{p}_2, \ldots, \mathbf{p}_{|\mathbf{P}|-1})$ are its vertices, while $\mathbf{p}_1$ is the direction of the first semi-infinite edge and $\mathbf{p}_{|\mathbf{P}|}$ is the direction of the last semi-infinite edge. The value of a flag *poly_stat* is returned as *closed* if Ω_i is closed, and *open* otherwise.

Algorithm 3.5. (Find the polygon Ω_i)

1. **Find** a direct neighbour $\mathbf{x}^{j_i}$ of $\mathbf{x}^i$. Let $\mathbf{P} = \emptyset$.

2. **Find** the point $\mathbf{q}$, the intersection of $[\mathbf{x}^i, \mathbf{x}^{j_i}]$ and e_{i,j_i}. Find also the direction $\mathbf{e} \in \mathbb{R}^2$ of e_{i,j_i} in the anticlockwise direction relative to $\mathbf{x}^i$. Let $\mathbf{r} := \mathbf{q}$, $\mathbf{d} := \mathbf{e}$, $j := j_i$ and $a := 1$.

3. **Let**
$$K_{i,j} = \{k \in \{1, \ldots, n\} : k \neq i, j, \ \mathbf{p}_{i,j,k} \text{ exists and } (\mathbf{p}_{i,j,k} - \mathbf{r}) \cdot \mathbf{d} > 0\}.$$

4. **If** $K_{i,j} = \emptyset$, $e_{i,j}$ is semi-infinite. Let $\mathbf{P} := (\mathbf{p}_1, \ldots, \mathbf{p}_{|\mathbf{P}|}, \mathbf{d})$.

4.a. If $a = 0$ **stop**. Let *poly_stat* := *open* and **output P** and *poly_stat*.

4.b. Else reverse the list, i.e. let $\mathbf{P} := (\mathbf{p}_{|\mathbf{P}|}, \ldots, \mathbf{p}_1)$, and let $\mathbf{r} := \mathbf{q}$, $\mathbf{d} := -\mathbf{e}$, $j := j_i$, $a := 0$ and goto 3.

5. **Else** let $k \in K_{i,j}$ such that

$$||\mathbf{p}_{i,j,k} - \mathbf{r}|| = \min_{l \in K_{i,j}} ||\mathbf{p}_{i,j,l} - \mathbf{r}||.$$

Then $\mathbf{p}_{i,j,k}$ is the end point of $e_{i,j}$ in the direction $\mathbf{d}$. Let $\mathbf{P} := (\mathbf{p}_1, \ldots, \mathbf{p}_{|\mathbf{P}|}, \mathbf{p}_{i,j,k})$.

5.a. If $k = j_i$, then **stop**. Let *poly_stat* := *closed* and **output P** and *poly_stat*.

5.b. Else let $\mathbf{r} := \mathbf{p}_{i,j,k}$ and $j := k$. Let $\mathbf{d}$ be the direction of the edge $e_{i,j}$ in the anticlockwise direction relative to $\mathbf{x}^i$ if $a = 1$, clockwise if $a = 0$. Then goto 3.

end.

Finally, we note that $\mathcal{L}$ can sometimes be used as the basis of a smoother convex interpolant to Hermite data.

Proposition 3.6. *Let us assume that $\mathcal{L}$ is well-associated with $\mathbf{X}$ and that the set of all edges obtained by connecting any two neighbours contains the boundary edges of $[\mathbf{X}]$. Then there exists a triangulation $\mathcal{T}$ of $\mathbf{X}$ such that the C^1 piecewise-quadratic interpolant F consisting of Powell-Sabin elements on the triangles of $\mathcal{T}$ is convex.*

Proof: Let $\mathcal{E}$ be the set of edges obtained by connecting every pair of neighbours in $\mathcal{L}$. The condition that $\mathcal{E}$ includes all edges of the boundary of $[\mathbf{X}]$, implies that the edges of $\mathcal{E}$ form a triangulation $\mathcal{T}$ (whose boundary is shared by $[\mathbf{X}]$) which, as a planar graph, is the dual of $\mathcal{L}$. The rest of the proof is essentially contained in [2]. ∎

We remark that though $\mathcal{T}^*$ is a likely candidate to be the dual of $\mathcal{L}$ in Proposition 3.6, in general $\mathcal{T} \neq \mathcal{T}^*$.

4 Numerical examples

In the examples, three strictly convex functions $f_1(x, y) = x^2 + y^2$, $f_2(x, y) = 1 - \sqrt{1 - x^2 - y^2}$, and $f_3(x, y) = e^{2x} + e^{2y}$ have been sampled on sets of random points in the unit disc. We note that f_1 is quadratic, f_2 a lower hemisphere, and f_3 an exponential function.

In particular, Lagrange and Hermite data were sampled at fifty points $\mathbf{X}_1$ and Figures 3 to 5 show the triangulations $\mathcal{T}^*$ and the tesselations $\mathcal{L}$ for f_1, f_2, and f_3. In the special case of f_1, $\mathcal{T}^*$ is a Delaunay triangulation and $\mathcal{L}$ is the Voronoi diagram. Figures 6 to 8 show the corresponding upper and lower bounds u and l. Figures 9 and 10 show shaded images and contour lines of the upper bound u of f_2.

A larger sample of 500 random points $\mathbf{X}_2$ was taken from the unit disc and Lagrange data sampled from f_2, the lower hemisphere. Figure 11 shows the upper bound u, the piecewise linear interpolant on $\mathcal{T}^*$. Meanwhile, in order to make a comparison, Figure 12 shows the piecewise linear interpolant on a Delaunay triangulation of $\mathbf{X}_2$. The evident lack of smoothness of the shading reflects the fact that it is not convex.

Research partially supported by the EU Project FAIRSHAPE, CHRX-CT94-0522. The first author was also partially supported by DGICYT PB93-0310 Research Grant.

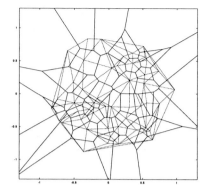

Fig. 3 Triangulation $\mathcal{T}^*$ and tesselation $\mathcal{L}$ of a sample from f_1

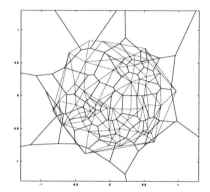

Fig. 4 Triangulation $\mathcal{T}^*$ and tesselation $\mathcal{L}$ of a sample from f_2

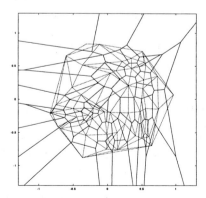

Fig. 5 Triangulation $\mathcal{T}^*$ and tesselation $\mathcal{L}$ of a sample from f_3

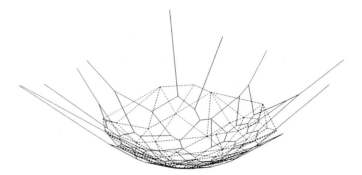

Fig. 6 Upper and lower bounds of a sample from f_1

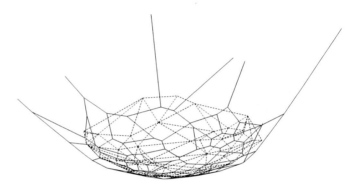

Fig. 7 Upper and lower bounds of a sample from f_2

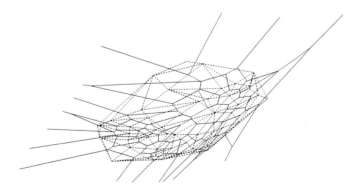

Fig. 8 Upper and lower bounds of a sample from f_3

Algorithms for convexity preserving interpolation of scattered data 183

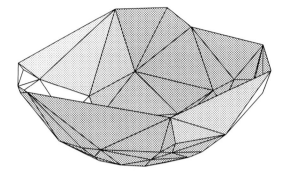

Fig. 9 Upper bound of a sample from f_2

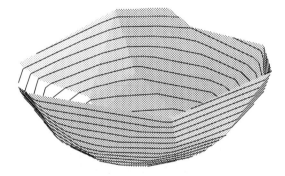

Fig. 10 Upper bound of a sample from f_2

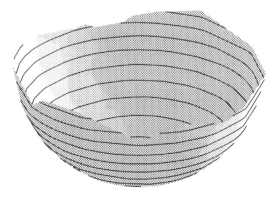

Fig. 11 Upper bound of a sample from f_2

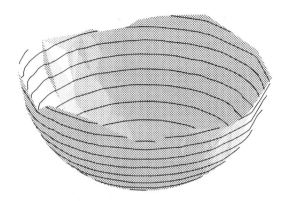

Fig. 12 Interpolant of a sample from f_2, on the Delaunay triangulation

References

[1] Carnicer, J. M.: *Multivariate convexity preserving interpolation by smooth functions.* Advances in Computational Mathematics **3** (1995), 395–404.

[2] Carnicer, J. M., Dahmen, W.: *Convexity preserving interpolation and Powell-Sabin elements.* Computer Aided Geometric Design **9** (1992), 279-289.

[3] Dahmen, W., Micchelli, C. A.: *Convexity of multivariate Bernstein polynomials and box spline surfaces.* Studia Scientiarum Mathematicorum Hungaricae, **23** (1988), 265-287.

[4] Mulansky, B.: *Interpolation of scattered data by a bivariate convex function I: Piecewise linear C^0-interpolation.* Memorandum no. 858, University of Twente, 1990.

[5] Powell, M. J. D., Sabin, M. A.: *Piecewise quadratic approximations on triangles.* ACM Transactions on Mathematical Software **3** (1977), 316-325.

[6] Scott, D. S.: *The complexity of interpolating given data in three-space with a convex function in two variables.* J. Approx. Theory **42** (1984), 52-63.

Authors

Prof. Dr. J.M. Carnicer
Dept. de Matematica Aplicada
Universidad de Zaragoza
Edificio de Matematicas
Ciudad Universitaria, Zaragoza
Spain
E–mail: carnicer@posta.unizar.es

Dr. Mikael Floater
Dept. de Matematica Aplicada
Universidad de Zaragoza
Ciudad Universitaria, Zaragoza
Spain
E–mail: floater@posta.unizar.es

Abstract schemes for functional shape–preserving interpolation

Paolo Costantini

Dipartimento di Matematica, Università di Siena

Abstract: Aim of this paper is to present a general and abstract method for the construction of constrained functions and describe its modifications and applications – developed in the recent years – in the field of functional shape–preserving interpolation.

1 Introduction

It is well known that constrained approximation or interpolation (a typical and important example being given by *shape–preserving interpolation*) has received considerable attention during the last years and several schemes with useful properties have been developed and analyzed from the points of view both of approximation theory and of CAGD. The reader is referred to the bibliography paper on this volume or to [16] for a large set of references.

A first subdivision of the ideas used in shape-preserving interpolation reveals three main categories. The first one is based on the variational approach where the function is defined and computed as the solution of a minimization problem with constraints given by shape-preserving and, eventually, interpolatory conditions. The resulting schemes are mathematically elegant but lead to global algorithms where the user has, in general, a limited local control on the shape of the curve. As far as the author knows, the number of papers on these variational methods is less conspicuous than the other two; here we limit ourselves to quote [13], [14], [18] and the references therein. The second one leads to the so called tension methods, where the pieces that form the piecewise interpolating function tend to the straight line passing through two interpolation points, in correspondence of limit values of proper tension variables. These variables are therefore used to control the shape, forcing the interpolant to be, in the limit, inherently co-monotone and/or co-convex. Tension methods will be discussed with more details in another paper on this volume.

The third category, perhaps the most studied, will constitute the argument of this paper. It follows a more natural approach, in the sense that a simple space of functions is chosen, and the interpolant is defined by a carefully computed sequence of *good* parameters. A closer inspection to this material reveals that, despite the different problems and goals examined, the greatest part of these papers is concerned with piecewise (polynomial, rational or exponential) functions and that the various approaches can be summarized in the following, common, way. First, a suitable set of piecewise functions (e.g. piecewise cubic C^1 polynomial spline) is chosen and a set of parameters (e.g. the derivatives at the knots) is selected; second, each piece of the function is expressed using these parameters (e.g. as cubic Hermite polynomials); third, the constraints (e.g. the monotonicity conditions) are also rewritten in terms of these parameters and a set of *admissible domains* is derived, and, fourth, a theory is developed for checking the feasibility of the problem and, eventually, an algorithm for computing a solution is provided.

In other words, as the examples and the references reported in the next section show, we are looking at this schizophrenic situation: domains of different form (and consequently different theorems and algorithms), correspond to the various problems, although the fundamental steps are practically the same.

A first attempt for overcoming this contradiction was presented some years ago in [23] and independently in [11], [6]. In these papers an abstract algorithm for the construction of univariate functions subject to separable constraints was developed, where "abstract" does not mean a generalization to abstract spaces and constraints, but, on the contrary, actually refers to a general purpose practical theory which can act as a theoretical and algorithmic frame where to collocate the various problems and goals. In fact, as shown in the following sections, its main attractive relies in the wideness of its applications; *any* piecewise 1-D function subject to piecewise constraints can in principle be constructed using this theory.

Although these abstract schemes have been used for several applications (see the references in the next section), it is the author's opinion they could provide further help both in classical shape–preserving interpolation / approximation theory and in CAGD, especially because of the possibility of "playing" with constraints, which can be changed to meet any possible request.

The purpose of this paper is therefore to make easier the usage of these tools, reviewing the fundamental ideas and the principal extensions and applications. The paper is divided in six sections. In the next one we will show with two simple examples the necessity of an unifying theory and we will present some basic results. In section three we will show how to improve the solution minimizing some general objective functional and in section four we will discuss how to insert boundary conditions. Section five is devoted to 2-D, tensor product extensions, and section six to final conclusions and remarks.

2   The basic scheme

Aim of this section is to introduce the basic scheme, which will be expanded in the next sections, and to show how its abstract form comprehends problems of apparently different nature. Let us start with a simple example.

Let the points

$$\{(x_0, y_0), (x_1, y_1), \ldots, (x_N, y_N)\} = \{(0,0), (1, 1/4), (2, 3/2), (3, 4), (4, 15/4)\}$$

be given; we want to construct a C^1 cubic piecewise polynomial function, interpolating the above data at the knots, which is increasing and convex in the intervals $[0,1]$, $[1,2]$ and concave in $[2,3]$, $[3,4]$. (Throughout the paper the shape constraints will be considered in a non strict sense). We can express the i-th, $i = 0, 1, 2, 3$ cubic piece in its Hermite form, that is:

$$p_i(x) = y_i H_0^{(i)}(x) + y_i' H_1^{(i)}(x) + y_{i+1} H_2^{(i)}(x) + y_{i+1}' H_3^{(i)}(x) \tag{1}$$

where $H_j^{(i)}$ are the Hermite basis functions, y_i, y_{i+1} are the given data and y_i', y_{i+1}' are the remaining free parameters. A simple differentiation shows that the shape constraints are satisfied if and only if

$$(y_0', y_1') \in D_0 \; ; \; (y_1', y_2') \in D_1 \; ; \; (y_2', y_3') \in D_2 \; ; \; (y_3', y_4') \in D_3 ,$$

where D_0, D_1 are the triangles with vertices

$$\{(1/4, 1/4), (0, 3/8), (0, 3/4)\}, \; \{(5/4, 5/4), (0, 15/8), (0, 15/4)\}$$

and D_2, D_3 the convex cones (in a generic uv-plane) of equations $v \geq -2u+3/2$, $v \leq -u/2+1$ and $v \geq -2u - 3/4$, $v \leq -u/2 - 3/8$. In other words our problem is solved if, and only if, we find a sequence $\{y_0', y_1', y_2', y_3', y_4', y_5'\}$ such that $(y_i', y_{i+1}') \in D_i \; ; \; i = 0, 1, 2, 3, 4$. We put in evidence that each of the internal variables y_1', y_2', y_3', y_4' is involved in two constraint domains.

Let us now consider another simple example. Let a histogram of classes $\{[0, 1), [1, 2), [2, 3), [3, 4]\}$ and frequencies $\phi_0 = 2, \phi_1 = 3, \phi_2 = 5, \phi_3 = 6$ be given; we want to construct a continuous piecewise linear function with knots at $\{0, 1, 2, 3, 4\}$ which is increasing and "interpolates" the histogram (a so-called *histospline*), that is

$$\int_i^{i+1} \ell_i(x) dx = \phi_i; \quad i = 0, 1, 2, 3,$$

where ℓ_i denotes the i-th linear segment. We may write $\ell_i(x) := y_i(x - i) + y_{i+1}(i + 1 - x)$, where y_0, y_1, y_2, y_3, y_4 are our free parameters, and we have immediately from the integral above $y_{i+1} = -y_i + 2\phi_i$, $i = 0, 1, 2, 3$. The increase of the function is immediately achieved if $y_i \leq y_{i+1}$ and, again, the problem of constructing an increasing, piecewise linear histospline, is equivalent to find a sequence $\{y_0, y_1, y_2, y_3, y_4\}$ such that $(y_i, y_{i+1}) \in D_i; \; i = 0, 1, 2, 3$ where $D_i := \{(u, v) \in \mathbb{R}^2 \text{ s.t. } v = -u + 2\phi_i \; ; \; u \leq v\}$.

The examples above show how two problems, at a first sight completely different, do on the contrary have the same structure. We are therefore driven toward a general formulation. Let the sequence of sets $D_i \subseteq \mathbb{R}^q \times \mathbb{R}^q \; ; \; D_i \neq \emptyset \; i = 0, 1, \ldots, N-1$, be given, let define

$$\mathbf{D} := \{(d_0, d_1, \ldots, d_N) \in \mathbb{R}^{q(N+1)} \text{ s.t.} (d_i, d_{i+1}) \in D_i \; ; \; i = 0, 1, \ldots, N-1\}$$

and state the following problems.

Problem P1[D]. Is $\mathbf{D}$ non-empty? In other words do there exist sequences $(d_0, d_1, \ldots, d_N)$ s.t. $(d_i, d_{i+1}) \in D_i \; ; \; i = 0, 1, \ldots, N-1$?

Problem P2[D]. If $\mathbf{D} \neq \emptyset$, how to compute a (the best) solution?

A solution for these problems can be obtained using a two-pass strategy, processing the data first *from left to right* (in algorithm A1), and then *from right to left* (in algorithm A2). We refer to [23] (where the term *staircase algorithm* was used) or to [6] for details and proofs. Let us denote with $\Pi_{u,v}^u : \mathbb{R}^q \times \mathbb{R}^q \to \mathbb{R}^q$ and $\Pi_{u,v}^v : \mathbb{R}^q \times \mathbb{R}^q \to \mathbb{R}^q$ the projection maps from the uv-"plane" onto the u-"axis" and v-"axis" respectively, and define the sets

$$B_i := \Pi_{u,v}^u(D_i); \; i = 0, 1, \ldots, N - 1; \quad B_N := \mathbb{R}^q. \tag{2}$$

We then introduce the following algorithm.

Algorithm A1[D].

1. Set $A_0 := B_0, J := N$
2. For $i = 1, \ldots, N$
   2.1. Set $A_i := \Pi_{u,v}^v(D_{i-1} \cap \{A_{i-1} \times B_i\})$.
   2.2. If $A_i = \emptyset$ set $J := i$ and stop.
3. Stop.

A geometrical interpretation of step 2.1, which is the kernel of the method, is shown in fig. 1; for a better comprehension of the figure we recall that the intervals B_i (on the i-axis) have been previously defined in (2) and that alg. A1 sweeps *from left to right*. So, in executing step 2.1, the intervals B_i, A_{i-1} and the set D_{i-1} are given, and the new interval A_i is computed. The following result holds.

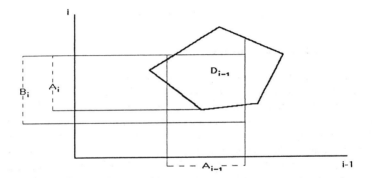

Figure 1. A graphical interpretation of step 2.1 of A1[D].

Theorem 1. *P1 has a solution if, and only if, $J=N$ that is $A_i \neq \emptyset$, $i = 0, 1, \ldots, N$. If $(d_0, d_1, \ldots, d_N)$ is a solution then*

$$d_i \in A_i \; ; \; i = 0, 1, \ldots, N. \tag{3}$$

We remark that, in general, a solution of P1 is not unique and that the necessary condition (3) is not sufficient. So, given the sequence of non empty sets $A_0, \ldots, A_N$ by algorithm A1, a first simple scheme for computing the sequence $(d_0, d_1, \ldots, d_N)$ is provided by the following algorithm and theorem.

Algorithm A2[D].

1. Choose any $d_N \in A_N$.
2. For $i = N-1, N-2, \ldots, 0$
   2.1. Choose any $d_i \in C_i(d_{i+1}) := \Pi_{u,v}^u(D_i \cap \{A_i \times \{d_{i+1}\}\})$
3. Stop.

Theorem 2. Let the sequence $A_0, A_1, \ldots, A_N$ be given by algorithm A1, with $A_i \neq \emptyset$ $i = 0, 1, \ldots, N$. Then algorithm A2 can be completed (that is sets $C(d_{i+1})$ are not empty) and any sequence $(d_0, d_1, \ldots, d_N)$ computed by algorithm A2 is a solution for problem P2.

The geometrical interpretation of step 2.1. is given in fig. 2.

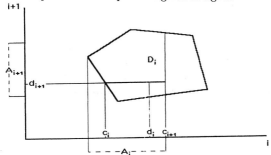

Figure 2. A graphical interpretation of step 2.1 of A2[D].

From a theoretical point of view algorithms A1 and A2 give a complete answer to P1 and P2. However, it could be observed that, in practical problems of constrained interpolation or constrained curve fitting, the user would have the best possible plot, whether or not P1 admits solutions. Let us suppose that alg. A1 applied to the complete set of $D_i; i = 0, 1, \ldots N - 1$, gives $J = j \neq N$, but that the problem is separately solvable *from 0 to j* and *from j to N*, that is there are solutions for the subsequences of constraints $D_0, \ldots, D_{j-1}$ and $D_j, \ldots, D_N$; if we reset $D_{j-1} := A_{j-1} \times B_j$ then A1, applied to this new family of sets, will produce $J = N$. We can, in other words, use the following revised version of alg. A1

Algorithm A1[D].

1. Set $A_0 := B_0$
2. For $i = 1, \ldots, N$
   2.1. Set $A_i := \Pi_{u,v}^v(D_{i-1} \cap \{A_{i-1} \times B_i\})$.
   2.2. If $A_i = \emptyset$ set $D_{i-1} := A_{i-1} \times B_i$; $A_i := B_i$.
3. Stop.

In the following, the term alg. A1[**D**] will refer to the above new version. It should be pointed out that in this way, we have cut off the $(j-1)$-th domain and so the sequence does not entirely satisfy the constraints. Other strategies, tailored for particular problems, can of course be used; we refer to [12] for a possible example.

This two passes scheme has turned out to be an useful tool in several problems, and we refer to the papers [2]–[6], [8]–[12], [17], [20]–[24] for details concerning the various applications and for the corresponding numerical and graphical examples. From all these papers and from the two simple examples at the beginning of this section, a reader could infer that the applicability of algs. A1–A2 is limited to the construction of *purely local* piecewise functions (for example C^1 piecewise cubic interpolants as opposed to C^2 cubic splines interpolating at the knots). Although this is surely the most common case (see also the *negative* remarks in the final section), from a theoretical point of view algs. A1–A2 do have a wider field of applications.

Just as an example (we refer to [4] for a complete exposition in the case of piecewise polynomial shape–preserving interpolation), let us consider the knots $x_0 < x_1 < \ldots < x_N$ and the set S_3^2 of cubic splines and let us denote with p_i the restriction of $s \in S_3^2$ in the interval $[x_i, x_{i+1}]$. Suppose in addition we are given a set of convex data $\{y_0, y_1, \ldots, y_N\}$ and our goal is to construct a convex spline interpolating the data at the knots. Let $d_i := (z_i, z_i^{(1)}, z_i^{(2)})$; any cubic polynomial p_i can be expressed as the solution of the (overdetermined) interpolation problem

$$p_i^{(j)}(x_i) = z_i^{(j)}, \ p_i^{(j)}(x_{i+1}) = z_{i+1}^{(j)}; \ j = 0, 1, 2 \ ,$$

where, of course, some of the above relations are linearly dependent. For example the last components of d_i and d_{i+1} are given by

$$\begin{aligned} z_i^{(2)} &= z_i D^{(2)} H_0^{(i)}(x_i) + z_i^{(1)} D^{(2)} H_1^{(i)}(x_i) + \\ &\quad z_{i+1} D^{(2)} H_2^{(i)}(x_i) + z_{i+1}^{(1)} D^{(2)} H_3^{(i)}(x_i) \\ z_{i+1}^{(2)} &= z_i D^{(2)} H_0^{(i)}(x_{i+1}) + z_i^{(1)} D^{(2)} H_1^{(i)}(x_{i+1}) + \\ &\quad z_{i+1} D^{(2)} H_2^{(i)}(x_{i+1}) + z_{i+1}^{(1)} D^{(2)} H_3^{(i)}(x_{i+1}) \end{aligned} \quad (4)$$

Now let define $E_i := \{(d_i, d_{i+1}) \in \mathbb{R}^3 \times \mathbb{R}^3 \text{ s.t. } (4) \text{ is satisfied}\}$, $F_i := \{(d_i, d_{i+1}) \in \mathbb{R}^3 \times \mathbb{R}^3 \text{ s.t. } z_i = y_i \ ; \ z_{i+1} = y_{i+1}\}$ and finally the convexity domains (easily obtainable differentiating p_i) $G_i := \{(d_i, d_{i+1}) \in \mathbb{R}^3 \times \mathbb{R}^3 \text{ s.t. } z_{i+1}^{(1)} \geq -2z_i^{(1)} + 3\sigma_i \ ; \ z_{i+1}^{(1)} \leq -z_i^{(1)}/2 + 3\sigma_i/2\}$, where $\sigma_i := (y_{i+1} - y_i)/(x_{i+1} - x_i)$. If we finally set

$$D_i := E_i \cap F_i \cap G_i \ ; \ i = 0, 1, \ldots, N-1,$$

our problem is once again restated in finding a sequence $(d_0, d_1, \ldots, d_N)$ such that $(d_i, d_{i+1}) \in D_i \ ; \ i = 0, 1, \ldots, N-1$.

We conclude observing that the results of this section have been proved in a very general setting and therefore hold even for $D_i \in I_i \times I_{i+1}$ where $I_0, \ldots, I_N$ are arbitrary, non-empty sets.

3 Optimal sequences

We have already noticed in the previous section that, given the sequence of sets A_i $i = 0, 1, \ldots, N$ from algorithm A1, it is possible to find an infinite number of sequences $(d_0, d_1, \ldots, d_N)$, because in alg. A2 the admissible sets $C_i(d_{i+1})$ defined in step 2.1 do not reduce, in general, to a single point. It is therefore a natural idea to seek for an optimal sequence; in other words we would like to find

$$\min_{(d_0, d_1, \ldots, d_N) \in \mathbf{D}} F(d_0, d_1, \ldots, d_N)$$

with e.g.,

$$F(d_0, d_1, \ldots, d_N) := \sum_{i=0}^{N} (d_i^\star - d_i)^2 \qquad (5)$$

or

$$F(d_0, d_1, \ldots, d_N) := \max_{0 \le i \le N} |d_i^\star - d_i|$$

where $\{d_0^\star, d_1^\star, \ldots, d_N^\star\}$ is, for some reason, an optimal target sequence, or, in the case we are calculating a constrained piecewise function s with knots $x_0 < x_1 < \ldots < x_N$:

$$F(d_0, d_1, \ldots, d_N) := \int_{x_0}^{x_N} (s''(x))^2 dx \ .$$

For this last functional some interesting results have been obtained using Fenchel's duality theory (see [21] and the references therein); however we prefer to present here an approach based on dynamic programming (DP), not because more efficient (it is well known DP is not, in general, the best optimization method) but because we are mainly concerned in the wideness of applications. In this connection DP is extremely flexible (many functionals and any kind of separable constraints can be processed using the same algorithmic structure) and, unlike other optimization methods, constraints play here a positive role, limiting the size of the *decision space*.

We refer to the classical book [1] for an introduction to DP and limit ourselves to say it can be used when the functional is obtained from the contribution of the i-th stages, $i = 0, 1, \ldots, N$, which do depend on the variables $d_i, d_{i-1}, \ldots, d_0$, and when the *Bellman's optimality principle* holds: "for any possible choice of the variable d_i (which is, for the moment, supposed fixed) the previous ones $d_{i-1}, d_{i-2}, \ldots, d_0$ must constitute an optimal choice for the i-th stage".

In the present case the functional recurrence relations of DP can be very efficiently linked with the constraints in alg. A2; below is reported a sketch of the new algorithm only for the choice (5), being possible to treat the other examples (as well as many other functionals) using exactly the same structure. We only advise that in Φ_i is stored the cost associate with the i-th stage (and therefore $\min_{(d_0,d_1,\ldots,d_N) \in \mathbf{D}} F(d_0, d_1, \ldots, d_N) = \min_{d_N} \Phi_N(d_N)$) and that in T_i is stored the optimal policy (therefore starting with the optimal d_N we obtain the optimal $d_{N-1} := T_{N-1}(d_N)$ and so on).

Algorithm A2DP[D].

1. For any $\delta_0 \in A_0$ compute $\Phi_0(\delta_0) := (d_0^\star - \delta_0)^2$
2. For $i = 1, 2, \ldots, N$
    2.1. For any $\delta_i \in A_i$ compute
    $$C_{i-1}(\delta_i) := \Pi_{u,v}^u(D_{i-1} \cap \{A_{i-1} \times \{\delta_i\}\})$$
    2.2. For any $\delta_i \in A_i$ compute
    $$\Phi_i(\delta_i) := (d_i^\star - \delta_i)^2 + \min_{\delta_{i-1} \in C_{i-1}(\delta_i)} \Phi_{i-1}(\delta_{i-1}) =$$
    $$= (d_i^\star - \delta_i)^2 + \Phi_{i-1}(T_{i-1}(\delta_i))$$
    and the corresponding optimizing value $T_{i-1}(\delta_i)$
3. Compute d_N such that $\Phi_N(d_N) = \min_{\delta_N \in A_N} \Phi_N(\delta_N)$
4. For $i = N - 1, \ldots, 0$
    4.1 $d_i := T_i(d_{i+1})$
5. Stop.

4 Boundary conditions

It is a standard practice both in interpolation theory and in CAGD to add end-conditions to the interpolating function; for example, if we are dealing with a spline s interpolating a given function f, we could ask for $s'(x_0) = f'(x_0)$; $s'(x_N) = f'(x_N))$ or, in the case f is periodic, $s'(x_0) = s'(x_N)$. In terms of our abstract scheme, conditions like these could be restated as $d_0 = \bar{d}_0$; $d_N = \bar{d}_N$ (or, more generally, in the case $\mathbb{R}^q = \mathbb{R}$, $d_0 \in [\bar{d}_0^{min}, \bar{d}_0^{max}]$ $d_N \in [\bar{d}_N^{min}, \bar{d}_N^{max}]$) and as $d_N = \beta(d_0)$ where β is any continuous function with continuous inverse. We want to find, among all the possible solutions of P1[D], those satisfying also one of the two above.

Boundary conditions of the first (local) form will be called *separable* (S) and those of the second form *non-separable* (NS). While S conditions can be very easily handled, the NS ones are by far more complicate because, relating directly the first and the last element of the sequence $(d_0, d_1, \ldots, d_N)$, destroy the sequential structure of our scheme.

Let us start with S conditions; we can simply look at them as new constraints which can be added to the corresponding domains; more specifically we can set

$$D_0^{(S)} := D_0 \cap \{(d_0, d_1) \in \mathbb{R}^q \times \mathbb{R}^q \text{ s.t. } d_0 = \bar{d}_0 \};$$
$$D_N^{(S)} := D_N \cap \{(d_{N-1})d_N) \in \mathbb{R}^q \times \mathbb{R}^q \text{ s.t. } d_N = \bar{d}_N \}$$

and, having defined,

$$\mathbf{D}^{(S)} := \{ (d_0, d_1, \ldots, d_N) \in \mathbb{R}^{q(N+1)} \text{ s.t. } (d_0, d_1) \in D_0^{(S)}$$
$$(d_i, d_{i+1}) \in D_i \text{ ; } i = 1, 2, \ldots, N - 2 \text{ , } (d_{N-1}, d_N) \in D_N^{(S)} \}$$

we can apply to $\mathbf{D}^{(S)}$ the theory developed in sections 2 and 3. We note parenthetically that local additional restrictions can be inserted in *any* domain D_i.

Let us now turn out our attention to NS boundary conditions and let us define

$$\mathbf{D}^{(NS)} := \{(d_0, d_1, \ldots, d_N) \in \mathbb{R}^{q(N+1)} \text{ s.t. } d_N = \beta(d_0)\} \text{ ;}$$

Functional shape–preserving interpolation

we can reformulate our problem of finding (if any) $(d_0, d_1, \ldots, d_N) \in \mathbf{D}^{(NS)}$ saying that we want to start with a $d_0 \in A_0$, then compute the elements $d_1, d_2, d_3, \ldots$ satisfying the constraints $D_0, D_1, D_2, \ldots$ and finally end with that precise $d_N \in A_N$ such that $d_N = \beta(d_0)$. In other words we must first check if there are, in the admissible interval A_0, some d_0 such that $\beta(d_0) \in A_N$ (from the theory of section two we know that for any feasible sequence it has $d_i \in A_i$) or, equivalently, check if $\beta(A_0) \cap A_N \neq \emptyset$. Then we must find a way to start with one of these d_N and go back ending in $d_0 = \beta^{-1}(d_N)$.

A full development of this problem, both for the existence theorems and for the constructive algorithms, requires some theory of set–valued maps which, for reasons of space, cannot be reported here. We refer to [8] for the theoretical aspects (stated in abstract Banach spaces) and to [9], [10] for an idea of some practical applications and limit ourselves to give a very intuitive sketch of the method.

Let $\delta_0 \in A_0$ be arbitrary but fixed and let us define

$$\Delta_0 := D_0 \cap \{(d_0, d_1) \in \mathbb{R}^q \times \mathbb{R}^q \text{ s.t. } d_0 = \delta_0\},$$
$$\Delta_i := D_i, \ i = 1, 2, \ldots, N-1$$

and

$$\boldsymbol{\Delta} := \{(d_0, d_1, \ldots, d_N) \in \mathbb{R}^{q(N+1)} \text{ s.t. } (d_i, d_{i+1}) \in \Delta_i\ ;\ i = 0, 1, \ldots, N-1\};$$

if we apply the forward algorithm A1[$\boldsymbol{\Delta}$] we obtain a sequence of sets $A_0(\boldsymbol{\Delta}), A_1(\boldsymbol{\Delta}), \ldots, A_N(\boldsymbol{\Delta})$ which are admissible for problem P1[$\boldsymbol{\Delta}$]; therefore, if we apply the backward algorithm A2[$\boldsymbol{\Delta}$] (or A2DP[$\boldsymbol{\Delta}$]) starting with any $d_N \in A_N(\boldsymbol{\Delta})$ we obtain a sequence $(d_0, d_1, \ldots, d_N)$ which satisfies all the constraints $\boldsymbol{\Delta}$, that is such that $d_0 = \delta_0$. Suppose now that $\beta(\delta_0) \in A_N(\boldsymbol{\Delta})$; we can start with $d_N = \beta(\delta_0)$ and the sequence $(d_0, d_1, \ldots, d_N)$ will surely satisfy our non-separable boundary conditions. So, the idea of the algorithms reported in [8], [9] is essentially a two steps strategy: in the first (using an iterative scheme based on Kakutani's fixed point theorem for set-valued functions) we find a $\delta_0 \in A_0(\boldsymbol{\Delta})$ such that $\beta(\delta_0) \in A_N(\boldsymbol{\Delta})$ and in the second we compute the sequence using A2[$\boldsymbol{\Delta}$] or A2DP[$\boldsymbol{\Delta}$] and starting with $d_N = \beta(\delta_0)$.

5 Tensor–product extensions

Repeating the scheme of the second section, let us start with an example. Suppose we are given a set of tensor–product data $(x_i, y_j, z_{i,j})$; $i = 0, 1, \ldots, N$, $j = 0, 1, \ldots, M$ which are increasing along x and y direction, that is $z_{i,j} < z_{i+1,j}$, $z_{i,j} < z_{i,j+1}$ and suppose we want to construct an "x-network" of increasing interpolating curves, that is a set of increasing functions $s_j(x)$, $j = 0, 1, \ldots, M$ such that $s_j(x_i) = z_{i,j}$. For simplicity, any s_j is supposed to be a C^1 piecewise cubic, expressed, in the generic interval $[x_i, x_{i+1}]$, in the Hermite form

$$p_i(x; j) = z_{i,j} H_0^{(i)}(x) + z_{i,j}^{(1,0)} H_1^{(i)}(x) + z_{i+1,j} H_2^{(i)}(x) + z_{i+1,j}^{(1,0)} H_3^{(i)}(x)$$

where $z_{-,-}^{(1,0)}$ denotes, as usual, the x-partials at interpolation points, which are the free parameters for the present problem.

If we restrict for the moment our attention to the rectangle $[x_i, x_{i+1}] \times [y_j, y_{j+1}]$ we have, from the well-known results reported in [15], that $p_i(x; r)$, $r = j, j+1$ is increasing with respect to x direction if $(z^{(1,0)}_{i,r}, z^{(1,0)}_{i+1,r}) \in DX_{i,r}$, where $DX_{i,r} := [0, 3\sigma_{i,r}] \times [0, 3\sigma_{i,r}]$ and $\sigma_{i,r} := (z_{i+1,j} - z_{i,j})/(x_{i+1} - x_i)$. We have now to consider increase with respect to y direction, that is we must impose, for any $x \in [x_i, x_{i+1}]$, $p_i(x,j) < p_i(x,j+1)$. From the Bezier representation of the cubic polynomials we have that the inequality is satisfied if $(z^{(1,0)}_{i,j}, z^{(1,0)}_{i,j+1}) \in DY^+_{i,j}$, $(z^{(1,0)}_{i+1,j}, z^{(1,0)}_{i+1,j+1}) \in DY^-_{i+1,j}$, where

$$DY^+_{i,j} := \{(u,v) \in \mathbb{R}^2 \text{ s.t. } v > u - 3(z_{i,j+1} - z_{i,j})/(x_{i+1} - x_i)\}$$
$$DY^-_{i+1,j} := \{(u,v) \in \mathbb{R}^2 \text{ s.t. } v < u + 3(z_{i+1,j+1} - z_{i+1,j})/(x_{i+1} - x_i)\}$$

The internal couples $(z^{(1,0)}_{i,j}, z^{(1,0)}_{i,j+1})$ are therefore subject to constraints from left and right, that is we must impose

$$(z^{(1,0)}_{i,j}, z^{(1,0)}_{i,j+1}) \in DY_{i,j} := DY^-_{i,j} \cap DY^+_{i,j},$$

while for the boundary ones we have simply $DY_{0,j} := DY^+_{0,j}$ and $DY_{N,j} := DY^-_{N,j}$. (Of course, more efficient constraints could have been derived for the problem considered in this example; but we are interested in introducing general ideas and prefer not to disturb the reading with technical details). In conclusion, our variables $z^{(1,0)}_{i,j}$ are simultaneously related by constraints along the x and y direction, as shown in fig. 3, and, since this example is clearly a sample from a wide class of problems, we are forced to derive another abstract formulation.

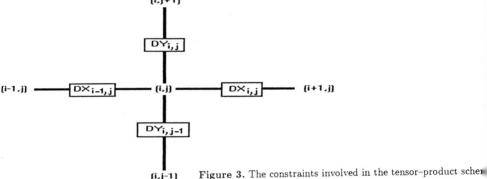

Figure 3. The constraints involved in the tensor-product scheme

Let the integers $i = 0, 1, \ldots, N$ and $j = 0, 1, \ldots, M$ be given, let $K = q(N+1)(M+1)$ and suppose we are given the following constraints:

$$DX_{i,j} \subset \mathbb{R}^q \times \mathbb{R}^q; \ DX_{i,j} \neq \emptyset; \ i = 0, 1, \ldots, N-1; \ j = 0, 1, \ldots, M,$$
$$DY_{i,j} \subset \mathbb{R}^q \times \mathbb{R}^q; \ DY_{i,j} \neq \emptyset; \ i = 0, 1, \ldots, N; \ \ \ \ j = 0, 1, \ldots, M-1.$$

In this section the domains $DX_{i,j}$ and $DY_{i,j}$ are supposed to be convex, an assumption which is always satisfied in shape-preserving problems. We want to find a matrix d, where

$$d := \{d_{i,j} \in \mathbb{R}^q \times \mathbb{R}^q \ ; \ i = 0, 1, \ldots, N \ ; \ j = 0, 1, \ldots, M\}$$

such that

$$(d_{i,j}, d_{i+1,j}) \in DX_{i,j} \ ; \ i = 0, 1, \ldots, N-1; j = 0, 1, \ldots, M, \quad (6.a)$$
$$(d_{i,j}, d_{i,j+1}) \in DY_{i,j} \ ; \ j = 0, 1, \ldots, M-1; i = 0, 1, \ldots, N. \quad (6.b)$$

Functional shape–preserving interpolation

We may define
$$\mathbf{DXY} := \{\, d \in \mathbb{R}^K \text{ s.t. (6) are satisfied}\,\}$$
and, having in mind the goals of checking if $\mathbf{DXY} \neq \emptyset$ and then of finding (if possible) a feasible solution $d \in \mathbf{DXY}$, we can introduce the following sets of constraints

$$\mathbf{DX_j} := \{(d_{0,j}, d_{1,j}, \ldots, d_{N,j}) \text{ s.t. } (d_{i,j}, d_{i+1,j}) \in DX_{i,j};$$
$$i = 0, 1, \ldots, N-1\,\};\ j = 0, 1, \ldots, M$$
$$\mathbf{DX} := \mathbf{DX_0} \times \mathbf{DX_1} \times \cdots \times \mathbf{DX_M}$$
$$\mathbf{DY_i} := \{(d_{i,0}, d_{i,1}, \ldots, d_{i,M}) \text{ s.t. } (d_{i,j}, d_{i,j+1}) \in DY_{i,j};$$
$$j = 0, 1, \ldots, M-1\,\}\ ;\ i = 0, 1, \ldots, N$$
$$\mathbf{DY} := \mathbf{DY_0} \times \mathbf{DY_1} \times \cdots \times \mathbf{DY_N}$$

and note that $\mathbf{DXY} = \mathbf{DX} \cap \mathbf{DY}$.

It is clear from the results of section 2 that we have no problem in finding $d \in \mathbf{DX}$ and, separately, $d \in \mathbf{DY}$ because both $\mathbf{DX}$ and $\mathbf{DY}$ are made up by independent one-dimensional constrained problems. The difficult point is to find (if any) an element in their intersection; the solution we are going to present is based on the *alternating projection* method, which is founded on the following famous result of Cheney and Goldstein [3]:

Theorem 3. *Let C_1, C_2 be convex sets of a finite dimensional Hilbert space H. Let $P_{C_1}(h)$, $P_{C_2}(h)$ be the best approximation to $h \in H$ from C_1, C_2. Let $x_0 \in H$ and $y_{n+1} := P_{C_2}(x_n)$, $x_{n+1} := P_{C_1}(y_n)$. Then*

$$\lim_{n \to \infty} x_{2n} = x \in C_1\ ;\ \lim_{n \to \infty} y_{2n+1} = y \in C_2$$

and

$$\|x - y\| = \inf_{\xi \in C_1, \eta \in C_2} \|\xi - \eta\|.$$

We can make use of the above result, because $\mathbf{DX}$ and $\mathbf{DY}$ are convex subset of $\mathbb{R}^K$, embedded with Euclidean norm. More specifically, suppose we are given $d^{(0)} \in \mathbb{R}^K$ and we want to find $P_{\mathbf{DX}}(d^{(0)})$ (analogous considerations hold obviously for $P_{\mathbf{DY}}(d^{(0)})$); we must compute

$$\min_{d \in \mathbf{DX}} \sum_{j=0}^{M} \sum_{i=0}^{N} (d_{i,j}^{(0)} - d_{i,j})^2 = \sum_{j=0}^{M} \min_{(d_{0,j},d_{1,j},\ldots,d_{N,j}) \in \mathbf{DX_j}} \sum_{i=0}^{N} (d_{i,j}^{(0)} - d_{i,j})^2\ ,$$

where the one-dimensional constrained approximations in the right hand side summation can be easily computed using the dynamic programming scheme of section two. The algorithm can be sketched as follows.

Algorithm.

1. Choose a suitable starting point $d^{(0)} \in \mathbb{R}^K$ and set $n = 0$.
2. While (*convergence test not satisfied*) do
   2.1 For $i = 0, 1, ..., N$
       2.1.1 Use A1[**DY$_i$**] and A2DP[**DY$_i$**] to compute $d_{i,0}^{(n+1)}, d_{i,1}^{(n+1)}, \ldots,$
       $d_{i,M}^{(n+1)}$ best approximation to $d^{(n)}$ from **DY$_i$**.
   2.2 Set $n = n + 1$.
   2.3 For $j = 0, 1, ..., M$
       2.3.1 Use A1[**DX$_j$**] and A2DP[**DX$_j$**] to compute $d_{0,j}^{(n+1)}, d_{1,j}^{(n+1)}, \ldots,$
       $d_{N,j}^{(n+1)}$ best approximation to $d^{(n)}$ from **DX$_j$**.
   2.4 Set $n = n + 1$.
3. Set $d := (d^{(n)} + d^{(n-1)})/2$.

Note that we have an output even if **DXY** $= \emptyset$ and in this case step 3 furnishes the best approximation to the constraints **DX** and **DY**. We note also that, assuming we are using the revised version of algorithm A1, we are implicitly supposing **DX** and **DY** non-empty, which, in turn, ensures the existence of best approximations in steps 2.1.1 and 2.3.1. From a practical point of view, the efficiency of the proposed scheme is closely related to that of the best approximations from the many sets **DX$_j$** and **DY$_i$**; fortunately, algorithm A2DP is very fast, its complexity being $O((N+1) \cdot Q)$ or $O((M+1) \cdot Q)$ where N and M give the number of x and y data and Q is the number of points in which we subdivide the constraint domains (see [10]). Some improvements are also possible. Suppose, for example, we have computed $d^{(n)} = \{d_{i,j}^{(n)} \; ; \; i = 0, \ldots, N \; ; \; j = 0, \ldots, M\} \in$ **DY** in step 2.1 and we are approaching the best approximations of step 2.3. It often happens that several of the **DX$_j$** constraints are satisfied by $d_{0,j}^m, d_{1,j}^m, \ldots, d_{N,j}^m$ and the corresponding computation of the best approximations (which would give zero distance) can therefore be skipped. Another improvement (which is too technical to be described here) consists in a preliminary reduction of the constraints, inserting the admissible domains A_- given by algorithm A1 along x direction in y constraints and vice-versa. The reader is referred to [7] for details.

6 Conclusions and remarks

It has already been observed that the main attractive of the abstract schemes developed in the previous sections is their generality, in the sense of the wide range of potential applications. This is indeed true, at least in theory; we have made no assumption on the subsets of $\mathbb{R}^q \times \mathbb{R}^q$ which are the basic elements of our constraints, and, on the contrary, we have been allowed to further generalize the mathematical settings: abstract sets (without algebraic structure) for sect. 2, Banach spaces for sect. 4, and Hilbert spaces for sect. 5.

However, a closer inspection to the several applications reported in the references, shows the main drawback of the method: with only two exceptions (as far as the author knows) the practical usage of these schemes has been confined to $q = 1$ that is to subsets of $\mathbb{R} \times \mathbb{R}$. The reason is in step 2.1 of algorithm A1, which is the kernel of all the modifications and improvements later developed:

$$A_i := \Pi_{u,v}^v(D_{i-1} \cap \{A_{i-1} \times B_i\}) \; ;$$

even in the *simple* case $D_{i-1} \subseteq \mathbb{R}^2 \times \mathbb{R}^2$ we should compute intersections and projections of arbitrary subsets of $\mathbb{R}^4$. Even if we supposed linear inequalities for the constraints (D_{i-1} would be a polytope of $\mathbb{R}^4$) we would obtain an (extremely difficult to implement) algorithm with complexity exponential in the number of points.

A solution for this problem has been presented in [20] for a particular, but important, case; however, at present, no completely satisfactory method *for the general case* has been proposed. We are going to explain a general idea (used, for example, in [19]) which provides a partial answer for our problem; this idea will, as usual, be preceded by an example which will also clarify why these extensions to spaces with higher dimension are so desirable.

Let $(t_0, f_0), (t_1, f_1), \ldots, (t_K, f_K)$ be a set of increasing and convex data points, let $x_0 < x_1 < \ldots < x_N$ be a set of knots; we want to construct an increasing and convex piecewise C^1 cubic, best approximation to data in the least square sense. We continue to express the i-th piece in its Hermite form (1), but, for notational purposes, we prefer to set $y_i = w_i$, $y'_i = z_i$ obtaining

$$p_i(x) = w_i H_0^{(i)}(x) + z_i H_1^{(i)}(x) + w_{i+1} H_2^{(i)}(x) + z_{i+1} H_3^{(i)}(x) ,$$

where, in the present case, $w_i, z_i, w_{i+1}, z_{i+1}$ are all free parameters. The increase and convexity of p_i is achieved if $(w_i, z_i, w_{i+1}, z_{i+1}) \in D_i$ where, having set $\sigma_i := (w_{i+1} - w_i)/(x_{i+1} - x_i)$,

$$D_i := \{(w_i, z_i, w_{i+1}, z_{i+1}) \in \mathbb{R}^4 \text{ s.t. } w_i \leq w_{i+1};$$
$$z_{i+1} \geq -2z_i + 3\sigma_i; \ z_{i+1} \leq -z_i/2 + 3\sigma_i/2.$$

Let $\Pi_{w,z}^w : \mathbb{R}^4 \mapsto \mathbb{R}^2$ the projective operator from the four dimensional $(w_i, z_i, w_{i+1}, z_{i+1})$-space onto the two dimensional (w_i, w_{i+1})-space, and let us define

$$DW_i := \Pi_{w,z}^w(D_i) \ ; \ i = 0, 1, \ldots, N-1 .$$

By definition, for any $(w_i, w_{i+1}) \in DW_i$ there are (z_i, z_{i+1}) such that $(w_i, z_i, w_{i+1}, z_{i+1}) \in D_i$, and so we have for the "inverse image"

$$DZ_i(w_i, w_{i+1}) := \{(z_i, z_{i+1}) \in \mathbb{R}^2 \text{ s.t. } (w_i, z_i, w_{i+1}, z_{i+1}) \in D_i\} \neq \emptyset .$$

Let us define the sets

$$\mathbf{DW} := \{(w_0, w_1, \ldots, w_N) \in \mathbb{R}^{N+1} \text{ s.t. } (w_i, w_{i+1}) \in DW_i,$$
$$i = 0, 1, \ldots, N-1 \},$$
$$\mathbf{DZ}(w_0, \ldots, w_N) := \{(z_0, z_1, \ldots, z_N) \in \mathbb{R}^{N+1} \text{ s.t. }$$
$$(z_i, z_{i+1}) \in DZ_i(w_i, w_{i+1}) \ ; \ i = 0, 1, \ldots, N-1\};$$

the proposed strategy can be sketched as follows.

- Apply algorithms A1[**DW**] and A2[**DW**] (or A2DP[**DW**]) to obtain a sequence $w_0, w_1, \ldots, w_N$ such that $(w_i, w_{i+1}) \in DW_i$; $i = 0, 1, \ldots, N-1$.
- Compute the sequence of non–empty sets $DZ_0(w_0, w_1), DZ_1(w_0, w_1), \ldots, DZ_{N-1}(w_{N-1}, w_N)$.
- Apply algorithms A1[**DZ**$(w_0, \ldots, w_N)$] and A2[**DZ**$(w_0, \ldots, w_N)$] (or A2DP[**DZ**$(w_0, \ldots, w_N)$]) to obtain a sequence $z_0, z_1, \ldots, z_N$ such that $(z_i, z_{i+1}) \in DZ_i(w_i, w_{i+1})$; $i = 0, 1, \ldots, N-1$.
- Form the sequence $d_i := (w_i, z_i, w_{i+1}, z_{i+1})$ such that $(d_i, d_{i+1}) \in D_i$; $i = 0, 1, \ldots, N-1$.

The obvious and important drawback of this approach is that we can loose feasible solutions because the set **DZ**$(w_0, \ldots, w_N)$, which does depend on the particular choice of the sequence $w_0, w_1, \ldots, w_N$, may result empty even if the "complete" set **D** is not. In this connection it is worthwhile to say that modifications of Dykstra's extension of alternating projection method seem to be somewhat promising; the corresponding research is under current work.

References

[1] Bellman, R. and S. Dreyfus: *Applied Dynamic Programming.* Princeton University Press, New York, 1962.

[2] Burmeister, W., W. Heß and J. W. Schmidt: *Convex spline interpolants with minimal curvature.* Computing **35** (1985), 219–229.

[3] Cheney, E.W, and A. Goldstein: *Proximity maps for convex sets.* Proc. Amer. Math. Soc. **10** (1959), 448–450.

[4] Costantini, P.: *Splines vincolate localmente ed interpolazione monotona e convessa.* In *Atti del Convegno di Analisi Numerica,* de Frede Editore, Napoli, 1985.

[5] Costantini, P.: *On monotone and convex spline interpolation.* Math. Comp., **46** (1986), 203–214.

[6] Costantini, P.: *An algorithm for computing shape-preserving interpolating splines of arbitrary degree.* J. Comp. Appl. Math., **22** (1988), 89–136.

[7] Costantini, P.: *On 2-D interpolation and highly separable constraints.* Università di Siena, Rapporto Matematico n. 242, 1992.

[8] Costantini, P.: *A general method for constrained curves with boundary conditions.* In *Multivariate Approximation: From CAGD to Wavelets,* K. Jetter and F.I. Utreras (eds.), World Scientific Publishing Co., Inc., Singapore, 1993.

[9] Costantini, P.: *Boundary valued shape-preserving interpolating splines.* Preprint, 1996, submitted to ACM Trans. Math. Software .

[10] Costantini, P.: *BVSPIS: a package for computing boundary valued shape-preserving interpolating splines.* Preprint, 1996, submitted to ACM Trans. Math. Software .

[11] Costantini, P. and R. Morandi: *Monotone and convex cubic spline interpolation.* CALCOLO, **21** (1984), 281–294.

[12] Costantini, P. and R. Morandi: *An algorithm for computing shape-preserving cubic spline interpolation to data.* CALCOLO, **21** (1984), 295–305.

[13] Ferguson, D.R.: *Construction of curves and surfaces using numerical optimization techniques.* Comput. Aided Design , **18** (1986), 15–21.

[14] Fiorot, J.C., Tabka, J.: *Shape–preserving C^2 cubic polynomial interpolating splines.* Math. Comp. , **57** (1991), 291–298.

[15] Fritsch, R.E. and R.E. Carlson: *Monotone piecewise cubic interpolation.* SIAM J. Numer. Anal. , **17** (1980), 238–246.

[16] Greiner, H.: *A survey on univariate data interpolation and approximation by splines of given shape.* Mathl. Comput. Modelling, **15** (1991), 97–106.

[17] Heß ,W. and J.W. Schmidt: *Convexity preserving interpolation with exponential splines.* Computing **36** (1986), 335–342.

[18] Micchelli, C.A. and F.I. Utreras: *Smoothing and interpolation in a convex subset of Hilbert space.* SIAM J. Sci. Stat. Comp. , **9** (1988), 728–746.

[19] Morandi, R. and P. Costantini: *Piecewise monotone quadratic histosplines.* SIAM J. Sci. Stat. Comp. **10** (1989), 397–406.

[20] Schmidt, J.W.: *Convex interval interpolation with cubic splines.* BIT **26** (1986), 377–387.

[21] Schmidt, J.W.: *On shape-preserving spline interpolation: existence theorems and determination of optimal splines. Approximation and Function Spaces*, Banach Center Publications, Volume 22, PWN-Polish Scientific Publishers, Warsaw, 1989.

[22] Schmidt, J.W.: *Staircase algorithm and construction of convex spline interpolants up to the continuity C^3.* Computer Math. Applic. **31** (1996), 67–79.

[23] Schmidt, J.W. and W. Heß: *Schwach verkoppelte Ungleichungssysteme und konvexe Spline-Interpolation.* El. Math. **39** (1984), 85–96.

[24] Schmidt, J.W. and W. Heß: *Quadratic and related exponential splines in shape-preserving interpolation.* J. Comp. Appl. Math. **18** (1987), 321–329.

Author

Paolo Costantini
Dipatimento di Matematica,
Università di Siena
via del Capitano 15, 53100 Siena
Italy
E–mail: costantini@unisi.it

Tensor Product Spline Interpolation subject to Piecewise Bilinear Lower and Upper Bounds

B. Mulansky, J. W. Schmidt, M. Walther

Technische Universität Dresden

Abstract: This paper is concerned with range restricted interpolation of gridded data by biquadratic and biquartic tensor product splines on refined rectangular grids. In particular, the given lower and upper bounds are assumed to be continuous and piecewise bilinear with respect to the original grid. Sufficient conditions for the fulfillment of the range restrictions are derived utilizing the tensor product structure as well as corresponding results for univariate quadratic and quartic splines with additional knots. The solvability of this system of sufficient conditions, hence the existence of interpolants meeting the constraints, can always be achieved for strictly compatible data by constructing the refined grid appropriately. The selection of a visually improved range restricted interpolant is based on a fit-and-modify approach or on the minimization of a bivariate Holladay functional.

1 Introduction

Depending on the practical background of the interpolation problem, it can be desirable or even required to satisfy given lower and upper bounds on the values of the interpolant. Such problems of *range restricted interpolation* could occur when approximating or designing a component which must fit within a restricted region or when avoiding the collision of an interpolated tool path with prescribed obstacles.

Univariate range restricted interpolation has been studied recently in [1], [4] based on the variational approach and in [12] using rational cubic splines. Some problems of spline interpolation subject to two-sided restrictions on the derivatives, which can be viewed as generalizations of monotonicity and convexity preserving interpolation, were considered earlier in [15].

A natural technique to apply univariate shape preserving interpolation methods to bivariate problems is the tensor product technique, see [6] for a survey. Our aim in this paper is to extend the univariate methods presented in [7], [10] using quadratic C^1 splines with additional knots to bivariate range restricted interpolation of gridded data. The interpolant is chosen from a space of biquadratic splines on a refined rectangular grid. Each rectangle of the original grid is divided into four subrectangles. The spline space is formed as the tensor product of univariate quadratic C^1 splines with additional knots. In [2], the same tensor product space of biquadratic splines has been used to develop a bivariate monotonicity preserving interpolation method.

We follow the approach in [9], where nonnegativity preserving interpolation by biquadratic tensor product splines has been considered. As mentioned there, the results can be straightforwardly extended to constant and even to piecewise constant lower and upper bounds, see also [18]. In these cases, the refinement of the rectangular grid can be fixed a priori independently of the given (compatible) data. Here, the range restrictions are assumed to be piecewise bilinear with respect to the original grid. Indeed, this type of restrictions is more attractive than piecewise constant bounds in view of potential applications.

Our derivation of sufficient conditions for the fulfillment of the constraints utilizes the tensor product structure of the bivariate spline spaces and of the bounds and avoids the technical details concerning the explicit representation of the bivariate splines.

In univariate interpolation by quadratic C^1 splines with additional knots subject to polygonal bounds, the existence of range restricted interpolants requires the appropriate choice of the additional knots depending on the data. A first proposal for the placement of the additional knots was presented in [7]. Since this proposal is not local, it is not suitable for the extension to a tensor product method. Hence, we start by proposing a new knot placement procedure, on which our tensor product method is based.

In this way, the solvability of the system of sufficient conditions, hence the existence of interpolants meeting the constraints, can always be achieved for strictly compatible data by constructing the refined grid appropriately.

Considering the refined rectangular grid as fixed, the derived sufficient range conditions establish a system of linear inequalities for the remaining parameters, namely the values of the first order partial derivatives and the second order mixed partial derivatives at each data site, which is separated with respect to the data sites. Since this system is solvable provided the refined grid has been constructed appropriately, in general there exists an infinite number of interpolants meeting the constraints. Here, the selection of a visually improved solution is based on a fit-and-modify approach or on the minimization of a fairness functional such as the bivariate Holladay functional.

Since C^1 smoothness is sometimes not sufficient for practical purposes, it is worthwhile to point out that the proposed methods can be immediately extended to interpolation by tensor products of quartic C^2 splines with additional knots subject to piecewise bilinear lower and upper bounds.

2 Univariate interpolation subject to polygonal range restrictions

2.1 Quadratic splines on refined grids

In this paragraph we introduce the needed spaces of univariate quadratic splines with additional knots. To prepare the consideration of the corresponding tensor products, the spline spaces are parameterized using suitable linear functionals.

Let be given a grid

$$\Delta : x^1 < x^2 < \cdots < x^n \tag{1}$$

of data sites, and denote its step sizes by $h^i := x^{i+1} - x^i$, $i = 1, \ldots, n-1$. We define a refined grid

$$\tilde{\Delta} : x^1 < \xi^1 < x^2 < \cdots < x^{n-1} < \xi^{n-1} < x^n \tag{2}$$

using the additional knots

$$\xi^i = \beta^i x^i + \alpha^i x^{i+1}, \quad i = 1, \ldots, n-1, \tag{3}$$

where the ratios $\alpha^i > 0$, $\beta^i > 0$, $\alpha^i + \beta^i = 1$ are arbitrary.

For fixed additional knots ξ^i, the linear space consisting of all piecewise quadratic polynomials on the refined grid $\tilde{\Delta}$ that are continuous on $[x^1, x^n]$ and differentiable at ξ^i, $i = 1, \ldots, n-1$, is denoted by $S_2^{1,0}(\tilde{\Delta})$. In usual terminology [3], $S_2^{1,0}(\tilde{\Delta})$ is the space of quadratic splines with *double* knots x^i and *simple* knots ξ^i.

The linear functionals λ^i, $i = 1, \ldots, 3n-2$, on $S_2^{1,0}(\tilde{\Delta})$ are defined by

$$\begin{aligned}
\lambda^i s &:= s(x^i), & i &= 1, \ldots, n, \\
\lambda^{n+i} s &:= s'(x^i + 0), & i &= 1, \ldots, n-1, \\
\lambda^{2n} s &:= s'(x^n - 0), & & \\
\lambda^{2n+i-1} s &:= s'(x^i - 0), & i &= 2, \ldots, n-1,
\end{aligned} \tag{4}$$

for all $s \in S_2^{1,0}(\tilde{\Delta})$. It follows from general results on spline interpolation [3] or can be easily verified directly (compare [10], [16]) that $S_2^{1,0}(\tilde{\Delta})$ is a $(3n-2)$-dimensional space and the $3n - 2$ linear functionals λ^i given in (4) are linearly independent.

The subspace $S_2^{1,1}(\tilde{\Delta}) := S_2^{1,0}(\tilde{\Delta}) \cap C^1[x^1, x^n]$ of quadratic C^1 splines on the refined grid $\tilde{\Delta}$ is obtained by considering the knots x^i as *simple* knots also. Obviously, the functionals λ^{2n+i-1} and λ^{n+i} are identical on $S_2^{1,1}(\tilde{\Delta})$, i.e.,

$$\lambda^{2n+i-1} s = \lambda^{n+i} s = s'(x^i), \quad i = 2, \ldots, n-1, \tag{5}$$

for all $s \in S_2^{1,1}(\tilde{\Delta})$. But each spline $s \in S_2^{1,1}(\tilde{\Delta})$ can be uniquely represented by the parameters

$$\begin{aligned}
z_i &= \lambda^i s = s(x^i), & i &= 1, \ldots, n, \\
p_i &= \lambda^{n+i} s = s'(x^i), & i &= 1, \ldots, n,
\end{aligned} \tag{6}$$

since the linear functionals λ^i, $i = 1, \ldots, 2n$, are linearly independent on the $2n$-dimensional space $S_2^{1,1}(\tilde{\Delta})$, see [10], [16].

2.2 Sufficient nonnegativity conditions

We want to deal with interpolation by splines $s \in S_2^{1,1}(\tilde{\Delta})$ subject to given lower and upper bounds polygonal (continuous piecewise linear) with respect to Δ. The fulfillment of the range restrictions is equivalent to the nonnegativity of the differences of the spline s and the bounds. Since these differences are from $S_2^{1,0}(\tilde{\Delta})$, we are interested in sufficient conditions for the nonnegativity of splines from $S_2^{1,0}(\tilde{\Delta})$. Again, these conditions are formulated using appropriate linear functionals to simplify the extension to tensor products.

The linear functionals μ^i, $i = 1, \ldots, 3n-2$, on $S_2^{1,0}(\tilde{\Delta})$ are defined by

$$\mu^i := \lambda^i, \qquad i = 1, \ldots, n,$$

$$\mu^{n+i} := \lambda^i + \frac{\alpha^i h^i}{2} \lambda^{n+i}, \qquad i = 1, \ldots, n-1,$$

$$\mu^{2n} := \lambda^n - \frac{\beta^{n-1} h^{n-1}}{2} \lambda^{2n},$$

$$\mu^{2n+i-1} := \lambda^i - \frac{\beta^{i-1} h^{i-1}}{2} \lambda^{2n+i-1}, \quad i = 2, \ldots, n-1.$$
(7)

The values μ^is deliver all the B-ordinates in the B-representation of the quadratic pieces of $s \in S_2^{1,0}(\tilde{\Delta})$ except the ordinates $s(\xi^i)$, which turn out to be convex combinations of the neighboring B-ordinates due to the C^1 conditions. Since the nonnegativity of all B-ordinates implies the nonnegativity of the quadratic pieces, the well-known sufficient nonnegativity conditions can be formulated in our notation as follows.

Proposition 1 *For $s \in S_2^{1,0}(\tilde{\Delta})$, the conditions*

$$\mu^i s \geq 0, \quad i = 1, \ldots, 3n-2,$$
(8)

imply

$$s(x) \geq 0 \quad \text{for all} \quad x \in [x^1, x^n].$$
(9)

The conditions (8) are separated with respect to the original knots x^i, since the functionals μ^i, μ^{n+i}, μ^{2n+i-1} involve only parameters corresponding to x^i, respectively.

2.3 Existence and construction of range restricted interpolants

Now we turn to the range restricted interpolation problem. For given data values z_i, $i = 1, \ldots, n$, and lower and upper bounds L, U we consider the problem:
Find $s \in S_2^{1,1}(\tilde{\Delta})$ such that the interpolation conditions

$$s(x^i) = z_i, \quad i = 1, \ldots, n,$$
(10)

and the range restrictions

$$L(x) \leq s(x) \leq U(x) \quad \text{for all} \quad x \in [x^1, x^n]$$
(11)

are satisfied.

The lower and upper bounds L, U are assumed to be polygonal (continuous piecewise linear) with respect to Δ, and are completely described by $L_i := L(x^i) = \mu^i L$, $U_i := U(x^i) = \mu^i U$, $i = 1, \ldots, n$. Obviously, the compatibility of the data values and the bounds, namely $L_i \leq z_i \leq U_i$, $i = 1, \ldots, n$, is a necessary condition for the solvability of the interpolation problem. Simple examples show that range restricted interpolants $s \in S_2^{1,1}(\tilde{\Delta})$ need *not* exist for compatible data, even if the additional knots ξ^i are variable. Therefore, we restrict our attention to strictly compatible data, i.e.,

$$L_i < z_i < U_i, \quad i = 1, \ldots, n. \tag{12}$$

If the refined grid is fixed independently of the data, even strictly compatible data can be constructed such that *no* range restricted interpolant $s \in S_2^{1,1}(\tilde{\Delta})$ exists [8]. However, the existence can be always achieved by choosing the additional knots appropriately depending on the data.

While the parameters z_i of the interpolant s are prescribed by the interpolation conditions (10), the remaining parameters p_i and the ratios α^i are at our disposal to fulfill the range restrictions. Applying Proposition 1 to the differences $s - L, U - s \in S_2^{1,0}(\tilde{\Delta})$, the conditions

$$\mu^{n+i} L \leq \mu^{n+i} s \leq \mu^{n+i} U, \quad i = 1, \ldots, 2n-2, \tag{13}$$

are seen to be sufficient for (11), which leads to the following result.

Proposition 2 *Let compatible data $z_i, L_i, U_i, i = 1, \ldots, n$, be given. If the relations*

$$L_i + \tfrac{1}{2}\alpha^i(L_{i+1} - L_i) \leq z_i + \tfrac{1}{2}\alpha^i h^i p_i \leq U_i + \tfrac{1}{2}\alpha^i(U_{i+1} - U_i),$$
$$i = 1, \ldots, n-1,$$
$$L_i - \tfrac{1}{2}\beta^{i-1}(L_i - L_{i-1}) \leq z_i - \tfrac{1}{2}\beta^{i-1} h^{i-1} p_i \leq U_i - \tfrac{1}{2}\beta^{i-1}(U_i - U_{i-1}),$$
$$i = 2, \ldots, n, \tag{14}$$

are satisfied, then the interpolant $s \in S_2^{1,1}(\tilde{\Delta})$ fulfills the range restrictions (11).

The existence of range restricted interpolants depends on the choice of the additional knots, i.e., on the ratios α^i. A first proposal for the determination of ratios such that the inequality system (14) is solvable for the parameters p_i was given in [7]. Since this method is not local, it is not suitable for the extension to tensor product interpolation. Therefore, we present another proposal here (see also [14], [18]), on which our tensor product method is based.

Proposition 3 *Let strictly compatible data $L_i < z_i < U_i, i = 1, \ldots, n$, be given, and define $\tau^i := (z_{i+1} - z_i)/h^i, i = 1, \ldots, n-1$. If the ratios α^i are chosen according to*

$$\alpha^1 \in (0,1), \quad \alpha^i \in (0,1), \quad 1/\alpha^i \geq M_i, \quad i = 2, \ldots, n-1, \tag{15}$$

where

$$M_i := \frac{1}{2} \max \left\{ \frac{L_{i+1} - L_i - h^i \tau^{i-1}}{z_i - L_i}, \frac{h^i \tau^{i-1} - U_{i+1} + U_i}{U_i - z_i} \right\},$$
$$i = 2, \ldots, n-1, \tag{16}$$

then the inequality system (14) is solvable for $p_i, i = 1, \ldots, n$, i.e. a range restricted interpolant $s \in S_2^{1,1}(\tilde{\Delta})$ exists.

It is easily verified that the particular choice $p_1 := \tau^1$, $p_i := \tau^{i-1}$, $i = 2,\ldots,n$, is feasible in this case; but, in general, the corresponding interpolant can be visually improved. Therefore, we propose a two-stage procedure. In the first stage, appropriate ratios α^i, $i = 1,\ldots,n-1$, are chosen according to Proposition 3. In the second stage, the refined grid $\tilde{\Delta}$ is considered as fixed. The parameters p_i, $i = 1,\ldots,n$, are determined by minimizing a suitable quadratic fairness functional, such as a fit-and-modify functional or a (weighted) Holladay functional, subject to the linear inequality constraints (14), see [10], [11].

2.4 Quartic splines

We briefly describe the extension of the results presented above to interpolation by quartic C^2 splines with additional knots subject to polygonal range restrictions.

For fixed additional knots ξ^i, the linear space consisting of all piecewise quartic polynomials on the refined grid $\tilde{\Delta}$ that are continuous on $[x^1, x^n]$ and twice differentiable at ξ^i, $i = 1,\ldots,n-1$, is denoted by $S_4^{2,0}(\tilde{\Delta})$. It is the space of quartic splines with *quadruple* knots x^i and *double* knots ξ^i. The subspace $S_4^{2,2}(\tilde{\Delta}) := S_4^{2,0}(\tilde{\Delta}) \cap C^2[x^1, x^n]$ of quartic C^2-splines on the refined grid $\tilde{\Delta}$ is obtained by considering the knots x^i as *double* knots also. Each spline $s \in S_4^{2,2}(\tilde{\Delta})$ is uniquely represented by the parameters

$$z_i = s(x^i), \quad p_i = s'(x^i), \quad P_i = s''(x^i), \quad i = 1,\ldots,n,$$
$$\Pi_i = s''(\xi^i), \quad i = 1,\ldots,n-1, \tag{17}$$

see [11].

Sufficient nonnegativity conditions for splines $s \in S_4^{2,0}(\tilde{\Delta})$ can be derived in the same way as described above [11]. Since the bounds are assumed to be polygonal on Δ, it suffices to consider the subspace of $S_4^{2,0}(\tilde{\Delta})$ that is defined by the conditions $s''(x^i \pm 0) = 0$, $i = 1,\ldots,n$, and $s''(\xi^i) = 0$, $i = 1,\ldots,n-1$. It turns out that the sufficient nonnegativity conditions for splines from this subspace are actually given by the inequality system (14) again [14]. Consequently, choosing the ratios α^i according to Proposition 3 also assures the existence of interpolating splines $s \in S_4^{2,2}(\tilde{\Delta})$ with $P_i = 0$, $i = 1,\ldots,n$, and $\Pi_i = 0$, $i = 1,\ldots,n-1$, fulfilling the range restrictions. Of course, the described two-stage procedure is also recommended to obtain more pleasant interpolants, see [11] for details.

We point out that the following considerations concerning the range restricted interpolation by biquadratic splines also apply to biquartic tensor product splines.

3 Bivariate tensor product interpolation subject to piecewise bilinear range restrictions

3.1 Biquadratic tensor product splines on refined grids

Now we describe the construction of the tensor products of the univariate spline spaces and the corresponding linear functionals.

The grid $\Delta = \Delta_x$ is considered as a given grid in the x-direction. To reflect this, the corresponding notations are supplemented by the subscript x.

Analogously, let $\Delta_y : y^1 < y^2 < \ldots < y^m$ be a given grid in the y-direction, which is refined to $\tilde{\Delta}_y : y^1 < \eta^1 < y^2 < \ldots < y^{m-1} < \eta^{m-1} < y^m$ by the additional knots $\eta^j = \beta_y^j y^j + \alpha_y^j y^{j+1}$, $j = 1, \ldots, m-1$, where $\alpha_y^j > 0$, $\beta_y^j > 0$, $\alpha_y^j + \beta_y^j = 1$. The linear functionals λ_y^j and μ_y^j, $j = 1, \ldots, 3m-2$, on $S_2^{1,0}(\tilde{\Delta}_y)$ are defined in the same way as in (4). For the justification of the following tensor product constructions we refer to [3] and [5]. The tensor product space $S_2^{1,0}(\tilde{\Delta}_x) \otimes S_2^{1,0}(\tilde{\Delta}_y)$ is built up from all finite linear combinations of functions of the form $s_x \otimes s_y$, $s_x \in S_2^{1,0}(\tilde{\Delta}_x)$, $s_y \in S_2^{1,0}(\tilde{\Delta}_y)$, which are defined by

$$(s_x \otimes s_y)(x,y) = s_x(x) \cdot s_y(y), \quad (x,y) \in [x^1, x^n] \times [y^1, y^m]. \tag{18}$$

The space $S_2^{1,0}(\tilde{\Delta}_x) \otimes S_2^{1,0}(\tilde{\Delta}_y)$ consists of biquadratic splines on the refined rectangular grid $\tilde{\Delta}_x \times \tilde{\Delta}_y$, which divides each rectangle of the grid $\Delta_x \times \Delta_y$ into four subrectangles. The splines from the subspace $S_2^{1,1}(\tilde{\Delta}_x) \otimes S_2^{1,1}(\tilde{\Delta}_y)$, defined analogously, are continuous and possess continuous first partial derivatives and continuous mixed second partial derivatives. The tensor product functional $\lambda_x \otimes \lambda_y$ on $S_2^{1,0}(\tilde{\Delta}_x) \otimes S_2^{1,0}(\tilde{\Delta}_y)$ of two linear functionals λ_x on $S_2^{1,0}(\tilde{\Delta}_x)$ and λ_y on $S_2^{1,0}(\tilde{\Delta}_y)$ is defined by linear extension from the basic rule

$$(\lambda_x \otimes \lambda_y)(s_x \otimes s_y) = \lambda_x s_x \cdot \lambda_y s_y, \quad s_x \in S_2^{1,0}(\tilde{\Delta}_x), \; s_y \in S_2^{1,0}(\tilde{\Delta}_y). \tag{19}$$

It follows from a general result in [3] that each spline s from the $(2n)(2m)$-dimensional tensor product space $S_2^{1,1}(\tilde{\Delta}_x) \otimes S_2^{1,1}(\tilde{\Delta}_y)$ can be uniquely represented by the parameters

$$\begin{aligned}
z_{i,j} &= (\lambda_x^i \otimes \lambda_y^j)s &&= s(x^i, y^j), & i &= 1,\ldots,n, \; j = 1,\ldots,m, \\
p_{i,j} &= (\lambda_x^{n+i} \otimes \lambda_y^j)s &&= \frac{\partial s}{\partial x}(x^i, y^j), & i &= 1,\ldots,n, \; j = 1,\ldots,m, \\
q_{i,j} &= (\lambda_x^i \otimes \lambda_y^{m+j})s &&= \frac{\partial s}{\partial y}(x^i, y^j), & i &= 1,\ldots,n, \; j = 1,\ldots,m, \\
r_{i,j} &= (\lambda_x^{n+i} \otimes \lambda_y^{m+j})s &&= \frac{\partial^2 s}{\partial x \partial y}(x^i, y^j), & i &= 1,\ldots,n, \; j = 1,\ldots,m,
\end{aligned} \tag{20}$$

see also [9].

3.2 Sufficient nonnegativity conditions

As in the univariate case, we are interested in sufficient conditions for the nonnegativity of splines from $S_2^{1,0}(\tilde{\Delta}_x) \otimes S_2^{1,0}(\tilde{\Delta}_y)$. The following result is obtained using an adaption of a theorem presented in [9].

Proposition 4 *For* $s \in S_2^{1,0}(\tilde{\Delta}_x) \otimes S_2^{1,0}(\tilde{\Delta}_y)$, *the conditions*

$$(\mu_x^i \otimes \mu_y^j)s \geq 0, \quad i = 1,\ldots,3n-2, \; j = 1,\ldots,3m-2, \tag{21}$$

imply

$$s(x,y) \geq 0 \quad \text{for all} \quad (x,y) \in [x^1, x^n] \times [y^1, y^m]. \tag{22}$$

We omit the detailed formulation of the conditions (21) and provide just one example for the representation of the occurring tensor product functionals, namely

$$(\mu_x^{n+i} \otimes \mu_y^{2m+j-1})s = \left(\lambda_x^i + \frac{\alpha_x^i h_x^i}{2}\lambda_x^{n+i}\right) \otimes \left(\lambda_y^j - \frac{\beta_y^{j-1} h_y^{j-1}}{2}\lambda_y^{2m+j-1}\right)s$$

$$= s(x^i, y^j) + \frac{\alpha_x^i h_x^i}{2} \cdot \frac{\partial s}{\partial x}(x^i + 0, y^j) \qquad (23)$$

$$- \frac{\beta_y^{j-1} h_y^{j-1}}{2} \cdot \frac{\partial s}{\partial y}(x^i, y^j - 0) - \frac{\alpha_x^i \beta_y^{j-1} h_x^i h_y^{j-1}}{4} \cdot \frac{\partial^2 s}{\partial x \partial y}(x^i + 0, y^j - 0),$$

where $i \in \{1, \ldots, n-1\}$, $j \in \{2, \ldots, m-1\}$ is assumed.

Again, the inequality system (21) is separated with respect to the original grid points (x^i, y^j). Nine conditions correspond to each interior grid point.

3.3 Existence and construction of range restricted interpolants

Now we are ready to consider the bivariate range restricted interpolation problem. For given data values $z_{i,j}$, $i = 1, \ldots, n$, $j = 1, \ldots, m$, and lower and upper bounds L, U we consider the problem:

Find $S_2^{1,1}(\tilde{\Delta}_x) \otimes S_2^{1,1}(\tilde{\Delta}_y)$ such that the interpolation conditions

$$s(x^i, y^j) = z_{i,j}, \quad i = 1, \ldots, n, \ j = 1, \ldots, m, \qquad (24)$$

and the range restrictions

$$L(x,y) \leq s(x,y) \leq U(x,y) \quad \text{for all} \quad (x,y) \in [x^1, x^n] \times [y^1, y^m] \qquad (25)$$

are satisfied.

The lower and upper bounds L, U are assumed to be continuous and piecewise bilinear with respect to $\Delta_x \times \Delta_y$, and are completely described by their values $L_{i,j} := L(x^i, y^j) = (\mu_x^i \otimes \mu_y^j)L$, $U_{i,j} := U(x^i, y^j) = (\mu_x^i \otimes \mu_y^j)U$, $i = 1, \ldots, n, \ j = 1, \ldots, m$.

For the same reasons as in the univariate case, we assume the given data to be strictly compatible, i.e.

$$L_{i,j} < z_{i,j} < U_{i,j}, \quad i = 1, \ldots, n, \ j = 1, \ldots, m. \qquad (26)$$

While the parameters $z_{i,j}$ of the interpolating tensor product spline s are determined by the interpolation conditions (24), the remaining parameters $p_{i,j}$, $q_{i,j}$, $r_{i,j}$ and the ratios α_x^i, α_y^j are at our disposal to fulfill the range restriction (25).

The application of Proposition 4 leads to the following sufficient conditions.

Proposition 5 *Let compatible data $z_{i,j}$, $L_{i,j}$, $U_{i,j}$, $i = 1, \ldots, n$, $j = 1, \ldots, m$, be given. If the relations*

$$(\mu_x^{n+i} \otimes \mu_y^j)L \leq (\mu_x^{n+i} \otimes \mu_y^j)s \leq (\mu_x^{n+i} \otimes \mu_y^j)U,$$
$$i = 1, \ldots, 2n-2, \ j = 1, \ldots, 3m-2,$$
$$(\mu_x^i \otimes \mu_y^{m+j})L \leq (\mu_x^i \otimes \mu_y^{m+j})s \leq (\mu_x^i \otimes \mu_y^{m+j})U, \qquad (27)$$
$$i = 1, \ldots, n, \ j = 1, \ldots, 2m-2,$$

are satisfied, then the interpolant $s \in S_2^{1,1}(\tilde{\Delta}_x) \otimes S_2^{1,1}(\tilde{\Delta}_y)$ fulfills the range restrictions (25).

A detailed description of the conditions (27) analogously to Proposition 2 is given in [18]. Based on Proposition 3, the following result concerning the choice of suitable ratios α_x^i, α_y^j is proved in detail also in [18].

Proposition 6 *Let strictly compatible data $L_{i,j} < z_{i,j} < U_{i,j}$, $i = 1, \ldots, n$, $j = 1, \ldots, m$, be given, and define*

$$\tau_x^{i,j} := (z_{i+1,j} - z_{i,j})/h_x^i, \qquad i = 1, \ldots, n-1, \ j = 1, \ldots, m,$$

$$\tau_y^{i,j} := (z_{i,j+1} - z_{i,j})/h_y^j, \qquad i = 1, \ldots, n, \ j = 1, \ldots, m-1, \qquad (28)$$

$$\tau^{i,j} := \frac{z_{i+1,j+1} - z_{i+1,j} - z_{i,j+1} + z_{i,j}}{h_x^i h_y^j}, \qquad i = 1, \ldots, n-1, \ j = 1, \ldots, m-1.$$

If the ratios α_x^i, α_y^j are chosen according to

$$\begin{aligned}
&\alpha_x^1 \in (0,1), \quad \alpha_x^i \in (0,1), \quad 1/\alpha_x^i \geq M_x^i/\epsilon, \quad i = 2, \ldots, n-1, \\
&\alpha_y^1 \in (0,1), \quad \alpha_y^j \in (0,1), \quad 1/\alpha_y^j \geq M_y^j/\epsilon, \quad j = 2, \ldots, m-1, \\
&1/(\alpha_x^i \alpha_y^j) \geq M^{i,j}/\delta, \quad i = 2, \ldots, n-1, \ j = 2, \ldots, m-1,
\end{aligned} \qquad (29)$$

where $\epsilon > 0$, $\delta > 0$, $2\epsilon + \delta = 1$,

$$\begin{aligned}
M_x^i &:= \frac{1}{2} \max_{j=1,\ldots,m} \max \left\{ \frac{L_{i+1,j} - L_{i,j} - h_x^i \tau_x^{i-1,j}}{z_{i,j} - L_{i,j}}, \frac{h_x^i \tau_x^{i-1,j} - U_{i+1,j} + U_{i,j}}{U_{i,j} - z_{i,j}} \right\}, \\
&\quad i = 2, \ldots, n-1, \\
M_y^j &:= \frac{1}{2} \max_{i=1,\ldots,n} \max \left\{ \frac{L_{i,j+1} - L_{i,j} - h_y^j \tau_y^{i,j-1}}{z_{i,j} - L_{i,j}}, \frac{h_y^j \tau_y^{i,j-1} - U_{i,j+1} + U_{i,j}}{U_{i,j} - z_{i,j}} \right\}, \\
&\quad j = 2, \ldots, m-1,
\end{aligned} \qquad (30)$$

and

$$\begin{aligned}
M^{i,j} &:= \frac{1}{4} \max \left\{ 0, \frac{L_{i+1,j+1} - L_{i+1,j} - L_{i,j+1} + L_{i,j} - h_x^i h_y^j \tau^{i-1,j-1}}{z_{i,j} - L_{i,j}}, \right. \\
&\qquad \left. \frac{h_x^i h_y^j \tau^{i-1,j-1} - U_{i+1,j+1} + U_{i+1,j} + U_{i,j+1} - U_{i,j}}{U_{i,j} - z_{i,j}} \right\}, \\
&\quad i = 2, \ldots, n-1, \ j = 2, \ldots, m-1,
\end{aligned} \qquad (31)$$

then the inequality system (27) is solvable for the parameters $p_{i,j}$, $q_{i,j}$, $r_{i,j}$, $i = 1, \ldots, n$, $j = 1, \ldots, m$, i.e., a range restricted interpolant $s \in S_2^{1,1}(\tilde{\Delta}_x) \otimes S_2^{1,1}(\tilde{\Delta}_y)$ exists.

Let us point out that the above conditions on the ratios α_x^i, α_y^j are also sufficient for the existence of range restricted interpolants from $S_4^{2,2}(\tilde{\Delta}_x) \otimes S_4^{2,2}(\tilde{\Delta}_y)$.

Also in the bivariate case we propose a two-stage procedure to compute a range restricted interpolant. In the first stage, suitable ratios are determined, e.g., by the algorithm

$$\alpha_x^1 := 1/2, \quad \alpha_y^1 := 1/2,$$

$$\frac{1}{\alpha_x^i} := \max\left\{2, \frac{M_x^i}{\epsilon}, \max_{j=2,\ldots,m-1} \sqrt{\frac{M^{i,j}}{\delta}}\right\}, \quad i = 2, \ldots, n-1, \qquad (32)$$

$$\frac{1}{\alpha_y^j} := \max\left\{2, \frac{M_y^j}{\epsilon}, \max_{i=2,\ldots,n-1} \sqrt{\frac{M^{i,j}}{\delta}}\right\}, \quad j = 2, \ldots, m-1.$$

The weights ϵ, δ have to be prescribed; we used $\epsilon = 0.4$, $\delta = 0.2$ in the examples presented below.

In the second stage, the refined grid $\tilde\Delta_x \times \tilde\Delta_y$ is considered as fixed. To obtain a visually improved solution we propose a fit-and-modify approach, which utilizes the separation of the conditions (27), see [9]. Initial estimates $p_{i,j}^0, q_{i,j}^0, r_{i,j}^0$ for the parameters, which define an initial interpolant s^0, are computed as usual approximations by divided differences, see [2], [18] for details. These estimates have to be modified as little as possible in order to fulfill the sufficient conditions for the range restrictions. This can be achieved by minimizing the functional [10]

$$\sum_{i\in\{1,n,n+1,\ldots,3n-2\}} \sum_{j\in\{1,m,m+1,\ldots,3m-2\}} \left[(\mu_x^i \otimes \mu_y^j)(s-s^0)\right]^2 \qquad (33)$$

subject to the constraints (27).

This problem reduces to $n \cdot m$ independent quadratic programs involving three variables only, which can be efficiently solved by standard methods of quadratic optimization.

Another proposal is to minimize the bivariate Holladay functional (see [5])

$$\int_{x^1}^{x^n} \int_{y^1}^{y^m} \left[\frac{\partial^4 s}{\partial^2 x \partial^2 y}(x,y)\right]^2 dxdy$$
$$+ \sigma \sum_{j=1}^{m} \int_{x^1}^{x^n} \left[\frac{\partial^2 s}{\partial x^2}(x,y^j)\right]^2 dx + \omega \sum_{i=1}^{n} \int_{y^1}^{y^m} \left[\frac{\partial^2 s}{\partial y^2}(x^i,y)\right]^2 dy \qquad (34)$$

subject to the linear inequality constraints (27).

The resulting quadratic optimization problem does not separate using this functional, and the special structure of the functional and the constraints should be utilized in a solution method, compare [7].

4 Numerical examples

The proposed interpolation methods were implemented in $MATLAB^{TM}$ using the Spline Toolbox by C. de Boor and the Optimization Toolbox by A. Grace. The programs have been written for testing and comparison purposes. In particular, the quadratic programs arising in the minimization of the bivariate Holladay functional are solved by standard methods.

Range Restricted Interpolation by Tensor Product Splines 211

In the example chosen for graphical presentation, the data are given at an equidistant rectangular grid on $[0,1] \times [0,1]$, where $n = m = 5$. The data values stem from a bivariate function similar to one of the functions proposed in [17], see also [10]. The given data are visualized in Fig. 1.

The lower and upper bounds L, U are prescribed by

$$L_{i,j} = 0, \quad U_{i,j} = z_{i,j} + 0.2, \quad i = 1, \ldots, 5, \ j = 1, \ldots, 5. \tag{35}$$

The application of algorithm (32) resulted in the ratios

$$\alpha_x^1 = \alpha_y^1 = 0.5, \quad \alpha_x^2 = \alpha_y^2 = 0.18, \quad \alpha_x^3 = \alpha_y^3 = 0.09, \quad \alpha_x^4 = \alpha_y^4 = 0.13. \tag{36}$$

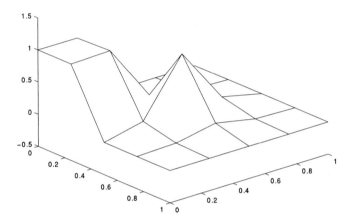

Fig. 1 Bilinear interpolant of the given data.

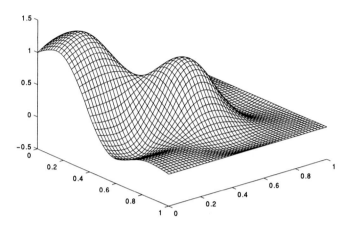

Fig. 2 Unconstrained interpolant minimizing the bivariate Holladay functional, ratios $\alpha_x^i = \alpha_y^j = 1/2$.

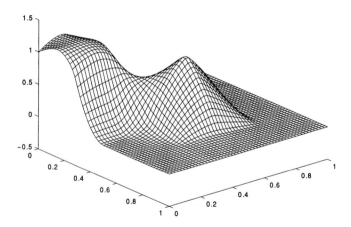

Fig. 3 Range restricted interpolant minimizing the bivariate Holladay functional, $\sigma = \omega = 10^6$, ratios given by (36).

For comparison, the unconstrained interpolant with all ratios $\alpha_x^i = \alpha_y^j = 1/2$ and the range restricted interpolant minimizing the bivariate Holladay functional are shown in Fig. 2 and Fig. 3. While the weights σ, ω do not influence the unconstrained interpolant, see [5], the weights $\sigma = \omega = 10^6$ for computing the range restricted interpolant were chosen experimentally.

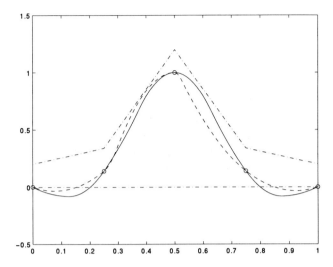

Fig. 4 Unconstrained interpolants, parallel cut.

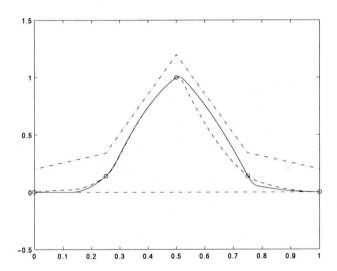

Fig. 5 Range restricted interpolants, parallel cut.

In the remaining figures, we present the cuts of certain unconstrained and range restricted interpolants along the line $y = 1/2$ parallel to the x-axis and along the diagonal $y = x$.

Here, the drawn curves have the following meanings:

solid: interpolant minimizing the bivariate Holladay functional, $\sigma = \omega = 10^6$,
dotted: interpolant minimizing the bivariate Holladay functional, $\sigma = \omega = 10^4$,
dashed: fit-and-modify interpolant,

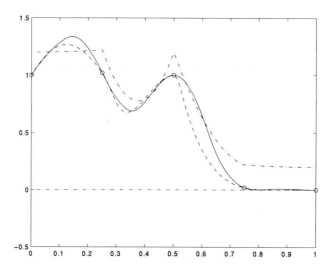

Fig. 6 Unconstrained interpolants, diagonal cut.

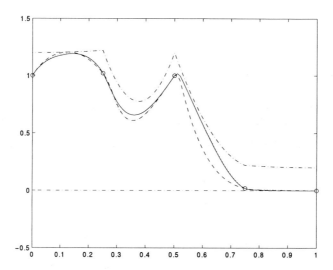

Fig. 7 Range restricted interpolants, diagonal cuts.

dashdotted: prescribed lower and upper bounds.
The given data points located in the cut are denoted by small circles.

In our opinion, the fit-and-modify method can be recommended in view of its simplicity and localness due to the separation of the optimization problems. In most cases, the resulting interpolants compete successfully with the interpolants obtained by minimizing the bivariate Holladay functional.

References

[1] Andersson, L.-E., Elfving, T.: *Best constrained approximation in Hilbert space and interpolation by cubic splines subject to obstacles.* SIAM Journal on Scientific Computing **16** (1995), 1209–1232.

[2] Asaturyan, S., Unsworth, K.: *A C^1 monotonicity preserving surface interpolation scheme.* In D. C. Handscomb (ed.): Mathematics of Surfaces III, Oxford University Press, Oxford (1989), 243–266.

[3] de Boor, C.: *A Practical Guide to Splines.* Springer 1978.

[4] Dontchev, A.L.: *Best interpolation in a strip.* Journal of Approximation Theory **73** (1993), 334–342.

[5] Ewald, S., Mulansky, B., Mühlig, H.: *Bivariate interpolating and smoothing tensor product splines.* In J. W. Schmidt and H. Späth (eds.): Splines in Numerical Analysis, Akademie-Verlag, Berlin (1989), 55–68.

[6] Fontanella, F.: *Shape preserving interpolation.* In W. Dahmen, M. Gasca, and C. A. Micchelli (eds.): Computation of Curves and Surfaces, Kluwer Academic Publ. (1990), 183–214.

[7] Herrmann, M., Mulansky, B., Schmidt, J. W.: *Scattered data interpolation subject to piecewise quadratic range restrictions.* Journal of Computational and Applied Mathematics (to appear).

[8] Mulansky, B., Neamtu, M.: *Interpolation and approximation from convex sets.* Preprint MATH-NM-02-1996, TU Dresden 1996.

[9] Mulansky, B., Schmidt, J. W.: *Nonnegative interpolation by biquadratic splines on refined rectangular grids.* In P.-J. Laurent, A. Le Méhauté, and L. L. Schumaker (eds.): Wavelets, Images and Surface Fitting, A K Peters, Wellesley (1994), 379–386.

[10] Mulansky, B., Schmidt, J. W.: *Powell-Sabin splines in range restricted interpolation of scattered data.* Computing **53** (1994), 137–154.

[11] Mulansky, B., Schmidt, J. W.: *Constructive methods in convex C^2 interpolation using quartic splines.* Numerical Algorithms **12** (1996), 111–124.

[12] Ong, B. H., Unsworth, K.: *On non-parametric constrained interpolation.* In T. Lyche and L. L. Schumaker (eds.): Mathematical Methods in Computer Aided Geometric Design II, Academic Press (1992), 419–430.

[13] Schmidt, J. W.: *Positive, monotone, and S-convex C^1-interpolation on rectangular grids.* Computing **48** (1992), 363–371.

[14] Schmidt, J. W.: *Strip interpolation using splines on refined grids.* Preprint MATH-NM-05-1996, TU Dresden 1996.

[15] Schmidt, J. W., Heß, W.: *Spline interpolation under two-sided restrictions on the derivatives.* Zeitschrift für Angewandte Mathematik und Mechanik **69** (1989), 353–365.

[16] Schumaker, L. L.: *On shape preserving quadratic spline interpolation.* SIAM Journal on Numerical Analysis **20** (1983), 854–864.

[17] Utreras, F. I.: *Positive thin plate splines*. Approximation Theory and its Applications **1** (1985), 77–108.

[18] Walther, M.: *Restringierte Interpolation mit bivariaten Splines*. Diplomarbeit, TU Dresden 1996.

Authors

Dr. Bernd Mulansky   mulansky@math.tu-dresden.de
Prof. Dr. Jochen W. Schmidt   jschmidt@math.tu-dresden.de
Dr. Marion Walther   walther@math.tu-dresden.de

Institut für Numerische Mathematik
Technische Universität Dresden
D–01062 Dresden
Germany

Construction of Surfaces by Shape Preserving Approximation of Contour Data

Bert Jüttler

University of Dundee

Abstract: We present a two–stage approach for the construction of surfaces from contour data. At first the contour curves are found using an algorithm for shape preserving least–square approximation of planar data by polynomial parametric spline curves. The obtained curves are then interpolated by tensor–product B–spline surfaces. The interpolation scheme used in the second step preserves the signs of the sectional curvature of the contours.

1 Introduction

Methods for the construction of surfaces from contour data are required in several applications of Computer Aided Geometric Design. These applications include the design of ship hulls, or also the reconstruction of bone surfaces in medical imaging. References to related literature can be found in the survey articles [13, 14]. As one of the possible approaches, such methods can be composed from two parts. At first, the contour curves of the desired surface are generated from the data. The contours are then interpolated in order to obtain the desired surface. We are particularly interested in methods which preserve the shape of the data as far as possible.

An algorithm for shape preserving least–square approximation of planar data by polynomial parametric spline curves is discussed in the first part of the present paper. It can be used for constructing the contour curves of the surface from the given data. Whereas shape preserving interpolation by parametric curves has been discussed in a number of papers (see the references cited in [9, pp. 103, 105, 114], see [7] for space curves), the corresponding approximation problem has not been considered so far. Even in the case of functional data, only the method proposed by Dierckx [2, 3] seems to be available. In Dierckx' method least–square approximation by cubic spline functions is formulated as a quadratic programming problem. Our approach generalizes this idea to the case of polynomial parametric spline curves. The desired shape of the approximating spline curve is specified with the help of a reference curve which is used in order to generate linear sufficient shape constraints. The control points of the approximating curve result by solving a quadratic programming problem.

After the contour curves have been obtained, the desired surface is to be found by interpolating them in the second step. In order to preserve the shape of the given data we use the notion of sectional curvature preserving interpolation which has been introduced by Kaklis and Ginnis [12]. Their approach to achieve this property is based on piecewise polynomial surfaces of non–uniform degree, whereby the degrees of the segments act as tension parameters. In contrast with this, we guarantee the desired shape with the help of appropriate linear constraints. The control points of the interpolating surface are then found by solving a simple optimization problem.

2 Approximation of planar data

In this section we describe a method for approximating a sequence of points by a B–spline curve satisfying some shape constraints. More precisely, the desired inflection points and the curvature signs of the spline segments in between can be specified. In order to be brief, we give an outline of the basic ideas of our approach only. For more details of the approximation scheme the reader is referred to [10].

2.1 Problem statement

Let a sequence of $P+1$ points $(\mathbf{p}_i)_{i=0,\ldots,P}$ in the plane $\mathbb{R}^2$ with associated parameter values $(t_i)_{i=0,\ldots,P}$ satisfying $t_i < t_{i+1}$ be given. If the parameters t_i are unknown yet, then they can be estimated from the data, cf. [9, pp. 201/202]. Additionally, the $K+1$ knots $(\tau_i)_{i=0,\ldots,K}$ of the approximating spline curve are assumed to be known ($\tau_i < \tau_{i+1}$, $\tau_0 = t_0$, $\tau_K = t_P$). Some of the knots are marked as the desired inflections of the spline curve. They are denoted by $(\tau_{w(i)})_{i=1,\ldots,W-1}$ where $w(i) < w(i+1)$. Additionally we set $w(0) = 0$ and $w(W) = K$. Finally, the curvature signs $\sigma_i \in \{+1, -1\}$ of the spline segment $t \in [\tau_{w(i)}, \tau_{w(i+1)}]$ have to be specified. Of course, the curvature signs of adjacent segments should be different: $\sigma_i \cdot \sigma_{i+1} = -1$.

The given data is to be approximated by a C^l B–spline curve ($l = 1, 2$) of degree d ($d > l$),

$$\mathbf{x}(t) = \sum_{j=0}^{D} \mathbf{d}_j \cdot N_j^d(t) \qquad t \in [\tau_0, \tau_K] \tag{1}$$

($D = d \cdot k - (k-1) \cdot l$) with the unknown control points $(\mathbf{d}_j)_{j=0,\ldots,D}$ in $\mathbb{R}^2$. The B–spline basis functions $(N_j^d(t))_{j=0,\ldots,D}$ are defined over the knot vector

$$(\underbrace{\tau_0, \ldots, \tau_0}_{d+1 \text{ times}}, \underbrace{\tau_1, \ldots, \tau_1}_{d-l \text{ times}}, \underbrace{\tau_2, \ldots, \tau_2}_{d-l \text{ times}}, \cdots \underbrace{\tau_{K-1}, \ldots, \tau_{K-1}}_{d-l \text{ times}}, \underbrace{\tau_K, \ldots, \tau_K}_{d+1 \text{ times}}). \tag{2}$$

For more information concerning B–spline curves we refer the reader to the textbooks [4, 9].

The construction described below will ensure that the approximating spline curve (1) possesses inflection points only for $t = \tau_{w(i)}$ ($i = 1, \ldots, W-1$). The curvature sign of the spline segments $t \in [\tau_{w(i)}, \tau_{w(i+1)}]$ in between will be equal to the specified value σ_i ($i = 1, \ldots, W$).

2.2 The method

The construction of the spline curve consists of three steps. At first, we construct a reference curve which possesses the specified inflection points and curvature signs. With the help of this curve we then generate an appropriate set of linear constraints ensuring the desired shape properties. Finally, the control points of the approximating curve are found by solving a constrained optimization problem.

(i) Construction of the reference curve

The reference curve $\mathbf{y}(t)$ is chosen as a C^l B–spline curve of degree $l+1$ which possesses the desired shape. Two possible construction of the reference curve are described in [10].

For the first construction we prescribe the direction vectors of the legs of the B–spline control polygon, whereas the lengths are found by minimizing the least–square sum. The directions of the legs which correspond to spline segments having constant curvature signs result from uniform rotations, This ensures the desired shape properties.

For the second construction we use a B–spline curve which is defined over an appropriate subset of the knot sequence $(\tau_i)_{i=0,\ldots,K}$. The curve is found by least–square approximation of the data. Its knots are chosen such that no undesired inflections occur. This is achieved by systematically deleting knots from the original knot sequence.

Whereas the first construction is guaranteed to yield a curve which has the desired shape properties, the second construction only ensures that no undesired inflections occur. But in most cases of practical interest the result of the second construction can be expected to have also the desired curvature signs. The second construction generally yields a better approximation of the data than the first one, and its result is more suitable for the following constructions.

In the case of C^2 spline curves ($l=2$), the construction of the reference curve additionally ensures that $\ddot{\mathbf{x}}(\tau_{w(i)}) = \vec{\mathbf{0}}$ holds for all inflection knots, $i = 1, \ldots, W-1$. This is necessary for the generation of the shape constraints in the second step.

In order to find a representation which is defined over the specified knot vector (2), the degree of the reference curve is elevated from $l+1$ to d.

(ii) Generation of linear shape constraints

The generation of the linear constraints ensuring the desired curvature distribution is based on the following observation:

Lemma. *Consider a Bézier curve* $\mathbf{y}(t)$ *of degree d with control points* $(\mathbf{b}_i)_{i=0,\ldots,d}$ *in* $\mathbf{R}^2$ *(see [9, p. 120]). If two vectors* $\mathbf{u}, \mathbf{v} \in \mathbf{R}^2 \setminus \{\vec{\mathbf{0}}\}$ *exist such that the $4d-2$ inequalities* [4]

$$\begin{array}{ll} [\vec{\mathbf{u}}, \Delta^1 \mathbf{b}_i] \geq 0 & [\Delta^1 \mathbf{b}_i, \vec{\mathbf{v}}] \geq 0 \qquad (i=0,\ldots,d-1) \\ [\vec{\mathbf{v}}, \Delta^2 \mathbf{b}_i] \geq 0 \ (\text{resp.} \leq 0) & [\Delta^2 \mathbf{b}_i, -\vec{\mathbf{u}}] \geq 0 \ (\text{resp.} \leq 0) \quad (i=0,\ldots,d-2) \end{array} \quad (3)$$

for the first and second differences vectors $\Delta^1 \mathbf{b}_i = \mathbf{b}_{i+1} - \mathbf{b}_i$ *and* $\Delta^2 \mathbf{b}_i = \Delta^1 \mathbf{b}_{i+1} - \Delta^1 \mathbf{b}_i$ *of the control points hold, then the curve segment $t \in [0,1]$ is convex and the curvature is non–negative (resp. non–positive).*

The conditions of the Lemma ensure that the wedges spanned by the first and second difference vectors of the control points can be separated as shown in Figure 1. The first and second derivatives of the curve are contained in these wedges, hence the curve has non–negative (resp. non–positive) curvature.

As shown in [10], any Bézier curve with non–positive (resp. non–negative) curvature can be subdivided into a finite number of segments such that the convexity of the segments can be

---
[4] The abbreviation $[.,.]$ denotes the exterior product $[\vec{\mathbf{s}}, \vec{\mathbf{t}}] = s_1 \cdot t_2 - s_2 \cdot t_1$ of two vectors $\vec{\mathbf{s}}, \vec{\mathbf{t}} \in \mathbf{R}^2$.

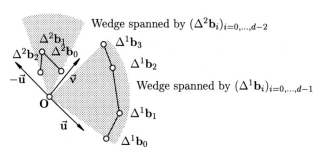

Fig. 1 Convexity conditions for a quartic Bézier curve.

guaranteed with the help of the above linear inequalities. This also applies to curves with vanishing curvature at the segment end points $t_0 = 0$ or 1, provided that the second derivative at the segment end points vanishes, $\ddot{\mathbf{y}}(t_0) = \vec{\mathbf{0}}$, and $[\dot{\mathbf{y}}(t_0), \dddot{\mathbf{y}}(t_0)] > 0$ (resp. < 0) holds. This is important for the generation of convexity constraints in the presence of desired inflection points.

Based on these observations, an algorithm has been formulated in [10] which generates a set of linear sufficient convexity constraints with the help of the reference curve. It yields two (possibly empty) sets of inequalities $\mathcal{I}$ and equalities $\mathcal{E}$ for the components of the unknown control points $(\mathbf{d}_i)_{i=0,\ldots,D}$. In order to generate these constraints, the reference curve is subdivided into an appropriate number of subsegments, and then the vectors $\vec{\mathbf{u}}$ and $\vec{\mathbf{v}}$ occurring in (3) are computed from the bounds of the wedges spanned by the first and second differences of the resulting control points.

(*iii*) **Computing the curve**

The unknown control points of the approximating B–spline curve (1) are found by minimizing the least–square sum

$$F(\mathbf{d}_0,\ldots,\mathbf{d}_D) = \sum_{i=0}^{P} \|\mathbf{x}(t_i) - \mathbf{p}_i\|^2 \qquad (4)$$

subject to the linear shape constraints $\mathcal{I}$ and $\mathcal{E}$. This quadratic programming problem (see [1]) is solved using the method of active set as described in [5]. The control points of the reference curve serve as initial point for the optimization.

As most algorithms for solving quadratic programming problems have difficulties with degenerate situations, it is of crucial importance to avoid redundancies of the sets of constraints as far as possible, see [10].

In most applications, the number of data points can be expected to be much bigger than the number D of B–spline control points, and also the distribution of the parameters t_i over the spline segments will be more or less uniform. Then the quadratic part of the function (4) is positively definite, and therefore the solution of the quadratic programming problem will be unique. Otherwise it is possible to add a "tension term" (cf. [9, p. 98] in order to achieve regularity.

Construction of Surfaces by Shape Preserving Approximation of Contour Data   221

If the approximating curve obtained from the quadratic programming is not satisfying, then the whole procedure can be iterated. New linear shape constraints can be generated by using the first curve as the new reference curve. This should yield constraints which are more suitable for the approximation of the given data.

2.3 Examples

In this section we will illustrate the capabilities of our method by an example. We computed two sets each consisting of 8 cubic C^2 B-spline curves (defined over appropriate knot vectors) which approximate the 296 data points (marked by crosses) of Figure 2. The curves in Figure

(a)

(b)

Fig. 2 Shape-preserving least-square approximation. The approximating curves have been computed (a) without constraints, (b) with shape constraints.

2a have been computed without any shape constraints. In contrast with this, the curves in Figure 2b have been obtained after two iterations of the above-described method. In order to analyze the shape of the curves, the normals of the spline curves have been drawn (thin black lines), whereby their magnitudes have been chosen proportional to the curvature at the curve points.

3 Interpolation of the contour curves

Now we outline a method for constructing a surface by sectional curvature-preserving interpolation of a given set of contour curves. A more detailed description of the scheme will be presented in [11].

3.1 Problem statement

Let a sequence of $C+1$ open contours $(\mathbf{x}_i(t))_{i=0,\ldots,C}$ in $\mathbb{R}^2$ with associated z–coordinates (heights) $z_0 < z_1 < \ldots < z_C$ be given. Each contour is described by a C^l B–spline curve ($l=1,2$)

$$\mathbf{x}_i(t) = \sum_{j=0}^{D_i} \mathbf{d}_{i,j} \cdot N_{i,j}^d(t) \qquad t \in [0,1], \quad i = 0,\ldots,C, \tag{5}$$

of degree d with $D_i + 1$ control points $(\mathbf{d}_{i,j})_{j=0,\ldots,D_i}$ in $\mathbb{R}^2$. The B–spline basis functions $(N_{i,j}^d(t))_{j=0,\ldots,D_i}$ of the i-th contour are defined over a certain knot vector $\mathcal{T}_i$. The knot vectors may be different, but they are assumed to have the $(d+1)$-fold boundary knots 0 and 1. The construction described below is based on the following three assumptions.

(1) The *matching problem* (cf. [14]) has already been solved; points on adjacent contours $\mathbf{x}_i(t)$, $\mathbf{x}_{i+1}(t)$ which share the parameter value t correspond to each other.

(2) If a contour curves possesses inflections, then the parameters of the inflection points are also knots of the spline curve. Moreover, in the C^2–case ($l=2$) we have $\ddot{\mathbf{x}}_i(t_{\text{infl}}) = \vec{0}$ at these points. (Note that this assumption is automatically fulfilled if the contour curves have been generated using the algorithm described in the first part of the paper.) No higher order flat points of the contour curves are assumed to exist.

(3) The shape of the contour curves corresponds with the shape of the control polygons. More precisely, if the curvature of a segment $t \in [t_a, t_b]$ of one of the contour curves $\mathbf{x}_i(t)$ has only non–negative (non–positive) values, then also the polygon formed by all control points acting on this segment has only non–negative (non–positive) angles between adjacent difference vectors of control points. Additionally, these angles are smaller [5] than $\pi/(d-1)$. The validity of this assumption can always be achieved by inserting additional knots.

We want to construct a C^l spline surface $\mathbf{y}(z,t)$ with $(t,z) \in [0,1] \times [z_0, z_C]$ which interpolates the given contour curves, i.e.,

$$\mathbf{y}(z_i, t) \equiv \mathbf{x}_i(t) \qquad \text{holds for} \qquad i = 0,\ldots,C. \tag{6}$$

The interpolating surface is to preserve the shape of the given contour curves. Consider two segments $t \in [t_a, t_b] \subseteq [0,1]$ of adjacent contours $\mathbf{y}(z_i, t) = \mathbf{x}_i(t)$ and $\mathbf{y}(z_{i+1}, t) = \mathbf{x}_{i+1}(t)$. If the curvature of both segments has only non–negative (resp. non–positive) values, then also the curvature of any interpolating contour segment $\mathbf{y}(z^*, t)$ in between ($z^* \in [z_i, z_{i+1}]$ is to be non–negative (resp. non–positive). If this assertion is true for all corresponding segments of adjacent contours, then the interpolating surface is said to preserve the sectional curvature of the data. This notion has been introduced by Kaklis and Ginnis [12].

---

[5] Under this assumption, the number of inflections of the curve is bounded by the number of inflections of the control polygon, see [6]. Of course, this also applies to any segment of the curves.

3.2 Surface definition and continuity constraints

The interpolating surface is composed from tensor–product B–spline surfaces of degree (d, n). Whereas d denotes the degree of the contour curves, the degree of the parameter lines $t = \text{const}$ is equal to $n \geq 2 \cdot l + 2$. For $z \in [z_{i-1}, z_i]$, the interpolating surface is described by

$$\mathbf{y}_i(z, t) = \sum_{j=0}^{\hat{D}_i} \hat{\mathbf{d}}_{i,j}(z) \cdot \hat{N}_{i,j}^d(t), \qquad (z, t) \in [z_{i-1}, z_i] \times [0, 1], \qquad i = 1, \ldots, C. \qquad (7)$$

The B–spline basis functions $(\hat{N}_{i,j}^d(t))_{j=0,\ldots,\hat{D}_i}$ are defined over the union of the knot vectors $\mathcal{T}_{i-1}$ and $\mathcal{T}_I$, see Figure 3. The $\hat{D}_i + 1$ control points of $(\hat{\mathbf{d}}_{i,j}(z))_{j=0,\ldots,\hat{D}_i}$ of the interpolating

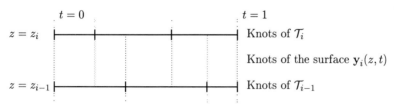

Fig. 3 The knots of interpolating surface.

contours depend on the height z. They run along Bezier curves of degree n with parameter z,

$$\hat{\mathbf{d}}_{i,j}(z) = \sum_{k=0}^{n} \mathbf{c}_{i,j,k} \cdot B_k^n\left(\frac{z - z_{i-1}}{z_i - z_{i-1}}\right) \qquad z \in [z_{i-1}, z_i],\; j = 0, \ldots, \hat{D}_i,\; i = 1, \ldots, C. \qquad (8)$$

Their control points $\mathbf{c}_{i,j,k} \in \mathbb{R}^2$ are unknown yet, whereas $B_i^n(u) = \binom{n}{i} u^i (1-u)^{n-i}$ are the Bernstein polynomials.

From the interpolation conditions (6) we immediately obtain the boundary control points $(\mathbf{c}_{i,j,0})_{j=0,\ldots,\hat{D}_i}$ and $(\mathbf{c}_{i,j,n})_{j=0,\ldots,\hat{D}_i}$ of the surface defined by (7) and (8) with the help of the knot insertion algorithm, cf. [9]. In addition, the surface has to fulfill the continuity constraints

$$\left(\frac{\partial}{\partial z}\right)^\lambda \mathbf{y}_i(z, t) \bigg|_{z=z_i} \equiv \left(\frac{\partial}{\partial z}\right)^\lambda \mathbf{y}_{i+1}(z, t) \bigg|_{z=z_i} \qquad \lambda = 1, \ldots, l,\; t \in [0, 1], \qquad (9)$$

$i = 1, \ldots, C - 1$. The continuity constraints leads to a set of linear equations involving the control points $(\mathbf{c}_{i,j,r})_{j=0,\ldots,\hat{D}_i}$ and $(\mathbf{c}_{i+1,j,n-r})_{j=0,\ldots,\hat{D}_{i+1}}$, $r = 0, \ldots, l$. Note that the knot vectors of adjacent spline surfaces $\mathbf{y}_i(z, t)$ may be different. Thus, the continuity constraints include certain not–a–knot–type conditions for the control points $\mathbf{c}_{i,j,k}$. Resulting from $n \geq 2 \cdot l + 2$, each coefficient $\mathbf{c}_{i,j,k}$ is subject to at most one set of constraints (9).

3.3 The method

The construction of the interpolating surface consists of three steps. At first we generate sufficient linear constraints for preserving the sectional curvature of the contour curves. Then we choose an appropriate quadratic objective function. In the last step we find an initial point for the optimization and construct the control points of the surface by solving a quadratic programming problem.

(i) Linear sufficient shape constraints

We consider one segment $\mathbf{y}_i(z,t)$ of the interpolating surface. Resulting from the assumptions formulated in (3), it is sufficient to consider the shape of the control polygons of the adjacent contour curves $\mathbf{x}_{i-1}(t) = \mathbf{y}_i(z_{i-1}, t)$ and $\mathbf{x}_i(t) = \mathbf{y}_i(z_i, t)$. If three successive contour control

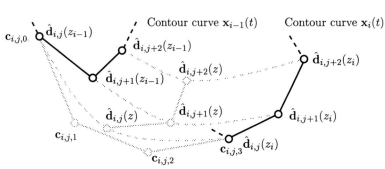

Fig. 4 Three adjacent contour control points ($n = 3$).

points $\hat{\mathbf{d}}_{i,j}(z)$, $\hat{\mathbf{d}}_{i,j+1}(z)$, $\hat{\mathbf{d}}_{i,j+2}(z)$ (cf. Figure 4) satisfy the condition

$$\left[\hat{\mathbf{d}}_{i,j+1}(z^*) - \hat{\mathbf{d}}_{i,j}(z^*),\ \hat{\mathbf{d}}_{i,j+2}(z^*) - \hat{\mathbf{d}}_{i,j+1}(z^*) \right] \geq 0 \quad (\text{resp. } \leq 0) \tag{10}$$

for $z^* = z_{i-1}$ and $z^* = z_i$, then we want to ensure that this inequality is true for all $z^* \in [z_{i-1}, z_i]$. (Note that the value of the left–hand side of the inequality for $z^* = z_{i-1}$ and $z^* = z_i$ results from the interpolation of the given contours, cf. (6).) In addition to (10), the contour control points are to fulfill

$$-\tfrac{\pi}{d-1} \leq \sphericalangle\left(\hat{\mathbf{d}}_{i,j+1}(z) - \hat{\mathbf{d}}_{i,j}(z),\ \hat{\mathbf{d}}_{i,j+2}(z) - \hat{\mathbf{d}}_{i,j+1}(z) \right) \leq \tfrac{\pi}{d-1} \text{ for } z \in [z_{i-1}, z_i] \tag{11}$$

because then the number of inflections of the contour curves is bounded by that of the control polygon, see [6]. Similar to the construction outlined in Section 2.2(ii), it is possible to derive linear sufficient constraints ensuring the inequalities (10) and (11), see [11].

(ii) An objective function

In addition to the constraints ensuring the desired shape of the contours we need an appropriate objective function in order to compute the unknown control points $\mathbf{c}_{i,j,k}$ of the interpolating surface. We want to achieve that the transition between control polygons of adjacent contours is as smooth as possible. So we construct the objective function by summing up the terms

$$\sum_{r=0}^{S} \left\| \left[\hat{\mathbf{d}}_{i,j+1}(\tfrac{S-r}{S} \cdot z_{i-1} + \tfrac{r}{S} \cdot z_i) - \hat{\mathbf{d}}_{i,j}(\tfrac{S-r}{S} \cdot z_{i-1} + \tfrac{r}{S} \cdot z_i) \right] - \mathbf{q}_{i,j,r} \right\|^2 \tag{12}$$

for each pair of adjacent contour control points $(\hat{\mathbf{d}}_{i,j}, \hat{\mathbf{d}}_{i,j+1})_{i=0,\ldots,\hat{D}_i-1}$ $(i = 1,\ldots,C)$, whereby the $S+1$ (S large) points $(\mathbf{q}_{i,j,r})_{r=0,\ldots,S}$ are sampled equidistantly from the spiral with center $\mathbf{O}$ which interpolates the difference vectors $\hat{\mathbf{d}}_{i,j+1}(z^*) - \hat{\mathbf{d}}_{i,j+1}(z^*)$ of adjacent contour control points for $z^* = z_{i-1}$ and $z^* = z_i$. For the trajectories of the first and last control point we add a "tension term" (cf. [9, p. 98]) like

$$\int_{z_{i-1}}^{z_i} \|\frac{d^2}{dz^2}\hat{\mathbf{d}}_{i,j_0}(z)\|^2 \, dz \quad (j_0 = 0, \hat{D}_i) \tag{13}$$

to the objective function. Note that without this term the objective function would depend only on difference vectors $\mathbf{c}_{i,j+1,k} - \mathbf{c}_{i,j,k}$ of control points, and hence the minimum would not be unique. The objective function is a quadratic functions of the components of the unknown control points $\mathbf{c}_{i,j,k}$.

(*iii*) **Computing the control points**

The control points $\mathbf{c}_{i,j,k}$ of the interpolating surface are found by minimizing the quadratic objective function subject to the linear shape constraints obtained from step (*i*). This quadratic programming problem is again solved with the help of the method of active set as described in [5]. The initial solution is found using the simplex algorithm. It is advantageous to start with an initial solution which is close to the minimum of the unconstrained objective function, see [11].

The linear shape constraints obtained from the first step are not guaranteed to provide feasible solutions. In most cases of practical interest the feasible region (in the linear space of control points $\mathbf{c}_{i,j,k}$) will be non–empty. Moreover it can be shown that feasible solutions always exist if the degree n of the parameter lines $t = \text{const}$ is chosen high enough [11].

3.4 An example

As an example we show in Figure 5 the interpolation of four given contour curves with equidistant heights $z_0,\ldots,z_3$. The four given contours are described by quadratic B-spline curves defined over different vectors. The degree of the three interpolating C^1 B–spline surface patches $\mathbf{y}_1(z,t),\ldots,\mathbf{y}_3(z,t)$ is equal to $(2,3)$.

The left figure shows the control polygons $\hat{\mathbf{d}}_{i,j}(z)$ of the interpolating contours and the control points $\mathbf{c}_{i,j,k}$ of their trajectories (in grey). The interpolating contour curves and the control polygons of the four given contours (in grey) have been drawn in Figure 5b. The thicker black lines in both figures indicate the given data.

References

[1] Boot, J. C. G., *Quadratic Programming*. Rand McNally, Chicago 1964.

[2] Dierckx, P., *An algorithm for cubic spline fitting with convexity constraints*. Computing 24 (1980), 349–371.

[3] Dierckx, P., *Curve and Surface Fitting with Splines*. Clarendon Press, Oxford 1993, 119–134.

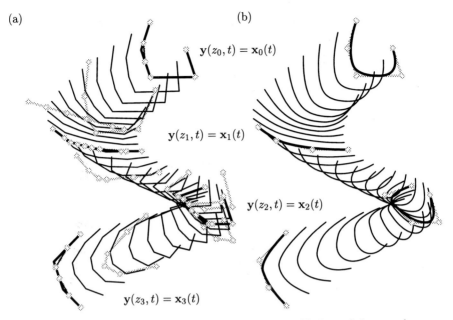

Fig. 5 Sectional curvature–preserving interpolation of four contours. The interpolating control polygons (a) and the contour curves (b).

[4] Farin, G., *Curves and Surfaces for Computer Aided Geometric Design*. Academic Press, Boston 1993.

[5] Fletcher, R., *Practical Methods of Optimization*. John Wiley & Sons, Chichester 1991 (2nd ed.).

[6] Goodman, T. N. T., *Inflections on curves in two and three dimensions*. Computer Aided Geometric Design 8 (1991), 37–50.

[7] Goodman, T. N. T., and Ong, B. H., *Shape preserving interpolation by curves in three dimensions*. These proceedings.

[8] Goodman, T. N. T., and Unsworth, K., *Shape preserving interpolation by curvature continuous parametric curves*. Computer Aided Geometric Design 5 (1988), 323–340.

[9] Hoschek, J. and Lasser, D., *Fundamentals of Computer Aided Geometric Design*. AK Peters, Wellesley MA 1993.

[10] Jüttler, B., *Shape preserving least–square approximation by polynomial parametric spline curves*. University of Dundee, Applied Analysis Report 965 (1996), submitted to Computer Aided Geometric Design.

[11] Jüttler, B., *Sectional curvature preserving approximation of contour lines*. in preparation.

[12] Kaklis, P. D., and Ginnis, A. I., *Sectional-Curvature Preserving Skinning Surfaces*. Technical Report, National Technical University of Athens, Dept. of Naval Architecture and Marine Engineering, 1994.

[13] Schumaker, L. L., *Reconstructing 3D objects from cross-sections*. in: Dahmen, W., Gasca, M., and Micchelli, C. A. (eds.), *Computation of Curves and Surfaces*. Kluwer, Dordrecht 1990, 275–309.

[14] Unsworth, K., *Recent developments in surface reconstruction from planar cross-sections*. University of Dundee, Computer Science Report 94/03 (1994), to appear in the proceedings of the Conference on CAGD held in Penang, July 1994 (to be published as a volume of the Annals of Numerical Mathematics).

Author

Dr. Bert Jüttler
Technische Hochschule Darmstadt
Fachbereich Mathematik
Schloßgartenstr. 7
D–64289 Darmstadt
E–mail: juettler@mathematik.th-darmstadt.de

B–Spline Approximation with Energy Constraints

Ulrich Dietz

Technische Hochschule Darmstadt

Abstract: This paper addresses the problem of reconstructing a free-form surface from measurement data. While the usual methods subdivide the point cloud and fit individual surfaces to these parts we fit a single integral tensor product B-spline surface to the entire point cloud. Holes in the point set, varying point densities, and free boundaries are handled. An effective algorithm is presented, which calculates a smooth approximation surface to a prescribed error tolerance with the help of energy terms.

1 Introduction

The data in the field of reverse engineering are often obtained from physical models (e.g., made of wood or clay). Typically these are large point clouds with measurement error (up to millions of points in the case of laser scanners). The point clouds are mostly scattered with varying densities and holes inside. Outliers may be present and in general there are no sharp or explicitly given boundaries. One class of reconstruction methods is based on the segmentation of the point cloud into polygonal (triangular, rectangular) regions. In current CAD systems this segmentation has to be done manually. To each region an individual surface is fitted. Problems arise in higher-order continuities and in the smoothness at the patch boundaries. Eck [3] recently published a fully automatic method solving the problem of surface reconstruction for objects with arbitrary topology. The other class of methods fit a single surface to all points. A solution has to be found how to handle non-rectangular bounded point sets (in the case of tensor product surface fitting) and holes in the point cloud. Referring to [20], the approximation of point sets with gaps is processed by an averaging method. In our paper the problem of satisfying the Schoenberg-Whitney conditions in gaps in areas of low point density and outside the point boundaries is bypassed by using fairness functionals. The appropriate weighting of the functional is calculated with respect to a prescribed error tolerance. The use of simple quadratic functionals leads to linear (and hence fast) approximation steps.

Another crucial point in parametric surface fitting is the determination of the point parametrization. Different strategies and heuristics for the assignment of good parameter values do exist in the literature. Hoschek [9] and Ma [13] propose parametrizations with the help of reference surfaces. These methods usually require a special point structure or some kind of user interaction. The basic idea behind the present method is to start an iteration process with the best fitting plane together with the corresponding parametrization. Successively, the plane together with the parametrization is transformed to the final surface with the optimal parametrization (orthogonal distance vectors). This is done by an automatic algorithm where fitting steps alternate with reparametrization steps.

2 Algorithm

The algorithm consists of the following three separate parts:

- initial parametrization
- fitting/reparametrization loop
- final surface trimming

The initial parameters are usually computed by projecting the points to the least squares fitting plane. The main part in the algorithm is the iteration loop in which linear fitting steps alternate with reparametrization steps. The data points are denoted by $\mathbf{P}_i, i = 0, \ldots, N$, the corresponding parameters by (u_i, v_i) and the approximation surface by $\mathbf{X}(u, v)$. If, further, $Q_{\text{dist}}(\mathbf{X})$ denotes the total error sum and $Q_{\text{fair}}(\mathbf{X})$ is a fairness measure, the fitting step can be written as

$$Q = Q_{\text{dist}}(\mathbf{X}) + \lambda\, Q_{\text{fair}}(\mathbf{X}) \xrightarrow{\mathbf{X}} \min. \tag{1}$$

The reparametrization or so-called parameter correction steps are given by

$$\|\mathbf{P}_i - \mathbf{X}(u_i, v_i)\|^2 \xrightarrow{(u_i, v_i)} \min, \qquad i = 0, \ldots, N. \tag{2}$$

In (2) the fairness measure doesn't occur as it is not dependent on the parameter values. While Laurent-Gengoux [12] deals with the full nonlinear approximation problem, the problem here is decoupled. Thus, we can take advantage of the geometric meaning and simple nature of the problem. The penalty factor λ in (1), controlling the influence of the fairness term, is determined automatically during the iteration loop. λ is initialized with a large value, forcing the approximation surface to be nearly planar. After each iteration step λ is reduced. This makes it possible for the surface to move towards the point cloud. The iteration ends if the prescribed error is reached. As an additional step the computed surface could be trimmed at the boundaries of the point cloud. The following sections describe the various steps of the algorithm in detail.

3 Parametrization and reparametrization

We assume a scattered point cloud is given and we want to calculate a parametric fitting surface in a least squares sense, i.e., $\mathbf{X}(u, v)$ with $\sum_i \|\mathbf{P}_i - \mathbf{X}(u_i, v_i)\|^2 \to \min$. But how can the parameters (u_i, v_i) be chosen? An optimal choice would be the one making the distance vectors $\mathbf{P}_i - \mathbf{X}(u_i, v_i)$ perpendicular to the surface. This means we are minimizing shortest distances. But as we don't know the surface $\mathbf{X}$ in advance the initial parametrization must be determined as a rough approximation of the optimal one. For unstructured point clouds several authors have proposed parametrizations with the aid of reference surfaces (base surfaces) [9, 13]. The points are projected orthogonally onto the reference surface. Parameter

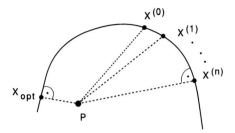

Fig. 1 A situation where parameter correction fails.

values can then be calculated in the parameter domain of this surface. The projection of the points must be a bijective mapping and the reference surface is expected to be close enough to the final approximation surface. In general, the construction of this reference surface requires user interaction or some additional information. Hoschek [9], for example, assumes that additional points on four boundaries are given. These points are then fitted by four curves defining a bilinear Coons patch.

After determining the parametrization a surface can be fitted to the data in a least squares sense. But for all parametrization strategies the distance vectors are generally not perpendicular to the surface. This means we have minimized an arbitrary error due to the parametrization. The approximation surface can be improved if we replace the old parameters by the parameter values of the perpendicular projections onto the surface and recalculate the approximation surface. This reparametrization step, or so-called parameter correction, has been used by various authors [6, 16, 18, 10, 19]. The task of finding the parameters of the orthogonal projection of a point $\mathbf{P}$ on a parametric surface $\mathbf{X}$ is equivalent to solving the nonlinear two-dimensional system

$$(\mathbf{P} - \mathbf{X})^T \mathbf{X}_u = 0$$
$$(\mathbf{P} - \mathbf{X})^T \mathbf{X}_v = 0. \tag{3}$$

Rogers [18] used a Gauss-Newton method, Hoschek [9] a simplified Newton method, and Sarkar [19] a Levenberg-Marquardt method. But as the first approximation could be far from the optimal situation and all mentioned methods are only locally convergent, parameter correction may fail (see Fig. 1). Peaks and wiggles can occur and may increase during the iterations.

A solution to this problem will be to start with a very smooth reference surface and to transform this surface together with the parametrization almost continuously towards the required approximation surface. This is done in an iterative process where fitting steps alternate with reparametrization steps. One Newton iteration per reparametrization has proven to be sufficient. The objective functional of the fitting step is a linear combination of the total error sum and a fairness measure. The penalty factor λ for the fairness functional is initialized with a large value and is reduced in every iteration step until a prescribed error tolerance is reached. This results in the largest possible value of λ or, equivalently, in the smoothest surface within the error tolerance. Contrary to the classic method of parameter correction where the reference surface is 'transformed' ad hoc to the approximation surface, this procedure is a much more stable one. The number of iterations and the stability of the process is controlled by the decreasing rate of λ. If this rate is low the iteration process is very stable but may be unnecessarily slow. On the other hand, if this rate is too high one could get the same problem as with the usual parameter correction. The optimal setup of λ and its optimal rate of decrease are subject to further investigation. A scaling factor of 0.5 for λ produced convincing results in all examined examples. As the calculated surfaces from one iteration step to another do not differ very much, each of the fitting problems can be solved by iterative methods (e.g., over-relaxation, conjugate gradient method [22, 24]) instead of direct methods. The advantage of these methods is the low memory requirement for sparse matrices (occurring when B-splines are used).

As a reference surface we use the least squares fitting plane in case no additional information on the point set is given, provided the orthogonal mapping is bijective. If four additional boundaries are given we can use the B-spline surface minimizing the fairness functional while keeping the boundaries as a reference surface.

4 B-spline approximation

In this section we shall summarize some important properties of l^2-approximation with tensor product B-spline surfaces. A general description of B-splines can be found in [4, 9] and, most recently, in [15].

We assume that $\mathbf{P}_i \in \mathbb{R}^3, i = 0, \ldots, N$ are given points with pairs of parameter values $(u_i, v_i), i = 0, \ldots, N$ assigned. The task of finding the tensor product B-spline surface

$$\mathbf{X}(u,v) = \sum_{i=0}^{n_u} \sum_{j=0}^{n_v} \mathbf{d}_{ij} N_{ik}(u) N_{jl}(v) \tag{4}$$

minimizing the total error sum

$$Q_{\text{dist}} = \sum_i \|\mathbf{P}_i - \mathbf{X}(u_i, v_i)\|^2 \longrightarrow \min \tag{5}$$

is a linear problem. Here the $\mathbf{d}_{ij}$'s denote the control points and N_{ik} denotes the i-th B-spline basis function of order k. Assuming the Schoenberg-Whitney conditions are satisfied, the control points can be uniquely determined either from the error equations

$$\sum_{ij} \mathbf{d}_{ij} N_{ik}(u_m) N_{jl}(v_m) = \mathbf{P}_m, \qquad m = 0, \ldots, N$$
$$\Longleftrightarrow \qquad \mathbf{M}\mathbf{d} = \mathbf{P} \tag{6}$$

with orthogonal factorization methods or from the normal equations

$$\mathbf{M}^T \mathbf{M} \mathbf{d} = \mathbf{M}^T \mathbf{P}. \tag{7}$$

The first method is numerically more stable than the second one since the condition number of $\mathbf{M}^T \mathbf{M}$ is the square of the condition number of $\mathbf{M}$ and $\mathbf{M}$ is usually badly conditioned (especially for high polynomial orders). Disadvantages when calculating the control points

B-Spline Approximation with Energy Constraints

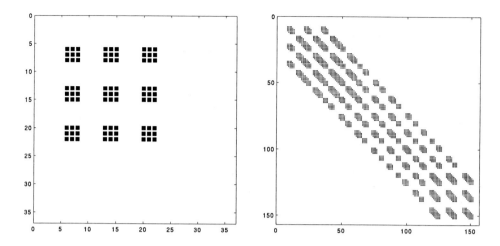

Fig. 2 left: Nonzero elements of single matrix $\mathbf{M}_i^T \mathbf{M}_i$ for a biquadratic B-spline surface; right: singular normal equation system for a point cloud not satisfying the Schoenberg-Whitney conditions

from the error equations are the computational costs of the orthogonal factorization methods and the enormous memory requirement for the matrix $\mathbf{M}$ (which is proportional to the number of data points).

In contrast, the matrix size of the normal equations does not depend on the number of data points. With respect to moderate polynomial order, de Boor [2] showed that the normal equations are relatively well conditioned. The matrix $\mathbf{M}^T \mathbf{M}$ can be calculated in a storage efficient manner as a cumulative sum of matrices $\mathbf{M}_i^T \mathbf{M}_i$ (see Fig. 2)

$$\mathbf{M}^T \mathbf{M} = \sum_{i=0}^{N} \mathbf{M}_i^T \mathbf{M}_i. \qquad (8)$$

These matrices are built as dyadic products of the i-th row of $\mathbf{M}$ with itself. They are of rank 1 and only $k^2 l^2$ elements are nonzero due to the local support of the basis functions. The resulting matrix $\mathbf{M}^T \mathbf{M}$ is symmetric, positive definite, and banded. The bandwidth depends on the control point order. Row-by-row order yields an upper bandwidth of $(k-1) + (l-1)(n_u + 1)$, column-by-column order yields $(l-1) + (k-1)(n_v + 1)$. These numbers may differ significantly (see Fig. 3), so the order resulting in the lower bandwidth should be chosen. The system can be solved stably and efficiently by a Cholesky decomposition taking the bandwidth into account. Iterative methods which only store the small fraction of non-vanishing elements can also be used. We, therefore, come to the conclusion that the B-spline basis with its local support results in a significantly reduced computational (and storage) expense in generating and solving the normal equations.

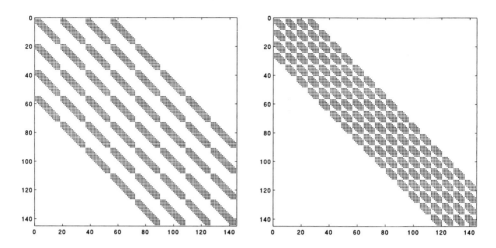

Fig. 3 Nonzero elements (5016 of 144^2) in the normal equation matrices for a B-spline surface with 15×5 bicubic patches; left side: row-by-row order with upper bandwidth 57; right side: column-by-column order with upper bandwidth 27

5 Fairness functionals

The use of the usual least squares surface approximation is limited in several ways. To obtain a unique approximation surface the Schoenberg-Whitney conditions must be satisfied. This requires well-distributed point sets without holes and with rectangularly structured boundaries. If noise in the data occurs the approximation surface will be noisy as well, due to the fact that only the total error sum is minimized. Most of these restrictions are overcome by introducing a new objective functional involving a fairness measure Q_{fair}:

$$Q = Q_{\text{dist}} + \lambda\, Q_{\text{fair}} \longrightarrow \min \tag{9}$$

λ is a penalty factor controlling the influence of the fairness functional; it is determined automatically during the presented iterative algorithm. Using this new objective functional we can handle point sets of varying density, point sets with holes, with arbitrary boundaries, and with measurement errors. A number of proposals for these fairness functionals can be found in the literature [1, 5, 14, 17]. Applications are to be found in [7, 8, 21, 23, 25]. Many of the functionals are related to energies. $\int (\kappa_1^2 + \kappa_2^2)\, dS$, for example, is the bending energy of a thin plate with small deflections (with κ_1, κ_2 principal curvatures). But this energy as many others depends highly nonlinearly on the unknown control points. If a fairness functional quadratic in the unknowns is chosen, the resulting system of equations is linear. The following energies are simplified measures of the area, bending energy, and change of curvature of a surface $\mathbf{X}$. They are all quadratic in the control points.

$$Q_1 = \iint (\mathbf{X}_u^2 + \mathbf{X}_v^2)\, du\, dv \tag{10}$$

$$Q_2 = \iint (\mathbf{X}_{uu} + \mathbf{X}_{vv})^2 - 2(1-\mu)(\mathbf{X}_{uu}\mathbf{X}_{vv} - \mathbf{X}_{uv}^2)\, du\, dv \tag{11}$$

$$Q_3 = \iint (\mathbf{X}_{uuu} + \mathbf{X}_{uvv})^2 + (\mathbf{X}_{uuv} + \mathbf{X}_{vvv})^2 \, du \, dv \tag{12}$$

These energies will be combined in a convex combination and used as our fairness functional

$$Q_{\text{fair}} = \sum_{p=1}^{3} \alpha_p Q_p, \quad \text{with} \quad \sum_{p=1}^{3} \alpha_p = 1. \tag{13}$$

The factors α_p control the influence of the individual energies and hence the shape of the resulting surface. These so-called design factors are to be set up by the user.
A necessary condition for the functional Q to be minimal is

$$\nabla Q \stackrel{!}{=} 0. \tag{14}$$

This leads to the linear system

$$\mathbf{Ad} = \mathbf{b} \quad \text{with} \quad \mathbf{A} = \mathbf{A}_{\text{dist}} + \lambda \mathbf{A}_{\text{fair}} \quad \text{and} \quad \mathbf{A}_{\text{fair}} = \sum_p \alpha_p \mathbf{A}^p. \tag{15}$$

Here $\mathbf{A}_{\text{dist}}$ is the matrix of the usual normal equations and $\mathbf{b}$ is the corresponding right-hand side. The vector $\mathbf{d}$ contains the unknown control points. Equation (15) is actually a system of equations with three right-hand sides (one for each coordinate). The matrices $\mathbf{A}^p$ correspond to the energies Q_p and do not depend on the point parametrization. Therefore, the matrix $\mathbf{A}_{\text{fair}}$ can be precalculated; only the matrix $\mathbf{A}_{\text{dist}}$ changes after reparametrization.

Using B-splines the matrices $\mathbf{A}^p$, $\mathbf{A}_{\text{fair}}$ are symmetric, positive semi-definite, and banded with the same bandwidth as the matrix $\mathbf{A}_{\text{dist}}$ for the usual least squares fit. Assuming that the B-spline surface is at least C^q-continuous ($q = \max\{p \,|\, \alpha_p > 0\}$), at least one $\alpha_p > 0$, and at least 8 non-collinear points are given, the matrix $\mathbf{A}$ is nonsingular. Thus, the approximation surface is uniquely determined.

The matrix $\mathbf{A}^1$ is given by

$$\begin{aligned}
\mathbf{A}^1 = \nabla^2 Q_1 &= \nabla^2 \iint (\mathbf{X}_u^2 + \mathbf{X}_v^2) \, du \, dv \\
&= \nabla^2 \iint (\mathbf{d}^T \mathbf{N}_u)(\mathbf{N}_u^T \mathbf{d}) + (\mathbf{d}^T \mathbf{N}_v)(\mathbf{N}_v^T \mathbf{d}) \, du \, dv \\
&= \iint (\mathbf{N}_u \mathbf{N}_u^T + \mathbf{N}_v \mathbf{N}_v^T) \, du \, dv
\end{aligned} \tag{16}$$

with $\mathbf{N}_u^T \mathbf{d} = \sum_{ij} N'_{ik}(u) N_{jl}(v) \, \mathbf{d}_{ij} = \mathbf{X}_u(u,v); \quad \mathbf{N}_v^T \mathbf{d}$ is defined analogously.

The integrals have to be understood element-wise. Since we use a tensor product surface the integration can be decoupled. For a single element A^1_{rs} of $\mathbf{A}^1$ we get

$$\begin{aligned}
A^1_{rs} &= \iint N'_{ik}(u) N_{jl}(v) N'_{gk}(u) N_{hl}(v) + N_{ik}(u) N'_{jl}(v) N_{gk}(u) N'_{hl}(v) \, du \, dv \\
&= \int N'_{ik} N'_{gk} \, du \int N_{jl} N_{hl} \, dv + \int N_{ik} N_{gk} \, du \int N'_{jl} N'_{hl} \, dv
\end{aligned} \tag{17}$$

with $r = i+j(n_u+1)$ and $s = g+h(n_u+1)$ (row-by-row order). Introducing the abbreviations

$$U^{\alpha\beta} = \int N_{ik}^{(\alpha)} N_{gk}^{(\beta)} \, du, \qquad V^{\alpha\beta} = \int N_{jl}^{(\alpha)} N_{hl}^{(\beta)} \, dv \qquad (18)$$

we can write

$$A_{rs}^1 = U^{11}V^{00} + U^{00}V^{11}. \qquad (19)$$

For the functionals Q_2 and Q_3 one obtains after analogous calculations

$$A_{rs}^2 = U^{22}V^{00} + \mu U^{20}V^{02} + \mu U^{02}V^{20} + U^{00}V^{22} + 2(1-\mu)U^{11}V^{11}, \qquad (20)$$

$$A_{rs}^3 = U^{33}V^{00} + U^{31}V^{02} + U^{13}V^{20} + U^{11}V^{22} + U^{22}V^{11} + U^{02}V^{31} + U^{20}V^{13} + U^{00}V^{33}. \qquad (21)$$

The integrals $U^{\alpha\beta}, V^{\alpha\beta}$ can be calculated exactly as the integrands are piecewise polynomials. For low polynomial orders the exact values can also be efficiently calculated with Gaussian quadrature. For the integration of a polynomial of order n one only needs $\lceil \frac{n}{2} \rceil$ evaluations of the polynomial. As already mentioned above, the matrices $\mathbf{A}^p$ are sparse and banded (see Fig. 3); thus, only a small fraction of the integrals has to be computed actually.

It is also possible to use the functionals proposed by Greiner [5] instead of the above-mentioned functionals since they are also quadratic. These functionals are based on reference surfaces and are independent of the actual surface parametrization. Usually they are better approximations to real energies and perhaps yield more pleasing surfaces. In the context of our algorithm the approximation surface of the actual iteration step can be used as a reference surface for the fairness functional in the next iteration step. The only drawback of Greiner's functionals are the high computational costs. The occurring integrals cannot be decoupled for the u- and v-directions. And as the functionals depend on a changing reference surface, the integral matrices cannot be precalculated. A way out of this is the use of the simpler quadratic functionals at the start of the algorithm; Greiner's functionals are then only used in some of the later iteration steps. An improvement of the surface quality is to be expected.

6 Surface trimming

The iterative approximation algorithm produces a smooth B-spline surface keeping a prescribed error tolerance. In general, the approximation surface will extend outside of the point cloud. These regions are then trimmed away by B-spline curves on the surface which are defined in the parametric domain of the surface. First of all the point cloud is projected orthogonally onto the surface and the corresponding parameters in the parametric domain are calculated (we actually could obtain these parameters from the last surface iteration step). Then the boundary points have to be determined, in case they are not already explicitly given. This can be done in the plane (but not in 3-space) by triangulating all parameter points (u_i, v_i). The parameters on the boundary of the triangulation are the required (and already ordered) boundary points. These points $(u_i, v_i), i = 0, \ldots, N_b$ in the parametric domain are approximated by a B-spline curve [11] with well-known least squares techniques together with parameter correction. While the approximation takes place in the parametric domain of the surface, the parameter correction works directly on the trimming curve on the

Fig. 4 Lense benchmark with 613 points, convex approximation surface with a single biquintic patch, $\delta_{max} = 0.0002\%$, $\delta_{avg} = 0.00005\%$, $t = 49$ secs. Isophotes on the left and mean curvature on the right side. The special characteristics of this bifocal lense can only be seen in the right figure.

surface. We get the required trimming curve by inserting the curve $\mathbf{C}(t)$ into the surface representation $\mathbf{X}$:

$$\mathbf{X}(u(t), v(t)) \tag{22}$$

7 Results

In this section we shall present approximation results for five different point clouds. The first three examples are approximation surfaces to the benchmark data sets. The other two examples are more complicated to show the efficiency of the algorithm. The best fitting plane is used as an initial reference surface in all examples. The thin grey lines in the figures mark the patch boundaries. The wide black lines shown in the right illustration of figure 4 are lines of equal mean curvature. In the right illustration of figure 6 they mark lines of equal Gaussian curvature. In all other figures they represent the isophotes visualizing the surface quality. All surfaces have been computed on an HP735 workstation (105MHz); for each example the computation time is given below.

8 Concluding remarks

An automatic algorithm for the reconstruction of smooth surfaces from scattered data has been presented. The use of a single tensor product B-spline surface makes the algorithm very efficient but restricts the method to the reconstruction of surfaces with simple topological type.

In the presented algorithm the penalty factor λ for the fairness functionals has a global effect over the whole surface. Further investigation will concentrate on providing the algorithm with a penalty factor λ with only local influence. During the iterations the factor λ will be lowered only in those surface regions where it is necessary, i.e., in regions where the error is greater than a given tolerance. This will improve the smoothness of the computed surfaces especially for surfaces with edges.

We can observe that during the early iterations the calculated surfaces are more or less flat. Thus, these surfaces do not need as high a number of patches as does the final surface. The algorithm can be speeded up if we begin the iterations with a single patch and increase the patch number during the iterations as needed.

9 Acknowledgment

The research work of the author is supported by the German *Bundesministerium für Bildung, Wissenschaft, Forschung und Technologie (BMBF)*.

References

[1] Bloor, M. I. G., Wilson, M. J. and Hagen, H.: *The smoothing properties of variational schemes for surface design.* Computer Aided Geometric Design **12** (1995), 381–394.

[2] de Boor, C.: *A Practical Guide to Splines.* Springer 1978.

[3] Eck, M., Hoppe, H.: *Automatic reconstruction of B-spline surfaces of arbitrary topological type.* Preprint 1800, Technische Hochschule Darmstadt, 1996.

[4] Farin, G.: *Curves and Surfaces for Computer Aided Geometric Design – A Practical Guide.* 3rd edition. Academic Press 1993.

[5] Greiner, G.: *Variational design and fairing of spline surfaces.* Computer Graphics Forum **13**:3 (1994), 143–154.

[6] Grossmann, M.: *Parametric curve fitting.* The Computer Journal **14** (1970), 169–172.

[7] Hadenfeld, J.: *Local energy fairing of B-spline surfaces.* In M. Dæhlen, T. Lyche, and L. L. Schumaker (eds.): Mathematical Methods in CAGD III, (1995), 203–212.

[8] Hagen, H., Schulze, G.: *Automatic smoothing with geometric surface patches.* Computer Aided Geometric Design **4** (1987), 231–235.

[9] Hoschek, J., Lasser, D.: *Fundamentals of Computer Aided Geometric Design.* A K Peters, 1993.

[10] Hoschek, J., Schneider, F.-J., Wassum, P.: *Optimal approximate conversion of spline surfaces.* Computer Aided Geometric Design **6** (1989), 293–306.

[11] Hoschek, J., Schneider, F.-J.: *Approximate spline conversion for integral and rational Bézier and B-spline surfaces.* In R. E. Barnhill (ed.): Geometry Processing for Design and Manufacturing, SIAM (1992), 45–86.

[12] Laurent-Gengoux, P., Mekhilef, M.: *Optimization of a NURBS representation.* Computer–Aided Design **25** (1993), 699–710.

[13] Ma, W., Kruth, J. P.: *Parametrization of randomly measured points for least squares fitting of B-Spline curves and surfaces.* Computer–Aided Design **27** (1995), 663–675.

[14] Moreton, H. P., Séquin, C. H.: *Functional optimization for fair surface design.* ACM Computer Graphics **26** (1992), 167–176.

[15] Piegl, L., Tiller, W.: *The NURBS book.* Springer 1995.

[16] Pratt, M. J.: *Smooth parametric surface approximations to discrete data.* Computer Aided Geometric Design **2** (1985), 165–171.

[17] Rando, T., Roulier, J.A.: *Designing faired parametric surfaces.* Computer–Aided Design **23** (1991), 492–497.

[18] Rogers, D. F., Fog, N. G.: *Constrained B-spline curve and surface fitting.* Computer–Aided Design **21** (1989), 641–648.

[19] Sarkar, B., Menq, C.-H.: *Parameter optimization in approximating curves and surfaces to measurement data.* Computer Aided Geometric Design **8** (1991), 267–290.

[20] Schmidt, R. M.: *Fitting scattered surface data with large gaps.* Surfaces in CAGD, R. E. Barnhill, W. Boehm (eds.), (1983), 185–189.

[21] Sinha, S. S., Schunck, B. G.: *A two-stage algorithm for discontinuity-preserving surface reconstruction.* IEEE Transactions on Pattern Analysis and Machine Intelligence **14** (1992), 36–55.

[22] Schwarz, H. R.: *Numerische Mathematik.* 3rd edition, Teubner 1993.

[23] Terzopoulos, D.: *Regularization of inverse visual problems involving discontinuities.* IEEE Transactions on Pattern Analysis and Machine Intelligence **8** (1986), 413–424.

[24] Törnig, W., Gipser, M., Kaspar, B.: *Numerische Lösung von partiellen Differentialgleichungen der Technik. Differenzenverfahren, finite Elemente und die Behandlung großer Gleichungssysteme.* 2nd edition, Teubner 1991.

[25] Welch, W., Witkin, A.: *Variational surface modeling.* ACM Computer Graphics **26** (1992), 157–166.

Author

Ulrich Dietz
Technische Hochschule Darmstadt
Fachbereich Mathematik
Schloßgartenstraße 7
D-64285 Darmstadt
E–mail: udietz@mathematik.th-darmstadt.de

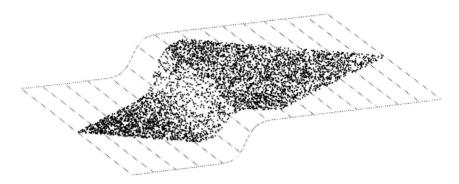

Fig. 5 Step benchmark with 5184 noisy points, nearly developable approximation surface with 1×15 patches of degree 1×4, $\delta_{max} = 0.001\%$, $\delta_{avg} = 0.00035\%$, $t = 59\,\text{secs}$, $|K| < 3 \cdot 10^{-9}$ (Gaussian curvature).

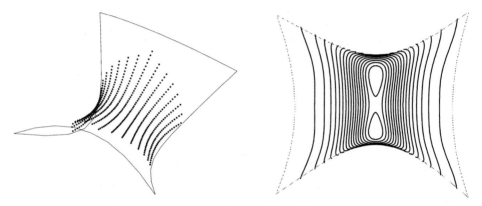

Fig. 6 Benchmark with 490 points from hyperbolic surface, approximation surface with a single bicubic patch (negative Gaussian curvature), $\delta_{max} = 0.00076\%$, $\delta_{avg} = 0.00022\%$, $t = 31\,\text{secs}$; points and patch boundaries on left side and Gaussian curvature on right side.

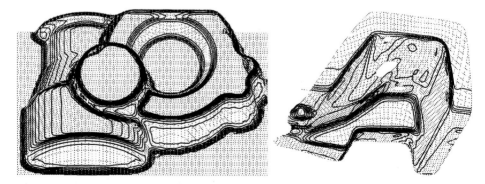

Fig. 7 Left side: approximation surface with 70×70 bicubic patches to 4101 points, $\delta_{max} = 0.6\%$, $\delta_{avg} = 0.03\%$, $t = 193\,\text{secs}$. Right side: approximation surface with 50×50 bicubic patches to 95986 points, $\delta_{max} = 1.2\%$, $\delta_{avg} = 0.02\%$, $t = 21\,\text{mins}$.

Curvature approximation with application to surface modeling

Günther Greiner

Universität Erlangen–Nürnberg
Am Weichselgarten 9
D–91058 Erlangen
E–mail: greiner@immd9.informatik.uni-erlangen.de

Abstract: When fair surfaces have to be designed, a uniformly distributed curvature is important. To achieve this designers try to minimize total curvature, or the variation of the curvature. This however is a very costly procedure. In the case of curves, the curvature can be easily approximated by quadratic functionals. This then allows optimization at interactive speed. The generalization of this method to surfaces is by no means straightforward.

In this note we describe how curvature can be approximated by quadratic functionals. Moreover, we apply the results to construct fairness functionals that are good approximations to the exact curvature functionals but are quadratic, hence relatively easy to minimize. We sketch how these functionals can be used to interpolate scattered data by fair tensor product B-spline functions.

1 Introduction

When fairing a parameterized curve $C(t)$ a standard procedure is the following (see e.g. Eck Hadenfeld [4])): Try to minimize the *total curvature* $\int_{[a,b]} \kappa(s)^2 \, ds$ or the variation of curvature $\int_{[a,b]} \kappa'(s)^2 \, ds$. Since the curvature is a complicated functional of the first and second order derivatives, one uses more simple functionals instead: $\int_{[a,b]} \|C''(t)\|^2 \, dt$ and $\int_{[a,b]} \|C'''(t)\|^2 \, dt$ for example. This can be justified as follows. If the parameterization C is chosen carefully, namely $\|C'(t)\|$ does not vary much, $\|C'(t)\| \approx \alpha = const$, then $\|C''(t)\|^2 \approx \alpha^4 \cdot \kappa(t)$ and $\|C'''(t)\|^2 \approx \alpha^6 (\kappa'^2(t) + \kappa^4(t))$.

The assumption that "$\|C'(t)\|$ does not vary much" can (theoretically) always be achieved. In fact, it is a well-known, that each curve can be reparameterized by arc length. In this case $\|C'(t)\| = 1$ and hence $\alpha = 1$.

Considering surfaces, and trying to approximate curvature in a similar way, one has to be aware of the following facts. In case the parameterization $F: \Omega \to \mathbb{R}^3$ is isometric, then $(F_{uu} + F_{vv})^2 = (\kappa_1 + \kappa_2)^2$ and $F_{uu}^2 + 2F_{uv}^2 + F_{vv}^2 = \kappa_1^2 + \kappa_2^2$ and $\langle F_{uu}|F_{vv}\rangle - F_{uv}^2 = \kappa_1 \cdot \kappa_2$. At a first glance these facts are quite promising. However, in contrast to the curve case, for most surfaces it is impossible to reparameterize the surface in an isometric manner. In fact, only for developable surfaces, or equivalently, for surfaces having zero Gaussian curvature an isometric reparameterization is possible. Thus, the simple quadratic expressions $(F_{uu}+F_{vv})^2$, $F_{uu}^2 + 2F_{uv}^2 + F_{vv}^2$ and $\langle F_{uu}|F_{vv}\rangle - F_{uv}^2$ respectively are good approximations to mean, total and Gaussian curvature only for surfaces having small Gaussian curvature. Such a surface can be reparameterized nearly isometrically (i.e. the first fundamental form is $\approx \begin{pmatrix} 1 & 0 \\ 0 & 1 \end{pmatrix}$). For most surfaces however, this is not possible and then, these expressions will be no good

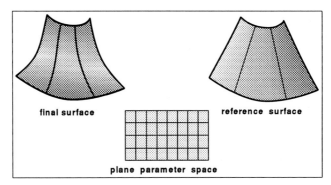

Fig. 1 To approximate the curvature of the "final surface", being parameterized over the rectangular parameter space, one uses a "reference surface" that has a similar geometry (shape)

approximations to the exact curvatures. Even reparameterization will not help much in this case!

In this paper we present quadratic approximations to curvature functionals, that are locally good approximations to the exact curvature, even for surfaces not being nearly isometrically parameterized. The basic idea is illustrated in Fig. 1. First one has to select a *reference surface*. Then one can define quadratic functionals that are good approximations to the (mean, total, Gaussian) curvature for all surfaces that are close to the chosen reference surface. These functionals are quadratic, hence a minimization process for functionals using these approximations of curvature is simple and can be performed relatively easy. Moreover, these functionals will be (locally) good approximations to the exact curvature functionals, hence the minimization process will yield surfaces of pleasant shape.

In Sec. 4 we will use these functionals to interpolate scattered data with tensor product B-spline functions. It is a variational approach. Similar approaches — for different modeling problems und using different fairness functionals — have been outlined by several authors, e.g. Celniker and Gossard[2], Welch and Witkin[13], Kallay[8], Brunett et al.[1], Moreton and Séquin[11, 12] Greiner[6, 7] and others. A variational approach to scattered data interpolation is due to Hoschek and Dietz (this volume).

2 Basics from differential geometry

2.1 Notation

First we want to introduce the notation used throughout this paper. We carefully have to distinguish between parameters $(u,v) \in \Omega \subset \mathbb{R}^2$ and vectors $p \in \mathbb{R}^3$. To simplify formulas we use index notation. Thus for parameters we write (u_1, u_2) instead of (u, v). Similarly, a vector p has components (p_1, p_2, p_3) instead of (p_x, p_y, p_z). The partial derivatives of a function $h : \Omega \to \mathbb{R}$ are denoted by $\partial_i h$. Thus $\partial_1 h = h_{u_1} = h_u$ and $\partial_2 h = h_{u_2} = h_v$. There are three second order partial derivatives of h: $\partial_1 \partial_1 h$, $\partial_2 \partial_2 h$ and $\partial_1 \partial_2 h = \partial_2 \partial_1 h$. Similarly, for a parameterized surface $S : \Omega \to \mathbb{R}^3$ we have two first order derivatives $\partial_i S = (\partial_i S_1, \partial_i S_2, \partial_i S_3)$ for $i = 1, 2$ and three second order partial derivatives $\partial_i \partial_j S$, $(i, j = 1, 2)$.

Curvature approximation with application to surface modeling

Throughout this section the *Einstein summation convention* is used. This convention states the following: if in an expression an index occurs twice, once as a super-index and once as a subindex, this expression is considered as the summation over this index. This means that $\sum_i \alpha_i^j x^i$ will be written briefly as $\alpha_i^j x^i$. Here the α_i^j denote some scalars, while the x^i may be scalars or vectors. For example, the directional derivative of a function $h : \Omega \to \mathbb{R}$ in direction $p = (p^1, p^2)$ is given as $p^j \partial_j h$. Similarly, the trace of a square matrix $A = (a_i^j)_{ij}$ is given as a_i^i. Note that the index of summation may be changed without affecting the result, for example $\alpha_i^j x^i = \alpha_p^j x^p$.

2.2 Fundamental forms and Christoffel symbols

When studying the geometry of surfaces, the first and the second fundamental form play an important role. We briefly review the definitions and some basic properties. More details can be found in advanced textbooks on differential geometry (e.g. [9, 10]).

Let $S : \Omega \to \mathbb{R}^3$ be a parameterized surface. We always assume that surfaces are differentiable, at least twice, and do not have singular points. That is, $\partial_1 S(u_1, u_2)$ and $\partial_2 S(u_1, u_2)$ are linearly independent for every parameter value (u_1, u_2). Thus they form a basis for the tangential vector space $T_u S$ of the surface. In this situation the *normal vector* $\vec{n}_S$ is uniquely defined at every point of the surface. It is given as $\vec{n}_S = \frac{\partial_1 S \times \partial_2 S}{\|\partial_1 S \times \partial_2 S\|}$.

The *first fundamental form* is the 2×2-matrix $\mathbf{I}_S = (g_{ij})$ with $g_{ij} = \langle \partial_i S | \partial_j S \rangle$. Here and further on, $\langle \cdot | \cdot \rangle$ denotes the usual inner product in 3D-space. The matrix $\mathbf{I}_S = (g_{ij})$ is invertible; its inverse is denoted by (g^{ij}). Thus we have

$$g^{ij} g_{jk} = \delta_k^i = \begin{cases} 1 & \text{if } i = k \\ 0 & \text{otherwise} \end{cases} . \tag{1}$$

The second partial derivatives $\partial_i \partial_j S$ will (in general) not be tangential to S. In any case, they can be represented in the coordinate system $\{\partial_1 S, \partial_2 S, \vec{n}_S\}$:

$$\partial_i \partial_j S = \Gamma_{ij}^k \partial_k S + h_{ij} \vec{n}_S \tag{2}$$

The coefficients can be determined from the following equations

$$h_{ij} = \langle \partial_i \partial_j S | \vec{n}_S \rangle \tag{3}$$
$$\Gamma_{ij}^k = g^{kl} \langle \partial_i \partial_j S | \partial_l S \rangle \tag{4}$$
$$= \tfrac{1}{2} g^{kl} (\partial_i g_{jl} + \partial_j g_{li} - \partial_l g_{ij}) .$$

The 2×2-matrix $\mathbf{II}_S = (h_{ij})$ is called the *second fundamental form*. The coefficients Γ_{ij}^k are called *Christoffel symbols (of the 2nd kind)*.

The first and second fundamental form carry all geometric information of the surface. In particular, curvature information of the surface S can be computed easily, once $\mathbf{I}_S$ and $\mathbf{II}_S$ are known. We summarize the results in a proposition.

Proposition 1

- *The principal curvatures κ_1 and κ_2 are the eigenvalues of the Weingarten map $\mathbf{I}_S^{-1} \mathbf{II}_S$. The corresponding eigenvectors are the principal curvature directions (with respect to the coordinate system $\{\partial_1 S, \partial_2 S\}$).*

- The mean curvature $H_S = \frac{1}{2}(\kappa_1 + \kappa_2)$ of S is obtained as $\frac{1}{2}\mathbf{trace}(\mathbf{I}_S^{-1}\mathbf{I\!I}_S)$.
- The Gaussian curvature $K_S = \kappa_1 \cdot \kappa_2$ of S is obtained by $\frac{\det(\mathbf{I\!I}_S)}{\det(\mathbf{I}_S)}$.

2.3 Differential calculus on surfaces

In this subsection the notions of covariant derivative and gradient are explained. Then the fundamental concept of the Hessian is derived.

First we briefly review the Euclidean situation. For a scalar function $h : \Omega \to \mathbb{R}$ the appropriate derivative is the *gradient*, given by $\mathbf{grad}(h) = \begin{pmatrix} \partial_1 h \\ \partial_2 h \end{pmatrix}$. Note that $\mathbf{grad}(h)$ is a vector field. Given a vector field $X = \begin{pmatrix} x^1 \\ x^2 \end{pmatrix}$ in the plane, its derivative is the Jacobian, $\nabla X = \begin{pmatrix} \partial_1 x^1 & \partial_2 x^1 \\ \partial_1 x^2 & \partial_2 x^2 \end{pmatrix}$. ∇X is considered a linear map in the plane. Combining these concepts one can define the second order derivative of a scalar function. This is called the Hessian and given by

$$\mathbf{Hess}(h) = \nabla(\mathbf{grad}(h)) = \begin{pmatrix} \partial_1 \partial_1 h & \partial_2 \partial_1 h \\ \partial_1 \partial_2 h & \partial_2 \partial_2 h \end{pmatrix}.$$

Now we want to consider scalar functions and tangential vector fields defined on a surface $S : \Omega \to \mathbb{R}^3$. There is a one-to-one correspondence between scalar functions $\tilde{h}$ defined on the surface point set $S(\Omega)$, and scalar functions h defined on the parameter space Ω of S. We have $h = \tilde{h} \circ S$. In other words, scalar functions defined on the parameter space Ω of S will be interpreted as functions defined on the surface in the following way. The function value $h(u_1, u_2)$ is assigned to the surface point $S(u_1, u_2)$. In a similar way, a 2-dimensional vector field $X : \Omega \to \mathbb{R}^2$ is considered a tangential vector field on the surface in the following way. The tangential vector $x^i(u_1, u_2)\partial_i S(u_1, u_2)$ is assigned to the surface point $S(u_1, u_2)$.

> In the following, all the functions and vector fields, formally defined over Ω are considered as functions and tangential vector fields on the surface S.

When differentiating a scalar function $h : \Omega \to \mathbb{R}$ we actually want to differentiate the corresponding function $\tilde{h} = h \circ S^{-1}$. In differential geometry there is a well developed calculus, which states how the derivatives can be computed.

We start with the *gradient* of a scalar function h.

$$\mathbf{grad}_S(h) = g^{jk}\partial_k h \partial_j S \tag{5}$$

Here the 2×2-matrix (g^{jk}) is the inverse of the first fundamental form (g_{ij}) of S.

The derivative $\nabla_S X$ of a vector field $X = (x^1, x^2) : \Omega \to \mathbb{R}^2$ is usually called *covariant derivative* in differential geometry. It is a linear map in the tangent space $T_u S$ defined as follows. Let $y = p^i \partial_i S$ be a vector in the tangent space $T_u S$. Then

$$\nabla_S X(p^i \partial_i S) = (\partial_j x^k + x^i \Gamma_{ij}^k) p^j \partial_k S \tag{6}$$

Thus the matrix representation of $\nabla_S X$ (with respect to the basis $\{\partial_1 S, \partial_2 S\}$) is given as

$$\nabla_S X = \begin{pmatrix} \partial_1 x^1 + x^i \Gamma_{i1}^1 & \partial_2 x^1 + x^i \Gamma_{i2}^1 \\ \partial_1 x^2 + x^i \Gamma_{i1}^2 & \partial_2 x^2 + x^i \Gamma_{i2}^2 \end{pmatrix}. \tag{7}$$

Combining gradient and covariant derivative, we can finally define the Hessian of a scalar function as follows.

$$\mathbf{Hess}_S(h) = \nabla_S\bigl(\mathbf{grad}_S(h)\bigr) \tag{8}$$

Some tedious calculations then lead to the following explicit representation of $\mathbf{Hess}_S(h)$ as 2×2-matrix.

$$\mathbf{Hess}_S(h) = \bigl(g^{kl}(\partial_j\partial_l h - \partial_i h \Gamma^i_{jl})\bigr)_{kj} \tag{9}$$

$$= \begin{pmatrix} g^{1l}(\partial_1\partial_l h - \partial_i h \Gamma^i_{1l}) & g^{1l}(\partial_2\partial_l h - \partial_i h \Gamma^i_{2l}) \\ g^{2l}(\partial_1\partial_l h - \partial_i h \Gamma^i_{1l}) & g^{2l}(\partial_2\partial_l h - \partial_i h \Gamma^i_{2l}) \end{pmatrix}.$$

2.4 Data dependent approximations to curvature functionals

In the following proposition, the curvature of a surface is expressed using the Hessian. This is fundamental for the derivation of the data dependent approximations that we are going to define after the proof of the proposition.

Proposition 2 *If $S = (S_1, S_2, S_3)$ is a twice differentiable function then*

a) $\sum_i \mathbf{trace}\bigl(\mathbf{Hess}_S(S_i)\bigr)^2 = (\kappa_1 + \kappa_2)^2$

b) $\sum_i \det\bigl(\mathbf{Hess}_S(S_i)\bigr) = \kappa_1 \cdot \kappa_2$

c) $\sum_i \mathbf{trace}\bigl(\mathbf{Hess}_S(S_i)^2\bigr) = \kappa_1^2 + \kappa_2^2$

Proof. By (9) we have $\mathbf{Hess}_S(S_i) = \bigl(g^{kl}(\partial_j\partial_l S_i - \partial_p S_i \Gamma^p_{jl})\bigr)_{kj}$. Using (2) we can replace $\partial_j\partial_l S_i$ and obtain

$$\mathbf{Hess}_S(S_i) = \bigl(g^{kl}(\Gamma^p_{jl}\partial_p S_i + h_{jl}[\vec{n}_S]_i - \partial_p S_i\,\Gamma^p_{jl})\bigr)_{kj} = \bigl(g^{kl} h_{jl}[\vec{n}_S]_i\bigr)_{kj} = \bigl(g^{kl} h_{jl}\bigr)_{jk} \cdot [\vec{n}_S]_i\,.$$

Now $g^{kl} h_{jl}$ is the (k,j)-th entry of the product of the matrices $\mathbf{I}_S^{-1} = (g^{ij})_{ij}$ and $\mathbf{I\!I}_S = (h_{ij})_{ij}$. Therefore we have

$$\mathbf{Hess}_S(S_i) = \mathbf{I}_S^{-1}\mathbf{I\!I}_S \cdot [\vec{n}_S]_i\,. \tag{10}$$

Since $\sum_i [\vec{n}_S]_i^2 = \|\vec{n}_S\|^2 = 1$, statement a) and b) now follow immediately from Proposition 1. Also c) is a consequence when additionally the following fact is taken into consideration. If A is a 2×2-matrix with eigenvalues λ_1 and λ_2, then for any scalar α the matrix $(\alpha A)^2$ has eigenvalues $\alpha^2\lambda_1^2$ and $\alpha^2\lambda_2^2$ and therefore $\alpha^2(\lambda_1^2 + \lambda_2^2) = \mathbf{trace}\bigl((\alpha A)^2\bigr)$. □

If we consider two surfaces F and S, defined over the same parameter space, which are close together (in the sense, that their fundamental forms do not differ very much), then from (9) we conclude that $\mathbf{Hess}_S(h) \approx \mathbf{Hess}_F(h)$ for functions h. Thus, the proposition above leads us to the following claim.

Claim. Assume that F and S have approximately the same shape then we have

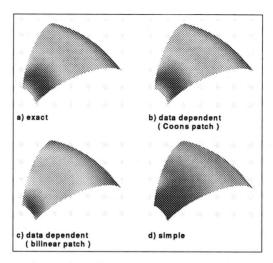

Fig. 2 A piece of a torus, color encoded with the exact mean curvature (upper left), two data dependent approximations (upper right and lower left) and the simple approximation (lower right). At the upper right picture, the reference surface is the linear Coons patch of the boundary curves. In the lower left part, the bilinear patch of the four vertices has been chosen.

- $\kappa_{F1}^2 + \kappa_{F2}^2 \approx \sum_i \operatorname{trace}\left(\operatorname{Hess}_S(F_i)^2\right)$;

- $K_F = \kappa_{F1} \cdot \kappa_{F2} \approx \sum_i \det\left(\operatorname{Hess}_S(F_i)\right)$;

- $4H_F^2 = (\kappa_1 + \kappa_2)^2 \approx \sum_i \operatorname{trace}\left(\operatorname{Hess}_S(F_i)\right)^2$.

The right hand sides are called *data dependent approximation* to the total curvature, the Gaussian curvature and the (square of the) mean curvature respectively.

In Fig. 2 and Fig. 3 several examples are given showing the exact curvature distribution and approximations to it. The surfaces are color encoded by the exact curvature, the data dependent approximation to the curvature and the simple approximation to the curvature. Here "simple approximation" refers to $F_{uu}^2 + 2F_{uv}^2 + F_{vv}^2$ for the total curvature, $\langle F_{uu}|F_{vv}\rangle - F_{uv}^2$ for the Gaussian curvature and $\frac{1}{4}(F_{uu} + F_{vv})^2$ for the (square of the) mean curvature.

3 Interpolating scattered data with tensor product B-spline functions

3.1 The general approach

The problem to be solved can be easily stated as follows.

- *Given data points $P_\rho = (x_\rho, y_\rho)$, $1 \leq \rho \leq r$ in the plane and corresponding data values z_ρ, $1 \leq \rho \leq r$.*

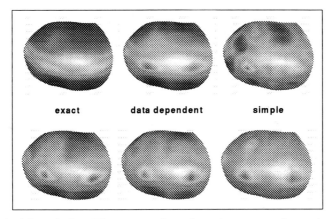

Fig. 3 A part of a deformed sphere. The upper row shows the total curvature and approximations to it. The lower row shows Gaussian curvature. As reference surface for the data dependent approximation of the curvatures, a fitting ellipsoid has been used.

- Find a bicubic TP B-spline function that interpolates the data, i.e. $f(x_\rho, y_\rho) = z_\rho$ for $1 \leq \rho \leq r$. In addition we require that (the graph) of f has a nice shape.

The approach to solve this problem is as follows:

1. Specify uniform knot vectors $\{t_{-2} < t_{-1} < t_0 < ... < t_n < t_{n+1} < t_{n+2}\}$ and $\{s_{-2} < s_{-1} < s_0 < ... < s_m < s_{m+1} < s_{m+2}\}$ such that all data points lie in the rectangle $\Omega := [t_0, t_n] \times [s_0, s_m]$. Let $\mathcal{S}$ be the class of TP B-spline functions defined over the rectangular grid $\{(t_i, s_j)\}$

2. Specify a fairness functional J, measuring somehow the fairness of surfaces defined over Ω.

3. Solve the constrained optimization problem

- $J(F) = \min$, $F(u,v) = (u, v, f(u,v))$ with $f \in \mathcal{S}$
- $f(x_\rho, y_\rho) = z_\rho$ for $1 \leq \rho \leq r$.

Of course, to ensure that interpolation with a TP B-spline function is possible the knot vectors $\{t_i\}$ and $\{s_j\}$ have to be chosen sufficiently fine. Then however there will be many interpolating TP B-spline functions. By the optimization in step 3 we single out the one with optimal fairness with respect to J. In order to keep the optimization process simple we restrict to quadratic fairness functionals as described in the following subsection.

Compared to other interpolation techniques (see [5]) the approach described above, has quite a few advantages.

- the resulting surface is given in a CAD-compatible format (TP B-spline);
- the interpolating surfaces are fair;
- the method can also handle large data sets.

3.2 Data dependent, quadratic fairness functional

We use the results of subsection 2.4 to approximate the exact thin plate energy $J_{exact}(F) = \int_\Omega \kappa_1^2 + \kappa_2^2 + 2\mu\kappa_1\kappa_2\, d\omega_F$ by a quadratic functional. It is given by The term "$\kappa_1^2 + \kappa_2^2$" in J_{exact} expresses bending energy while the remaining term is energy occurring from shearing (see [3]). μ is a constant depending on the material, $0 \leq \mu \leq 1$. The simple quadratic approximation to J_{exact} is given by $J_{simple}(F) = \int_\Omega (F_{uu} + F_{vv})^2 - 2(1-\mu)(\langle F_{uu}|F_{vv}\rangle - F_{uv}^2)\, dudv$.

Better quadratic approximations to J_{exact} will be obtained when reference surfaces $S: \Omega \to \mathbb{R}^3$ are used. Such an S is obtained by a least square fit to the data. Again we use TP B-spline functions. This time over a coarse grid $\{\tilde{t}_{-2} < \tilde{t}_{-1} < \tilde{t}_0 < ... < \tilde{t}_{\tilde{n}} < \tilde{t}_{\tilde{n}+1} < \tilde{t}_{\tilde{n}+2}\}$ ($\tilde{t}_0 = t_0$, $\tilde{t}_{\tilde{n}} = t_n$) and $\{\tilde{s}_{-2} < \tilde{s}_{-1} < \tilde{s}_0 < ... < \tilde{s}_{\tilde{m}} < \tilde{s}_{\tilde{m}+1} < \tilde{s}_{\tilde{m}+2}\}$, ($\tilde{s}_0 = s_0$, $\tilde{s}_{\tilde{m}} = s_m$). The grid size is chosen such that $\tilde{n} \cdot \tilde{m} < \frac{r}{2}$, ($r$ the number of data points). The graph of the least square fit $s: \Omega \to \mathbb{R}$ is used as reference surface S, i.e. $S(u,v) = (u, v, s(u,v))$.

The fairness functional J_S, measuring the fairness of parameterized surfaces having Ω as parameter space is defined as follows.

$$J_S : F \mapsto \sum_i \int_\Omega \text{trace}\left(\text{Hess}_S(F_i)^2\right) + 2\mu \det(\text{Hess}_S(F_i))\, d\omega_S \tag{11}$$

The functional J_S has some interesting properties, which show that it can be treated numerically much simpler then J_{exact}. These properties are listed in the following theorem. We will discuss the consequences more detailed after the theorem is proved. Note that the case of J_{simple} is included, in fact, for $S(u,v) = (u,v,0)$ we have $J_S = J_{simple}$.

Theorem 1 *For a two times differentiable parameterized surface S, the functional J_S defined in (11) has the following properties.*

a) *J_S is a quadratic functional.*

b) *J_S is a positive semi-definite functional. For $\mu < 1$ the null space has dimension ≤ 9.*

c) *$J_S(S) = J_{exact}(S)$.*

d) *J_S does not depend on the specific parameterization of the surface S.*

<u>Supplement to b)</u>. *If S is <u>not</u> a developable surface (and $\mu < 1$) then the null space of J_S has dimension $= 3$.*

Proof. We only consider the case $\mu = 0$.
a) From (9) it follows that for fixed S $\text{Hess}_S(F_i)$ linearly depends on the first and second order partial derivatives of F_i. Thus $\text{Hess}_S(F_i)^2$ quadratically depends on these derivatives and so does its trace.

b) Semi-definiteness is clear from the definition of J_S (note that the trace of the square of a symmetric matrix A is always ≥ 0.) Moreover, assuming that $J_S(F) = 0$, then $\text{Hess}_S(F_i) = 0$. This means that the covariant derivative of the vector field $\text{grad}_S(F_i)$ vanishes. Therefore the vector field $\text{grad}_S(F_i)$ is composed by parallel displacements of a unique tangent vector on S. Such a vector can be specified by two coordinates. Finally, when integrating the gradient

$\mathbf{grad}_S(F_i)$ to obtain F_i, there is an additional constant, the value of F_i at a single point. Thus for each component F_i we have three degrees of freedom.

Proof of the supplement: Non-zero parallel vector fields only exist on developable surfaces (see [10]). Hence if S is not developable and $J_F(S) = 0$ the conclusion above implies that $\mathbf{grad}_S(F_i)$ vanishes identically ($i = 1, 2, 3$). Thus $F(u, v) = P = const$ for all $(u, v) \in \Omega$

c) follows from assertion d) of Proposition 2.

d) Both, the gradient and the covariant derivative are *geometric invariants* (see Lemma 4.1.5 in [9]), that is, they do not depend on a specific parameterization of S. Since our construction only uses these notions, J_S is independent of the parameterization of S as well. □

Remark. For $\mu = 1$ we have $J_{exact}(F) = \int_\Omega (\kappa_1 + \kappa_2)^2 \, d\omega_F = 4 \int_\Omega H_F^2 \, d\omega_F$ and $J_S(F) =$
$\sum_i \int_\Omega \mathbf{trace}\left(\mathbf{Hess}_S(F_i)^2\right) + 2 \det(\mathbf{Hess}_S(F_i)) \, d\omega_S = \sum_i \int_\Omega \left(\mathbf{trace}(\mathbf{Hess}_S(F_i))\right)^2 d\omega_S =$
$= \int_\Omega \left(\Delta_S(F)\right)^2 d\omega_S$ where Δ_S is the Laplacian on the surface S (often called Laplace-Beltrami-operator). In this case, both for J_{exact} and J_S the null space will not be finite dimensional. In fact, to any (reasonable) boundary condition there is a minimal surface and also a harmonic function. Minimal surfaces have zero mean curvature thus being in the null space of J_{exact}. Harmonic functions satisfy $\Delta_S(F) = 0$ which shows that they are in the null space of J_S.

Discussion and consequences of the theorem.

- By a) J_S has a representation

$$J_S(F) = \sum_{1 \le |\alpha|, |\beta| \le 2} \int_\Omega w_{\alpha\beta}(u, v) \langle D^\alpha F | D^\beta F \rangle \, du dv \qquad (12)$$

with certain weight functions $w_{\alpha\beta}$ depending on the (first fundamental form) of the reference surface. At a first glance J_S looks quite complicated. However, since it is quadratic, it can be minimized very easily, much faster than J_{exact}. In fact, the minimum of such a functional can be obtained as the solution of a linear system.

- A consequence to b) is that for $\mu < 1$ only few constraints are necessary in order to guarantee a unique solution to a constraint optimization problem for J_S. In general, specifying 3 (vector) constraints, will lead to a positive definite problem. For example, in our situation, if at least 3 data are given that do not lie on a line, the problem will be positive definite. For a non developable reference surface even one interpolation condition ensures positive definiteness.

- Assuming that the least square fit is already a good approximation to the interpolating surface, assertion c) implies that J_S is (locally) a good approximation to J_{exact}.

3.3 Solving the optimization problem

Let S denote the bicubic TP B-spline functions over the grid $\{(t_i, s_j)\}$. And let $\{N_i^u : 0 \le i \le n\}$ be the (univariate, cubic) B-spline basis functions subordinated to the knot vector $\{t_i\}$. The support of N_i^u is the interval $[t_{i-2}, t_{i+2}]$. Similarly $\{N_j^v : 0 \le j \le m\}$ are the basis functions with respect to the knot vector $\{s_j\}$.

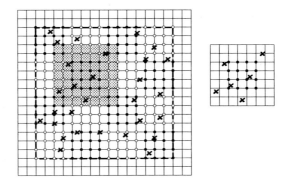

Fig. 4 The grid points are subdivided in subsets of 4 by 4. To such a subset those data points are associated that lie in the union of the supports (e.g. gray region). The corresponding *local interpolation problem* is sketched on the right (6 data points, 16 control points).

A function $f \in S$ is given by $f(u,v) = \sum_{ij} d_{ij} N_i^u(u) N_j^v(v)$. The control points $\{d_{ij}\}$ will be determined by solving the constrained optimization problem

- $J_S(F) = \min$, $F(u,v) = (u,v,f(u,v))$ with $f \in S$
- $f(x_\rho, y_\rho) = z_\rho$ for $1 \leq \rho \leq r$.

We use the method of Lagrangian multipliers. This leads to a linear system of size $(nm + r) \times (nm + r)$ for the control points $\mathbf{d} = \{d_{ij}\}$ and r Lagrangian multipliers $\boldsymbol{\lambda} = (\lambda_1, ..., \lambda_r)$

$$\begin{pmatrix} \mathbf{A} & \mathbf{P}^t \\ \mathbf{P} & \mathbf{0} \end{pmatrix} \begin{pmatrix} \mathbf{d} \\ \boldsymbol{\lambda} \end{pmatrix} = \begin{pmatrix} \mathbf{0} \\ \mathbf{z} \end{pmatrix}$$

The entries of the $nm \times nm$ matrix $\mathbf{A}$ are given by

$$\mathbf{A}_{ij\,pq} = \sum_{1 \leq |\alpha|, |\beta| \leq 2} \int_\Omega w_{\alpha\beta}(u,v) N_i^{u(\alpha_1)}(u) N_j^{v(\alpha_2)}(v) N_p^{u(\beta_1)}(u) N_q^{v(\beta_2)}(v) \, du dv$$

and will be determined by Gaussian quadrature. $\mathbf{P}$ is a $r \times nm$ matrix, $\mathbf{P}_{ij\,\rho} = N_i^u(x_\rho) N_j^v(y_\rho)$. On the right hand side the data values $\mathbf{z} = (z_1, ..., z_r)$ appear.

Due to the local support of the basis functions the linear system is sparse. We use a block SOR method to solve this system. Each block is formed by a collection of 4 by 4 control points and the interpolation conditions corresponding to these control points (see Fig. 4) In each single iteration step one has to solve one of these *local interpolation problems*.

In Fig. 5 an example is shown. The Ritchie surface has been sampled at 100 randomly chosen points. The picture on the left hand side gives two views of the original surface and show the distribution of the sample points. The picture in the middle and on the right show the surfaces obtained by interpolating the 100 data. Both are bicubic TP B-spline functions over a uniform grid of knots. Each surface has 32×32 control points. In the middle the functional J_{simple} (for $\mu = 0$) has been minimized. The interpolating surface obtained by minimizing J_S ($\mu = 0$) is shown on the right. As reference surface S a least square fit to the data has been used. It the bicubic TP B-spline with 7×7 control points.

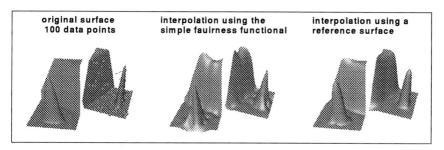

Fig. 5 One of the standard test examples: the Ritchie surface interpolated at 100 sample points.

References

[1] G. Brunnett, H. Hagen, and P. Santarelli. Variational design of curves and surfaces. *Surv. Math. Industry*, 3:1–27, 1993.

[2] G. Celniker and D. Gossard. Deformable curve and surface finite–element for free–form shape design. *ACM Computer Graphics*, 25:257–266, 1991.

[3] R. Courant and D. Hilbert. *Methods of Mathematical Physics, Vol. 1*. Wiley, New York, N.Y., 1953.

[4] M. Eck and J. Hadenfeld. Local energy fairing of B-spline curves. In G. Farin, H. Hagen, and H. Noltemeier, editors, *Computing Supplementum* **10**. Springer Verlag, 1995.

[5] R. Franke and G. M. Nielson. Scattered data interpolation and applications: A tutorial and survey. In H. Hagen and D. Roller, editors, *Geometric Modelling, Methods and Applications*, pages 131–160. Springer Verlag, 1991.

[6] G. Greiner. Surface construction based on variational principles. In P. J. Laurent, A. LeMéhauté, and L. L. Schumaker, editors, *Wavelets, Images, and Surface Fitting*, pages 277–286, Wellesley MA, 1994. AKPeters.

[7] G. Greiner. Variational design and fairing of spline surfaces. *Computer Graphics Forum*, 13(3):143–154, 1994.

[8] M. Kallay. Constrained optimization in surface design. In B. Falcidieno and T. L. Kunii, editors, *Modeling in Computer Graphics*, Berlin, 1993. Springer–Verlag.

[9] W. Klingenberg. *A Course in Differential Geometry*. Springer–Verlag, Berlin Heidelberg, 1978.

[10] S. Kobayashi and K. Nomizu. *Foundations of differential geometry (Vol. I)*. Interscience Publ., New York, 1963.

[11] H. P. Moreton and C. Séquin. Functional optimization for fair surface design. *ACM Computer Graphics*, 26:167–176, 1992.

[12] H. P. Moreton and C. H. Séquin. Scale-invariant minimum-cost curves: Fair and robust design implements. *Computer Graphics Forum*, 12(3):473–484, 1993.

[13] W. Welch and A. Witkin. Variational surface modeling. *ACM Computer Graphics*, 26:157–166, 1992.

Scattered Data Approximation with Triangular B-Splines

Ron Pfeifle, Hans-Peter Seidel

Universität Erlangen

Abstract: Scattered data is, by definition, irregularly spaced. Uniform surface schemes are generally not well adapted to the locally varying nature of such data. Conversely, triangular B-spline surfaces are more flexible in that they can be built over arbitrary triangulations and thus can be adapted to the scattered data.

This paper discusses the use of DMS spline surfaces for approximation of scattered data. A method is provided for automatically triangulating the domain containing the points and generating basis functions over this triangulation. A surface approximating the data is then found by a combination of least squares and bending energy minimization. This combination serves both to generate a smooth surface and to accommodate for gaps in the data. Examples are presented which demonstrate the effectiveness of the technique for mathematical, geographical and other data sets.

1 Introduction

Scattered data fitting problems come from many different areas, including scientific and engineering data visualization, mining and mapping, geographic information systems, meteorology and other fields [15]. Often we would like to visualize the geometry of the surface that corresponds to a set of scattered data in order to better understand the underlying data.

The problem we address is the fitting of a functional surface $F(x,y)$ to a collection of scattered functional data $\{(x_i, y_i, z_i(x_i, y_i))\}$. Our goal is to find a smooth surface F that is a reasonable approximation to the data.

The kind of functional surface we choose is the DMS spline formulation [4]. The DMS spline surfaces have numerous positive characteristics that make them appropriate for this data fitting problem, including their automatic smoothness properties, the ability to define a surface over an arbitrary triangulation (which can be adapted to the local density of sampled data) and their "completeness", in that all piecewise polynomials of a particular degree over a given triangulation can be represented by a DMS spline of the same degree.

Schumaker [18] first introduces the problem of scattered data approximation. Auerbach *et al.* [1] examine the functional data approximation problem for a slightly different simplex spline space [12, 3]. We find that they do not fully address the problem of producing a good triangulation of a data set, nor do they describe how to deal with data sets that cause difficulties when using the least squares method.

The following describes the organization of the paper. In Section 2 we review the definition of simplex splines and that of the DMS spline scheme. In Section 3, we examine how a triangulation of the domain may be formed that adapts to the distribution of sample data. In Section 4, we look at the minimization problem used to find the approximating surface, namely the *least squares method* and describe how it can be augmented with a smoothing term in order to overcome certain difficulties. Lastly, we present our conclusions and suggestions for further work in Section 5. Examples fitting quadratic DMS splines (with C^1 continuity) to various data sets appear as figures throughout the paper.

2   DMS Splines

A DMS spline surface is formed as a linear combination of basis functions. These basis functions are defined over a triangulation of the plane. This section explains how these basis functions are constructed [4, 19].

Bivariate simplex splines form the individual basis functions for the DMS spline surface. Simplex splines are defined *via* affine combinations of points. Formally, consider a point u and a set of points $V = \{t_0, \ldots, t_{n+2}\}$, called *knots*, from $\mathbb{R}^2$. The degree n simplex spline $M(u|V)$ is defined recursively as follows [14]:

- For $n > 0$, with $V = \{t_0, \ldots, t_{n+2}\}$, select three points $W = \{t_{i_0}, t_{i_1}, t_{i_2}\}$ from V, such that W is affinely independent. Then

$$M(u|V) = \sum_{j=0}^{2} \lambda_j(u|W) \, M(u|V \setminus \{t_{i_j}\}) \qquad (1)$$

where $u = \sum_{j=0}^{2} \lambda_j(u|W) t_{i_j}$ and $\sum_{j=0}^{2} \lambda_j(u|W) = 1$.

- For $n = 0$, with $V = \{t_0, t_1, t_2\}$,

$$M(u|t_0, t_1, t_2) = \frac{\chi_{[t_0,t_1,t_2]}(u)}{2|\Delta(t_0, t_1, t_2)|} \quad \text{where} \quad \chi_{[t_0,t_1,t_2]} = \begin{cases} 1 & \text{if } u \in [t_0, t_1, t_2) \\ 0 & \text{otherwise} \end{cases} \qquad (2)$$

with $[t_0, t_1, t_2)$ being the *half-open convex hull* of points t_0, t_1 and t_2.

Remark: u is in $[t_0, t_1, t_2)$ if the set $\{u + s\eta + t\xi \mid s, t > 0, \, s + t < \epsilon\}$ is contained within the convex hull of $\{t_0, t_1, t_2\}$, for some $\epsilon > 0$, ξ being the horizontal unit vector in $\mathbb{R}^2$ and η a vector with positive slope [19].

A slightly modified version of Equation 1 can be used to find the derivative of $M(u|V)$ with respect to a parametric vector v:

$$D_v M(u|V) = n \sum_{j=0}^{2} \mu_j(v|W) M(u|V \setminus \{t_{i_j}\}) \qquad (3)$$

with $v = \sum_{j=0}^{2} \mu_j(v|W) t_{i_j}$ and $\sum_{j=0}^{2} \mu_j(v|W) = 0$.

Simplex splines possess a number of interesting properties. They are piecewise polynomial of degree n, are zero outside the convex hull of the knots V and non-negative within it, and are smooth, in the sense that if the knots of V are in general position, then $M(u|V)$ exhibits C^{n-1} continuity.

Simplex splines can now be combined to form a spline surface [4]. Let T be an arbitrary proper triangulation of $\mathbb{R}^2$ or some bounded domain $D \subset \mathbb{R}^2$. "Proper" means that every pair of domain triangles I, J are disjoint, or share exactly one edge, or exactly one vertex.

To each vertex t_i of the triangulation T, we assign a *knot cloud*, which is a sequence of points (*knots*) $t_{i,0}, \ldots, t_{i,n}$, where $t_{i,0} \equiv t_i$. For each triangle $\Delta = (t_0, t_1, t_2) \in T$, we require that $(t_{0,i}, t_{1,j}, t_{2,k})$ always form a proper triangle. We then define, for each Δ and $i + j + k = n$, the knot sets

$$V_{i,j,k}^{\Delta} = \{t_{0,0}, \ldots, t_{0,i}, t_{1,0}, \ldots, t_{1,j}, t_{2,0}, \ldots, t_{2,k}\}$$

which yields $\binom{n+2}{n}$ simplex splines $M(u|V_{i,j,k}^{\Delta})$.

The *normalized B-splines* are then defined as $N_{i,j,k}^{\Delta}(u) = d_{i,j,k}^{\Delta} M(u|V_{i,j,k}^{\Delta})$, where $d_{i,j,k}^{\Delta}$ is defined to be twice the area of $\Delta(t_{0_i}, t_{1_j}, t_{2_k})$. This normalization ensures that the basis functions sum to one.

A functional surface F of degree n over the triangulation T with *knot net* $\mathcal{K} = \{t_{i,l} | i \in \mathbb{Z}, l = 0, \ldots, n\}$ is then defined as

$$F(u) = \sum_{\Delta \in T} \sum_{i+j+k=n} c_{i,j,k}^{\Delta} N_{i,j,k}^{\Delta}(u),$$

with coefficients $c_{i,j,k}^{\Delta} \in \mathbb{R}$. These normalized B-spline surfaces demonstrate properties similar to B-spline curves, in that they are affine invariant, have local control, and exhibit the Convex Hull Property. Moreover, every degree n piecewise polynomial over a triangulation T can be represented as a normalized B-spline surface.

3 Finding a Triangulation

DMS spline basis functions are defined with respect to a triangulation of the domain, so our first task is to generate an appropriate triangulation of the portion of the domain containing our sampled data. Once a set of basis functions has been determined, an approximating surface F in the span of these functions can be found by functional minimization, as described in Section 4.

3.1 Properties of a Good Triangulation

The triangulation we form should possess the following four properties:

1. All sample points must be contained in some triangle of the triangulation.
2. No triangle has too many or too few sample points within it, and points within each triangle are distributed as uniformly as possible.
3. Triangles are not too elongated.
4. Neighbouring triangles are roughly comparable in size.

Property One ensures that all of the data has an influence on the final surface. Property Two ensures that each data point is well-represented by the surface. Properties Three and Four ensure that we get a "good" set of basis functions. Very thin triangles could cause numerical problems in evaluation.

Unfortunately, in extreme cases, our goals may not be consistent with one another. For example, if all our sample points are nearly collinear, then to satisfy Property Two we should probably have elongated triangles in our triangulation. This would violate Property Three.

It is clear that some other means of triangulating the data points must be used that adapts the size of triangles used to the local density of the data. The initial temptation is to use the sample (x, y) values as vertices of a Delaunay triangulation [8]. This is, of course, ridiculous, as this choice results in an explosion of triangles. Another adaptive data structure commonly used for 2D domains is the quadtree, which we will use in the next section in order to build a triangulation.

3.2 Quadtree Division of the Domain

We now describe a triangulation scheme that satisfies Properties One, Three and Four and partially satisfies Property Two. We will rely on the minimization technique of Section 4.2 to ameliorate the remaining difficulties with the triangulation scheme.

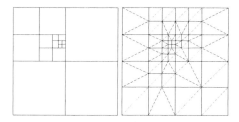

Fig. 1 An unbalanced quadtree with large leaf nodes adjacent to small ones, and the corresponding balanced quadtree, with triangulation edges added.

We begin by finding a bounding box around the data, and generate a quadtree partition of this box such that no leaf node contains more than a specified number of sample points. The corners of leaf nodes generated by the quadtree division can be used as vertices of (relatively) equally sized triangles.

This simple quadtree violates Property Three for data sets that contain tight clusters of points. Point clusters can cause finely-subdivided areas to be adjacent to far more coarsely subdivided areas, which, in turn, generates long thin triangles. This can be remedied by requiring that the quadtree be *balanced* [17]. A quadtree is balanced if the depths of two adjacent leaf nodes differ by at most one.

A potential problem exists when the bounding box is a poor fit to the data, that is, when large areas within the bounding box contain no data. In this case, one could accept user input to decide whether to exclude leaf nodes containing no sample points from the triangulation.

3.3 Assigning Knot Clouds

Once a triangulation has been formed, knot clouds are added to vertices of the triangulation. The knot clouds are chosen so as to avoid collinearity of knots associated with a particular triangle. The method used is similar to the one given in Auerbach et al. [1, page 81]. The selection of the knot clouds then defines the basis functions for the surface.

Knots can also be selected to promote collinearity along certain edges of the triangulation. If $k+2$ of the knots used to define a particular DMS basis function are placed collinearly, then the continuity of the surface along that parametric line will be reduced by k. Reduced continuity can be introduced into the surface using this method.

Once a triangulation and its knot clouds have been defined, linear combinations of the corresponding DMS basis functions can be used to form surfaces. In the next section we discuss the problem of selecting, from among this set of surfaces, some surface F that is a good approximation to the data.

4 Least Squares Fitting and Fairing by Minimizing a Functional

In this section, we assume that a fixed set of basis functions exists, and that a surface F in the span of those functions should be used to approximate the data.

4.1 Least Squares Minimization

We use the least squares method [13, 5] for approximating a set of functional scattered data. Given a surface $F(x, y)$, the unweighted least squares functional $LS(F) = \sum_l (F(x_l, y_l) - z_l)^2$ provides a measure of how well F approximates the data. If we minimize this sum, we will obtain a good approximation to the data. Since F is the linear combination of some fixed set of basis functions, the functional LS is quadratic in their coefficients, and so the minimization problem can be expressed as a linear system in the coefficients of those basis functions. Standard solution techniques form either the *observation equations* or *normal equations* for a given data set [5]. Both kinds of systems can be solved using standard techniques.

In order to form the individual entries of either of these matrices, each basis function (in our case, each DMS basis function $N_{i,j,k}^{\triangle}$) must be evaluated at each (x_l, y_l) location. This poses no particular difficulty, as evaluation algorithms for DMS splines based on Equation 1 are readily available [7, 16].

Formulating the approximation problem as a least squares problem has a number of advantages and disadvantages. The main advantages of the least squares method are that it is simple to understand and relatively easy to implement. The chief disadvantage is that the solution is very sensitive to the location of data points with respect to the given set of basis functions. A particular basis function may have no influence on the value of the least squares functional (because it contains no data points within its region of support), or a number of basis functions may be linearly dependent with respect to the given data. Moreover, the surface determined by the least squares approach may very well lie close to data points, but may not be very smooth.

In order to deal with these problems we modify the least squares functional to take surface smoothness into account. The smoothness factor can be used to assign reasonable values to coefficients undetermined by the data, by choosing values that make the surface as smooth as possible.

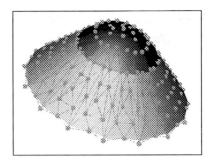

Fig. 2 Capping a truncated cone by minimizing thin-plate energy

4.2 Smoothing by Minimizing Thin Plate Energy

"Fairing" is the process of reducing irregularities in a surface in order to make it smoother. The typical fairing process proceeds by defining a fairness functional for a given set of surfaces, and finding the minimum surface F with respect to that functional. Different functionals have been defined and used for this purpose [9, 10].

If we choose a functional that is quadratic in the basis function coefficients, then as is the case with the least squares functional, it is possible to express the minimization problem as a linear system that can be solved using matrix techniques.

Since we are dealing with *functional* DMS surfaces, an appropriate smoothing functional is the *linearized thin plate energy* functional $J(F(x,y))$ [9, 10, 2, 11], defined as

$$J(F) = \int_\Omega F_{xx}^2 + 2F_{xy}^2 + F_{yy}^2 \, dx \, dy.$$

The region Ω can be used to restrict the functional to only part of the surface in question, thereby localizing the smoothing effect. Figure 2 shows a truncated cone which has been capped in such a way as to minimize the thin plate energy of the capping surface and form a C^1 join with the truncated cone.

When actually minimizing this functional, we form a linear system similar to the normal equations used for least squares minimization. If we define $N_I = N_{i,j,k}^\Delta$ and $N_J = N_{l,m,n}^{\Delta'}$ to be two DMS basis functions (not necessarily distinct) then the individual entries $E_{I,J}$ of this matrix are of the form:

$$\int_\Omega N_{I_{xx}} N_{J_{xx}} + 2 N_{I_{xy}} N_{J_{xy}} + N_{I_{yy}} N_{J_{yy}} \, dx \, dy.$$

In order to find this matrix, we must evaluate this integral. We can use Equation 3 to express the second derivative of each DMS basis function in terms of simplex splines of lower order. Once expressions for the second derivatives of two basis functions have been obtained, we can find the integral of their product. It is conceivable to do this numerically, but we would like to represent these second derivatives symbolically, so that the integral can be found exactly.

If the DMS spline is of degree two, then each of its second derivatives is piecewise constant. It follows from Equation 3 that each second derivative can be expressed as the sum of many

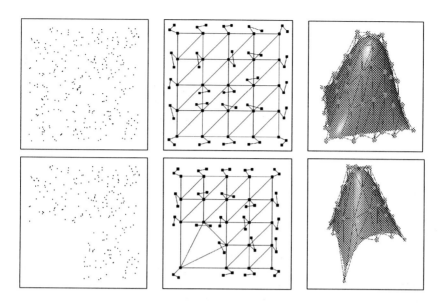

Fig. 3 Top Row: *Left*, the locations in $[0,1] \times [0,1]$ of data points where the function $\sin(\pi x)\sin(\pi y)$ is randomly sampled. *Middle*, quadtree-based triangulation of the region, showing the knot cloud associated with each vertex. *Right*, surface fit using pure least squares, with control points. Bottom Row: *Left*, as above, but with data missing from the region $[0, 0.5] \times [0, 0.5]$. *Middle*, corresponding triangulation with knot clouds. Triangles in the lower left corner contain no data. *Right*, surface fit using a combination of least squares and thin plate energy minimization. The front corner of the surface corresponds to the domain triangles that have no data. Because of the missing data, this data set cannot be fit without the assistance of a smoothing term.

special piecewise constant functions, where each special function is a non-zero constant within a (half-open) triangular region of the domain and vanishes outside it. Let $S_I = \sum_i T_i$ and $S_J = \sum_j T_j$ be second derivatives of N_I and N_J, respectively, represented as sums of special piecewise constant functions. The integral of their product can be rewritten as:

$$\int_\Omega S_I S_J \, dx \, dy = \int_\Omega (\sum_i T_i)(\sum_j T_j) \, dx \, dy$$
$$= \sum_i \sum_j \int_\Omega T_i T_j \, dx \, dy$$

Integration of the product of two special piecewise constant functions can be performed by clipping the domain triangle of one against the domain triangle of the other. The resulting polygon is then clipped against the Ω region and the area of the result is multiplied by the heights of both piecewise constants (in Traas [20], the intersection of two triangles is found using linear programming). The partial results are then summed to form the final integral. Since the regions of support of most DMS basis functions will be disjoint from one another, we can accelerate the calculation if we only compute the product of special functions that lie in the intersection of the bounding boxes of the two basis functions N_I and N_J. This

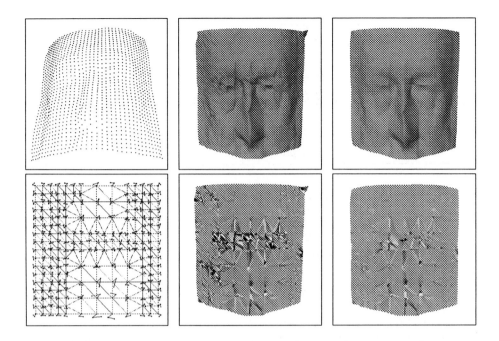

Fig. 4 Fitting a surface to data taken from a bust of Victor Hugo (data courtesy the University of Waterloo Computer Graphics Lab). *Top Left*, the data points in 3D. *Bottom Left*, corresponding triangulation with knot clouds. *Top Middle*, surface fit using pure least squares. *Top Right*, surface fit using mixed least squares and thin plate energy minimization ($\alpha = 0.25$) *Bottom Middle/Right*, corresponding plots of Gaussian curvature. Darker regions have negative curvature, lighter regions have positive curvature. The mixed fit surface exhibits far less variation in curvature.

method deals with the integration problem uniformly for degree two surfaces, no matter how the knots of any particular basis function are configured.

The smoothing functional cannot be used on its own for scattered data fitting. However, it is easy to combine thin plate energy with least squares and obtain a functional that, when minimized, tries to smooth the surface while fitting the surface to the scattered data.

4.3 Combining Least Squares and Smoothing

One way of combining a smoothing factor with our least squares functional is to form a linear combination of the original least squares functional and the thin plate energy functional:

$$LSJ(F) = (1 - \alpha) LS(F) + \alpha J(F), \quad 0 \leq \alpha \leq 1$$

By changing the value of α, we can vary the relative strengths of the least squares approximation and the thin plate energy terms. The least squares term ensures that the surface approximates the given data points, while the smoothness term ensures that the surface maintain a certain degree of smoothness, and that that basis functions that are either under-

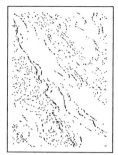 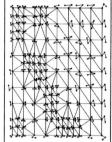 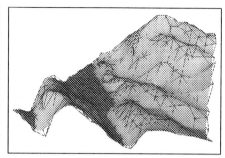

Fig. 5 Geographic elevation data for a portion of Île D'Orléans, downstream from Québec City (US Geological Survey data set *quebec-e* [EC85]). *Left,* data sampled at 21m contour levels (East is to the top). *Middle,* the quadtree-based triangulation with knot clouds. *Right,* side view of the surface fitted using a combination of least squares and thin plate energy minimization, with control net (surface patches lying substantially in the St. Lawrence River are darkly coloured). The vertical scale has been exaggerated 40 times to enhance detail.

determined or linearly dependent with respect to the data are assigned values minimizing bending energy.

The process of finding an approximating surface can now be summarized. First, the data points are used to define a triangulation, based on the quadtree approach outlined above. Knot clouds are then assigned to each vertex of the triangulation. This defines the DMS basis functions, which gives us a space of functions in which to find our approximating surface F.

In both the purely least squares and mixed least squares thin plate energy cases, we evaluate the set of basis functions at each sample (x, y) location. When minimizing a purely least squares problem, these values are assembled into an observation matrix and the basis function coefficients are calculated using linear algebra techniques.

If we wish to solve a least squares with smoothing problem, then the second derivatives of the basis functions must also be evaluated symbolically as outlined in Section 4.2 using the bounding box of the quadtree as the Ω region. The results of integration are then placed into a matrix, combined with the normal equations assembled for the purely least squares solution (taking the α factor into account), and the basis function coefficients are found using linear algebra techniques.

5 Conclusions

We have presented a way of using DMS spline surfaces for functional scattered data approximation. The method generates a triangulation of the parameter domain which, in turn, defines a set of basis functions adapted to the local density of the data. A surface is then found using either a pure least squares method or a combination of least squares and thin-plate energy minimization. The combined method allows data sets to be fit which cannot be fit using the least squares method alone.

Further work should be done in the following areas. By placing certain knots collinearly it is possible to have some of a DMS spline surface's triangular patches meet with lower-than-maximal continuity. This is useful for modeling known discontinuities in the data. This scheme should be extended to incorporate into the triangulation parametric edges where lower continuity is desired.

The possibility of finding a different triangulation method that completely addresses our triangulation goals should be more fully examined. The question of finding a "fair" triangulation (in the sense of Property Two) should be more completely explored, as it is likely that it could be used in a wider setting to characterize attributes of a data set.

References

[1] S. Auerbach, R. H. J. Gmelig Meyling, M. Neamtu, and H. Schaeben. Approximation and geometric modeling with simplex B-splines associated with irregular triangles. *Computer Aided Geometric Design*, 8:67–87, 1991.

[2] G. Celniker and D. Gossard. Deformable curve and surface finite-elements for free-form shape design. In *Proceedings of SIGGRAPH '91*, pages 257–265. ACM SIGGRAPH, acm Press, 1991.

[3] W. Dahmen and C. A. Micchelli. On the linear independence of multivariate B-splines, I. triangulations of simploids. *SIAM Journal on Numerical Analysis*, 19(5):993–1012, October 1982.

[4] W. Dahmen, C. A. Micchelli, and H.-P. Seidel. Blossoming begets B-spline bases built better by B-patches. *Mathematics of Computation*, 59(199):97–115, July 1992.

[5] P. Dierckx. *Curve and Surface Fitting with Splines*. Monographs on Numerical Analysis. Clarendon Press, 1993.

[6] A. A. Elassal and V. M. Caruso. *Digital Elevation Models*. USGS Digital Cartographic Data Standards. U.S. Geological Survey, 1985.

[7] P. Fong and H.-P. Seidel. An implementation of triangular B-spline surfaces over arbitrary triangulations. *Computer Aided Geometric Design*, 10:267–275, 1993.

[8] S. Fortune. Voronoi diagrams and Delaunay triangulations. In D.-Z. Du and F.K. Hwang, editors, *Computing in Euclidean Geometry*, pages 193–233. World Scientific Publishing, 1994.

[9] G. Greiner. Surface construction based on variational principles. In P. J. Laurent, A. Le Méhauté, and L. L. Schumaker, editors, *Wavelets, Images and Surface Fitting*. A K Peters, 1994.

[10] G. Greiner. Variational design and fairing of spline surfaces. In *Proceedings of EUROGRAPHICS '94*, pages C.143–C.154. Eurographics Association, Blackwell, 1994.

[11] M. Halstead, M. Kass, and T. DeRose. Efficient, fair interpolation using Catmull-Clark surfaces. In *Proceedings of SIGGRAPH '93*, pages 35–44. ACM SIGGRAPH, acm Press, 1993.

[12] K. Höllig. Multivariate splines. *SIAM Journal on Numerical Analysis*, 19(5):1013–1031, October 1982.

[13] J. Hoschek and D. Lasser. *Grundlagen der geometrischen Datenverarbeitung.* Teubner, Stuttgart, 1989.

[14] C. A. Micchelli. On a numerically efficient method for computing multivariate B-splines. In W. Schempp and K. Zeller, editors, *Multivariate Approximation Theory.* Birkhäuser, Basel, 1979.

[15] G. M. Nielson. Scattered data modeling. *IEEE Computer Graphics & Applications*, 13(1):60–70, January 1993.

[16] R. Pfeifle. *Approximation and Interpolation using Quadratic Triangular B-Splines.* PhD thesis, Universität Erlangen, Computer Graphics Group, 1996.

[17] H. Samet. *Design and Analysis of Spatial Data Structures.* Addison-Wesley, 1990.

[18] L. L. Schumaker. Fitting surfaces to scattered data. In G. G. Lorentz, C. K. Chui, and L. L. Schumaker, editors, *Approximation Theory.* Academic Press, New York, 1976.

[19] H.-P. Seidel. Polar forms and triangular B-Spline surfaces. In *Blossoming: The New Polar-Form Approach to Spline Curves and Surfaces SIGGRAPH '91 Course Notes # 26.* ACM SIGGRAPH, 1991.

[20] C. R. Traas. Practice of bivariate quadratic simplicial splines. In W. Dahmen, M. Gasca, and C. A. Micchelli, editors, *Computation of Curves and Surfaces*, NATO ASI Series, pages 383–422. Kluwer Academic Publishers, Dordrecht, 1990.

Authors

Prof. Dr. Hans-Peter Seidel
Lehrstuhl für Informatik
Friedrich–Alexander Universität
Am Weichselgarten 9
D–91058 Erlangen
E–mail: seidel@informatik.uni–erlangen.de

Ron Pfeifle
Lehrstuhl für Informatik
Friedrich–Alexander Universität
Am Weichselgarten 9
D–91058 Erlangen

Benchmarks

Benchmarking in the Area of Planar Shape-Preserving Interpolation

Panagiotis D. Kaklis and Nikolaos C. Gabrielides

National Technical University of Athens
Department of Naval Architecture and Marine Engineering,
Ship-Design Laboratory, CAGD Group.

Abstract: This paper contains a representative sample of the benchmarking activity of FAIRSHAPE in the area of Shape-Preserving Interpolation. Four (4) partners of the project, namely the Universities of Dundee (UK), Florence (IT) and Zaragoza (SP), as well as the National Technical University of Athens (GR), have tested their algorithms for planar shape-preserving interpolation against two (2) academic and ten (10) industrial data sets. The outcomes of the algorithms are presented, in tabular and graphical form, and compared on the basis of a set of fairness measures, which were agreed upon at the 2nd Internal Workshop of FAIRSHAPE (Berlin, August 1995).

1 Introduction

The numerical results, presented in this paper, are provided by the following four (4) partners of FAIRSHAPE: the University of **DUN**dee (senior research scientist: T.N.T. Goodman), the University of **FLO**rence (senior research scientists: F. Fontanella and P. Costantini), the University of **ZAR**agoza (senior research scientists: M. Gasca and J. Carnicer) and the National Technical University of **A**thens (**NTUA**) (senior research scientist: P.D. Kaklis). The shape-preserving interpolation methods used for obtaining the DUN, FLO, ZAR and NTUA results are described in [1] [2], [5], [3] and [4], respectively.

The fairness measures for two-dimensional curves $\mathbf{Q}(u)$, $u \in I_u \subset \mathbb{R}$, fixed by the 2nd Internal Workshop of FAIRSHAPE, have as follows:

1. Analytical type of the curve (*type* for short),
2. Continuity of the curve (*conti/ty* for short),
3. Maximum absolute curvature $|\kappa|_{max} = \max_{u \in I_u} |\kappa(u)|$, with $\kappa(u)$ denoting the curvature of $\mathbf{Q}(u)$ at the parametric point u,
4. Integral of the square of the curvature $E_\kappa = \int_\mathcal{L} |\kappa(s)|^2 ds$, with s denoting the curve element of $\mathbf{Q}(u)$ and $\mathcal{L}$ its length,
5. Integral of the square of the arc-length derivative of the curvature $E_{\kappa'} = \int_\mathcal{L} |d\kappa(s)/ds|^2 ds$,
6. The maximum absolute discontinuity of $d\kappa/ds$, and
7. The sum of all absolute discontinuities of $d\kappa/ds$.

Due to lack of output data, fairness measures 6 and 7 will not be considered in this work.

2 The Benchmarking Output

The collected benchmarking output is associated with twelve planar data sets, namely the *academic* data sets: *PRUESS_DATA, NTUA_45* and the *industrial* data sets: *HOOD_j, j=1,3,5,7,9* and *ROOF_i, i=1,2,3,4,5*, stemming from the *hood* and the *roof* of a a car respectively. The industrial data sets are provided by Mercedes-Benz AG, the one of our two industrial partners. The co-ordinates of these data sets can be obtained by anonymous ftp from the directory: *pub/deslab/CAGD/FAIRSHAPE_bench_SPIN* of the ftp-site: *ftp.deslab.naval.ntua.gr* of NTUA.

The benchmarking output is documented, for each data set, with a table and pairs of figures, containing curves along with their curvature distributions. Tables collect the available numerical output, i.e., the values of the fairness measures of the shape-preserving interpolants provided by the participating partners (see, e.g., Table 1 for *PRUESS_DATA*), while figures provide a representative part of the corresponding graphical output; for example, in the case of *PRUESS_DATA*, Figs. 1-2 and 3-4 depict the graphical output provided by FLO and NTUA, respectively. All figures are collected at the homonymous last section (§4) of the paper.

Regarding the notation used in the tables, the symbols "pol./rat.(=m)" and "pol./rat.($\leq n$)" signify that the interpolant is a polynomiali/rational spline of (resp.: up to) degree m (resp.: degree n). The latter case arises when the interpolant is a polynomial spline of non-uniform degree of the type developed in [4].

PRUESS_DATA					
partner	type	conti/ty	$|\kappa|_{max}$	E_κ	$E_{\kappa'}$
FLO	pol.(=6)	C^2	$1.5 < |\kappa|_{max} < 2.0$	1.68×10^{-6}	1.51×10^{-7}
ZAR	pol.(2 or 3)	C^1	200.8812	381.631	-
NTUA	pol.(≤ 8)	C^2	0.001712	4.41×10^{-5}	1.56×10^{-6}

TABLE 1

NTUA_45					
partner	type	conti/ty	$|\kappa|_{max}$	E_κ	$E_{\kappa'}$
FLO	pol.(=6)	C^2	$4.5 < |\kappa|_{max} < 5.0$	0.57270	25.9153
DUN	rat.(=3)	G^2	1.31108	4.99754	32.3207
NTUA	pol.(≤ 5)	C^2	1.19064	4.50098	7.13512

TABLE 2

HOOD_1					
partner	type	conti/ty	$|\kappa|_{max}$	E_κ	$E_{\kappa'}$
FLO	pol.(=6)	C^2	$1.2 < |\kappa|_{max} < 1.4$	0.07108	1.18655
ZAR	pol.(2 or 3)	C^1	0.45834	0.17206	-
NTUA	pol.(≤ 5)	C^2	0.34680	0.13543	0.24644

TABLE 3.1

HOOD_3					
partner	type	conti/ty	$\|\kappa\|_{max}$	E_κ	$E_{\kappa'}$
FLO	pol.(=6)	C^2	$0.8 < \|\kappa\|_{max} < 0.9$	0.74909	2.34350
ZAR	pol.(2 or 3)	C^1	0.54465	0.19363	-
NTUA	pol.(≤ 6)	C^2	0.47629	0.19352	0.30024

TABLE 3.2

HOOD_5					
partner	type	conti/ty	$\|\kappa\|_{max}$	E_κ	$E_{\kappa'}$
FLO	pol.(=6)	C^2	$1.0 < \|\kappa\|_{max} < 1.1$	0.56052	3.66958
ZAR	pol.(2 or 3)	C^1	0.42984	0.20713	-
NTUA	pol.(≤ 5)	C^2	0.46492	0.16145	0.33766

TABLE 3.3

HOOD_7					
partner	type	conti/ty	$\|\kappa\|_{max}$	E_κ	$E_{\kappa'}$
FLO	pol.(=6)	C^2	$0.5 < \|\kappa\|_{max} < 0.6$	0.24973	2.49218
ZAR	pol.(2 or 3)	C^1	0.10661	0.03956	-
NTUA	pol.(≤ 4)	C^2	0.08438	0.03475	0.00746

TABLE 3.4

HOOD_9					
partner	type	conti/ty	$\|\kappa\|_{max}$	E_κ	$E_{\kappa'}$
FLO	pol.(=6)	C^2	$0.3 < \|\kappa\|_{max} < 0.35$	0.10603	0.17156
ZAR	pol.(2 or 3)	C^1	0.14413	0.04989	-
NTUA	pol.(≤ 5)	C^2	0.10903	0.04516	0.00835

TABLE 3.5

ROOF_1					
partner	type	conti/ty	$\|\kappa\|_{max}$	E_κ	$E_{\kappa'}$
FLO	pol.(=6)	C^2	$0.8 < \|\kappa\|_{max} < 0.9$	0.12324	0.31254
ZAR	pol.(2 or 3)	C^1	0.01678	0.00178	-
NTUA	pol.(≤ 7)	C^2	0.17678	0.02120	0.07706

TABLE 4.1

ROOF_2					
partner	type	conti/ty	$\|\kappa\|_{max}$	E_κ	$E_{\kappa'}$
FLO	pol.(=6)	C^2	$0.9 < \|\kappa\|_{max} < 1.0$	0.28873	0.88422
ZAR	pol.(2 or 3)	C^1	0.04833	0.01074	-
NTUA	pol.(≤ 5)	C^2	0.54279	0.20737	0.44090

TABLE 4.2

ROOF_3					
partner	type	conti/ty	$\|\kappa\|_{max}$	E_κ	$E_{\kappa'}$
FLO	pol.(=6)	C^2	$0.8 < \|\kappa\|_{max} < 1.0$	0.20628	1.28174
ZAR	pol.(2 or 3)	C^1	0.07865	0.02219	-
NTUA	pol.(≤ 9)	C^2	3.00228	0.33286	0.28058

TABLE 4.3

ROOF_4					
partner	type	conti/ty	$\|\kappa\|_{max}$	E_κ	$E_{\kappa'}$
FLO	pol.(=6)	C^2	$1.2 < \|\kappa\|_{max} < 1.4$	0.93347	14.8625
ZAR	pol.(2 or 3)	C^1	0.06194	0.01682	-
NTUA	pol.(≤ 5)	C^2	0.62543	0.37858	0.27909

TABLE 4.4

ROOF_5					
partner	type	conti/ty	$\|\kappa\|_{max}$	E_κ	$E_{\kappa'}$
FLO	pol.(=6)	C^2	$0.8 < \|\kappa\|_{max} < 0.9$	0.02595	0.30776
ZAR	pol.(2 or 3)	C^1	0.04272	0.01405	-
NTUA	pol.(≤ 13)	C^2	0.44457	0.16640	0.33242

TABLE 4.5

3 Comparative Comments

On the basis of the numerical output collected in Tables 1-4, we are lead to the following general remarks regarding the performance of the employed algorithms:

1. The smallest-degree shape-preserving interpolants are obtained by the ZAR algorithm which, however, ensures up to C^1 continuity. Second-order continuity, G^2 or C^2, can be achieved by appealing to the DUN or FLO/NTUA algorithms, respectively.

2. The degree of the FLO polynomial-spline interpolants is fixed, equal to 6, whereas the degree of NTUA interpolants varies from segment to segment. In the latter case, the maximum degree (=13) occured for the *ROOF_5* data, while for eight (8) data sets, namely *NTUA_45, HOOD_5, HOOD_7, HOOD_9, ROOF_2, ROOF_4* and *ROOF_5*, the maximum degree was less or equal to that of the FLO interpolant.
3. The smallest value of the maximum absolute curvature $|\kappa|_{max}$ is achieved by the ZAR algorithm for seven (7) data sets and the NTUA algorithm for the remaining five (5) data sets.
4. The relative performance of the algorithms with respect to the integral measure E_κ is more or less equilibrated. More specifically, the smallest value of E_κ is achieved by the FLO algorithm for three (3) data sets, the NTUA algorithm for four (4) data sets and the ZAR algorithm for five (5) data sets.
5. The smallest value of the integral measure $E_{\kappa'}$ is achieved by the NTUA algorithm for eleven (11) data sets and the FLO algorithm for the remaining one (1) data set.

4 Figures

4.1 Figures for the *PRUESS_DATA*

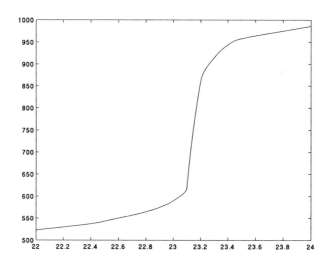

Fig. 1 The FLO shape-preserving interpolant for the *PRUESS_DATA*.

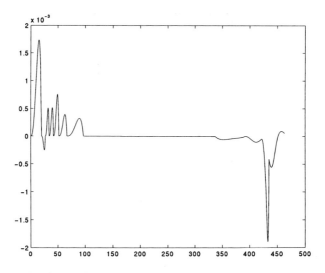

Fig. 2 The curvature distribution of the curve in Fig. 1.

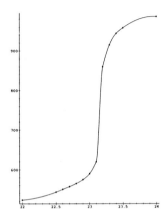

Fig. 3 The ZAR shape-preserving interpolant for the PRUESS_DATA; interpolation points are denoted by bullets.

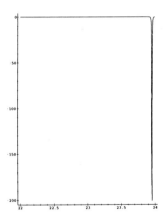

Fig. 4 The curvature distribution of the curve in Fig. 3.

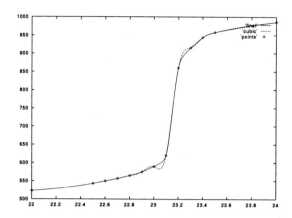

Fig. 5 The C^2 cubic spline (dotted line) and the NTUA shape-preserving interpolant (solid line) for the PRUESS_DATA; interpolation points are denoted by rhombuses.

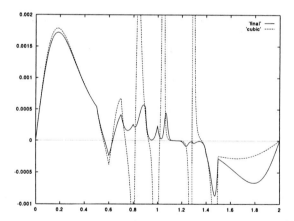

Fig. 6 The curvature distribution of the curves in Fig. 5.

4.2 Figures for the *NTUA_45* Data

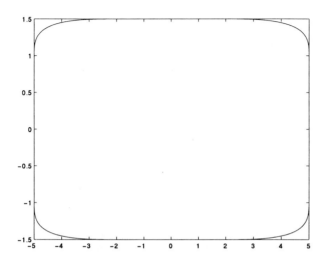

Fig. 7 The FLO shape-preserving interpolant for the *NTUA_45* data.

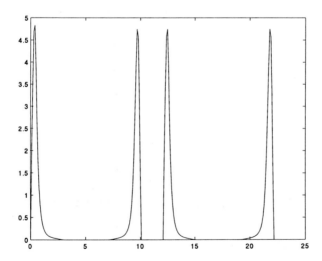

Fig. 8 The curvature distribution of the curve in Fig. 7.

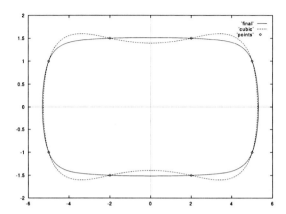

Fig. 9 The C^2 cubic spline (dotted line) and the NTUA shape-preserving interpolant (solid line) for the $NTUA\_45$ data; interpolation points are denoted by rhombuses.

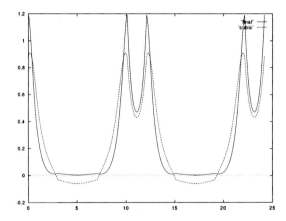

Fig. 10 The curvature distribution of the curves in Fig. 9.

4.3 Figures for the *HOOD_5* Data

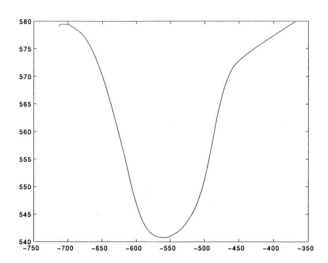

Fig. 11 The FLO shape-preserving interpolant for the *HOOD_5* data.

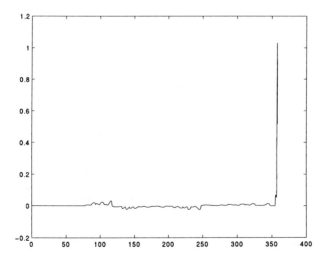

Fig. 12 The curvature distribution of the curve in Fig. 11.

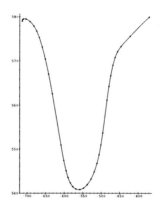

Fig. 13 The ZAR shape-preserving interpolant for the HOOD_5 data; interpolation points are denoted by bullets.

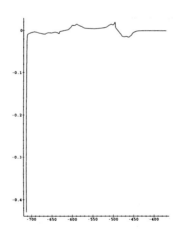

Fig. 14 The curvature distribution of the curve in Fig. 13.

Benchmarking in the Area of Planar Shape-Preserving Interpolation 277

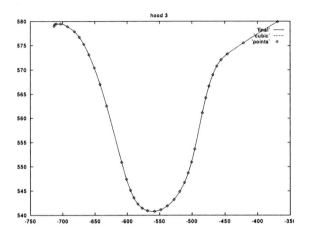

Fig. 15 The C^2 cubic spline (dotted line) and the NTUA shape-preserving interpolant (solid line) for the HOOD_5 data; interpolation points are denoted by rhombuses.

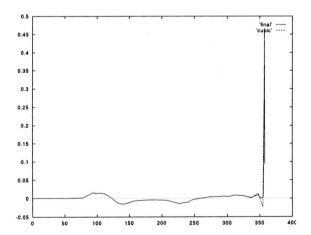

Fig. 16 The curvature distribution of the curves in Fig. 15.

4.4 Figures for the ROOF_1 Data

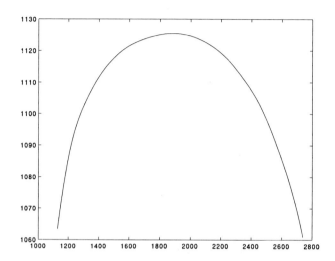

Fig. 17 The FLO shape-preserving interpolant for the ROOF_1 data.

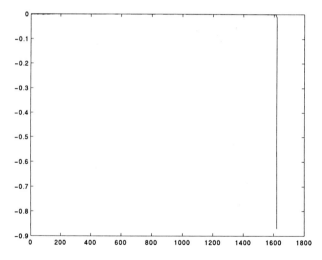

Fig. 18 The curvature distribution of the curve in Fig. 17.

Benchmarking in the Area of Planar Shape-Preserving Interpolation   279

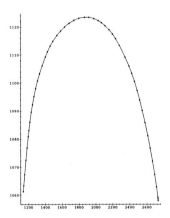

Fig. 19 The ZAR shape-preserving interpolant for the *ROOF_1* data; interpolation points are denoted by bullets.

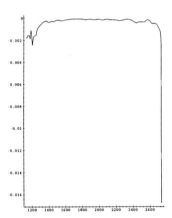

Fig. 20 The curvature distribution of the curve in Fig. 19.

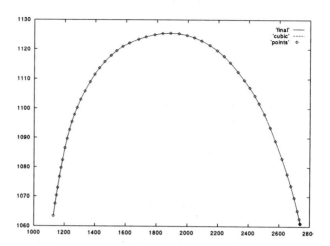

Fig. 21 The C^2 cubic spline (dotted line) and the NTUA shape-preserving interpolant (solid line) for the ROOF_1 data; interpolation points are denoted by rhombuses.

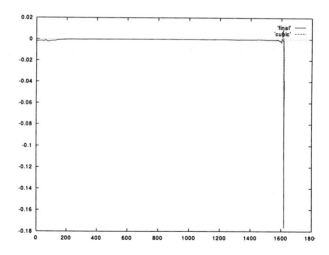

Fig. 22 The curvature distribution of the curves in Fig. 21.

5 Acknowledgements

The first of the authors would like to thank S.Asaturyan, J. Carnicer, P. Costantini and K. Unsworth, whose contributions with output data made the materialization of this paper possible.

References

[1] T.N.T. Goodman: *Shape preserving interpolation by parametric rational cubic splines.* In International Series of Numerical Mathematics, **86**, Birkhauser Verlag, Basel (1988), 149–158.

[2] T.N.T. Goodman and K. Unsworth: *An algorithm for generating shape preserving parametric interpolating curves using rational cubic splines,* University of Dundee Report CS 89/01, October 1989.

[3] J. Carnicer and M. Floater: *Coconvexity Preserving Curve Interpolation.* In these Proceedings.

[4] P.D. Kaklis and N.S. Sapidis: *Convexity-Preserving Interpolatory Parametric Splines of Non-Uniform Polynomial Degree.* CAGD, **12** (1995), 1-26.

[5] P. Costantini: *BVSPIS: A Package for Computing Boundary Valued Shape-Preserving Interpolating Splines.* University of Siena, Technical Report (1995), submitted for publication in Transactions of Mathematical Software.

Authors

Prof. Dr. Panagiotis D. Kaklis
National Technical University of Athens
Dept. of Naval Architecture and
Marine Engineering
Ship–Design Laboratory, CAGD Group
E–mail: kaklis@naval.ntua.gr

Nikolaos C. Gabrielides
National Technical University of Athens
Dept. of Naval Architecture and
Marine Engineering
Ship–Design Laboratory, CAGD Group
E–mail: kaklis@naval.ntua.gr

Benchmark Process in the Area of Shape-Constrained Approximation

Steffen Wahl

Mercedes-Benz AG

Introduction

On the following pages one can find a description of the benchmark examples in the area of shape-constrained approximation. The data files are available at the ftp side of the Technische Hochschule Darmstadt.

1 Contents

1.1 Benchmark Examples

1.1.1 Curve Approximation

Technische Hochschule Darmstadt	Helix
University of Leeds	Airfoil Sections
Mercedes-Benz AG	Formlinie

1.1.2 Surface Approximation

Universität Erlangen	Lens (K>0)
Technische Hochschule Darmstadt	Waist (K<0)
Technische Hochschule Darmstadt	Developable (K=0)
Kockums Computer Systems Ltd.	Water Lines
Mercedes-Benz AG	Boundary Curves

1.2 Solutions

2 Benchmark Examples

2.1 Curve Approximation

2.1.1 Helix

The given 101 points are disturbed points from a helix.
Approximate this set of points by a curve of

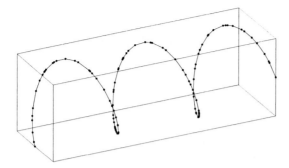

Fig. 1 Helix

- nearly constant curvature and
- nearly constant torsion

by B-splines and distance error as less as possible.

Constraints:

- nearly constant curvature
- nearly constant torsion

2.1.2 Airfoil Sections

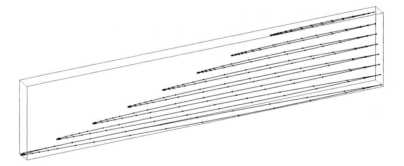

Fig. 2 Airfoil Sections

Successive airfoil sections taken from a wing. Airfoils given as point sets (39 points per set):

- x coordinate (approx) parallel to wing chord
- y coordinate (approx) parallel to wing thickness (increased by a factor of 10)
- z coordinate is the span wise direction

Constraints:

- convex curves

Additional Constraints:

- simultaneous approximation in order to create a surface out of the approximating curves

2.1.3 Formlinie

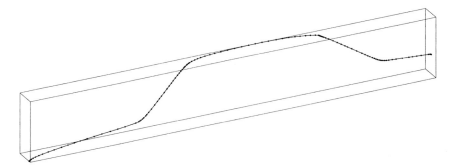

Fig. 3 Formlinie

'Formlinien' of the side of a upper class car model (119 points). 'Formlinien' are styling lines and true 3D. They are used to describe the styling of the shape of the car model.

Constraints:

- not more than four inflection points in side view
- number of monotonic segments in curvature distribution as low as possible
- no unnecessary torsion

2.2 Surface Approximation

2.2.1 Lens (K>0)

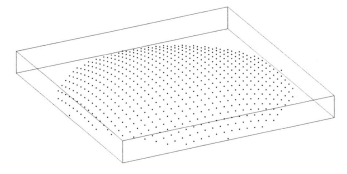

Fig. 4 Lens (K>0)

The given dataset (613 points) should be approximated by a convex (B-spline)-surface. The maximal distance error E should be as low as possible (as example E = 0.05). The data volume should be as low as possible. The reflection lines or the isophotes should be well distributed. The Mean Curvature (or their derivatives) should be as stable as possible.

Constraints:

- convex surface
- Mean Curvature as stable as possible (isolines)
- reflection lines or isophotes well distributed

2.2.2 Waist (K<0)

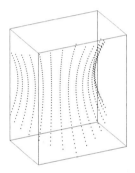

Fig. 5 Waist (K<0)

The given dataset (490 points) should be approximated by a B-spline-Surface with negative Gaussian Curvature. The maximal distance error E should be as low as possible (as example E = 0.05). The data volume should be as low as possible. The reflection lines or the isophotes should be well distributed. The Gaussian Curvature (or their derivatives) should be as stable as possible.

Constraints:

- negative Gaussian Curvature
- Gaussian Curvature as stable as possible
- reflection lines or isophotes well distributed
- symmetry

2.2.3 Developable (K=0)

The given dataset (5189 points) should be approximated by a developable (B-spline)-Surface (Gaussian curvature equal zero!) The maximal distance error E should be as low as possible (as example E = 0.05). The data volume should be as low as possible. The reflection lines or the isophotes should be well distributed.

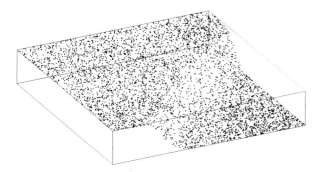

Fig. 6 Developable (K=0)

Constraints:

- developable surface
- reflection lines or isophotes well distributed

2.2.4 Water Lines

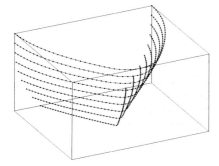

Fig. 7 Water Lines

The data supplied is in the end-surface region at the Fore part of the ship. Each Water Line consists of 60 points. Water Lines are intersections with parallel planes with constant z coordinate. The created surface should be a bi-cubic B-Spline surface and have only one sign of Gaussian Curvature.

Constraints:

- positive Gaussian Curvature (e.g. porcupine approach)
- tangent constraints (symmetry)

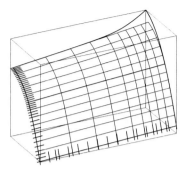

Fig. 8 Boundary Curves

2.2.5 Boundary Curves

The bi-cubic B-spline surface should match the new boundary curves and hold the tangent planes given as normals on the left and lower bound. The approximation algorithm should also keep the characteristic of the surface as good as possible.

Constraints:

- almost no modification in the lower and left part
- nice reflection lines:

– no waves
– regularly varying distances

- tolerance about 0.1mm

3 Solutions

The solutions for the above benchmark examples should be placed at the same ftp side. So, everyone could compare the resulting curves and surfaces.

Author

Steffen Wahl
Mercedes–Benz AG
Abt. EP/CDF
D–71059 Sindelfingen
E–mail: Steffen_Wahl@ep.mbag.sifi.daimlerbenz.com